Control of Centrifugal Compressors

An Independent Learning Module
from the
Instrument Society of America

CONTROL OF CENTRIFUGAL COMPRESSORS

By Ralph L. Moore

INSTRUMENT SOCIETY OF AMERICA

Copyright© Instrument Society of American 1989

All rights reserved

Printed in the United States of America

No part of this publication may be reproduced, stored in a retrieval system, or transmitted, in any form or by any means, electronic, mechanical, photocopying, recording or otherwise, without the prior written permission of the publisher.

INSTRUMENT SOCIETY OF AMERICA
67 Alexander Drive
P.O. Box 12277
Research Triangle Park
North Carolina 27709

Library of Congress Cataloging-in-Publication Data

Moore, Ralph L.
  Control of centrifugal compressors.

  Bibliography: p.
  1. Compressors. I. Title.
TJ990.M66 1989   621.5'1   88-35767
ISBN 1-55617-171-4

*To those from whom I have learned so much
about that amazing machine, the centrifugal compressor—*
    Robert C. Dean, Turbonetics
    Mike Rohacek, DuPont
    George Shanks, DuPont
    Naum Staroselsky, Compressor Controls
    Bernd Wagner, Atlas-Copco
    Guy Williamson, Dresser Rand

TABLE OF CONTENTS

Preface   ix

UNIT 1   **Introduction and Overview**
- 1-1. Purpose   3
- 1-2. Audience and Prerequisites   3
- 1-3. Study Material   4
- 1-4. Organization and Sequence   4
- 1-5. Course Objectives   5
- 1-6. Course Length   5

UNIT 2   **Centrifugal Pumps**
- 2-1. Introduction   9
- 2-2. Pump Casings   10
- 2-3. Pump Performance Curves   13
- 2-4. Load Curves   17
- 2-5. Series or Parallel Operation   18
- 2-6. Net Positive Suction Head (NPSH)   20
- 2-7. Capacity Regulation   20
- 2-8. Variable Speed Drives   21

UNIT 3   **Thermodynamics of Compression**
- 3-1. Basic Laws of Compression   29
- 3-2. Thermodynamic Compression Processes   32
- 3-3. Compressor Cooling   37
- 3-4. Head   39
- 3-5. Compressor Map   40
- 3-6. Surge   43
- 3-7. System Load Curve   47
- 3-8. Compressor Map Parameters   50
- 3-9. Use of the Compressor Map   55

UNIT 4   **Prime Movers and Ancillary Systems**
- 4-1. Prime Movers   63
- 4-2. Prime Mover Characteristics   68
- 4-3. Speed Governors   78
- 4-4. Turbine Overspeed Protection   83
- 4-5. Oil Supply Systems   83
- 4-6. Bearing Temperatures   86
- 4-7. Vibration Monitoring   86

UNIT 5   **Capacity Control**
- 5-1. Typical Control of an Air Compressor   91
- 5-2. Overload Protection Control   102
- 5-3. Variable Speed Compressors   108
- 5-4. Multistage Compressors   112

UNIT 6   **Surge Control**
- 6-1. General   123
- 6-2. Constant Speed Compressor with Inlet Throttling   125

	6-3.	Surge Control Based on Fan Laws	132
	6-4.	Surge Control Based on Other Variable Pairs	135
	6-5.	Simplified Surge Control System	137
	6-6.	Manual Surge Control for Start-Up	147
	6-7.	Sophisticated Surge Control System	149
	6-8.	Surge Control for Complex Systems	150
	6-9.	Surge Control System Components	153
UNIT 7		**Control System Interaction**	
	7-1.	General	167
	7-2.	Block Diagrams	170
	7-3.	Matrix Block Diagrams	173
	7-4.	Relative Gain Array	176
	7-5.	Dynamic Interaction	184
	7-6.	Interaction Compensation	187
UNIT 8		**Coordinated Compressor Control**	
	8-1.	Introduction	195
	8-2.	Incipient Surge	199
	8-3.	Experimental Performance Testing	201
	8-4.	Integrated Surge Control	204
	8-5.	Integrated Surge Controller	211
	8-6.	Integrated Capacity Control	213
	8-7.	Integrated Multistage Surge Control	216
	8-8.	Automatic Start-Up and Shutdown	221
UNIT 9		**Multiple Compressor Systems**	
	9-1.	General	227
	9-2.	Specific Power Consumption	236
	9-3.	Multiple Compressors in Series	240
	9-4.	Multiple Compressors in Parallel	244
UNIT 10		**Optimization of Compressor Operation**	
	10-1.	General	256
	10-2.	Optimized Surge Control Line Adaption	257
	10-3.	Compressor Optimization with Multiple Loads	258
	10-4.	Optimization by Improved Control Response	260
	10-5.	Optimization of Compressors with Turboexpanders	263
	10-6.	Optimization of Multiple Compressors	267
	10-7.	Optimization of Pumps	269
	10-8.	Optimization by Pressure Modulation	270
	10-9.	Optimization of Pipeline Compressors	273
	10-10.	Power House Optimization	274
APPENDIX A:		**Nomenclature**	281
APPENDIX B:		**Suggested Readings and Study Materials**	285
APPENDIX C:		**Solutions to All Exercises**	289
Index			299

PREFACE

ISA'S Independent Learning Modules

This is an Independent Learning Module (ILM) on the subject of the control of centrifugal pumps and compressors; it is part of the ISA series of modules on Unit Process and Unit Operation Control.

The ILMs are the principal components of a major educational system designed primarily for independent self-study. This comprehensive learning system has been custom designed and created for ISA to more fully educate people in the basic theories and technologies associated with applied instrumentation and control.

The ILM system is divided into several distinct sets of modules on closely related topics; such a set of individually related modules is called a series. The ILM System is composed of:

- The ISA Series of Modules on Control Principles and Techniques
- The ISA Series of Modules on Fundamental Instrumentation
- The ISA Series of Modules on Unit Process and Unit Operation Control
- The ISA Series of Modules for Professional Development

The principal components of the series are the individual ILMs such as this one. They are especially designed for independent self-study; no other texts or references are required. The unique format, style, and teaching techniques in the ILMs make them a powerful addition to any library.

Comments About this Volume

This ILM, *Control of Centrifugal Compressors*, is designed as an introduction to the basic operating principles of centrifugal pumps and compressors and as a compilation of instrument application practices in the control of these rotating machines. The pumping process and the control configuration interact intimately. The theoretical relationships and the graphical data supplied by manufacturers are reviewed as background for control system design.

Emphasis is given to the extremely large financial rewards resulting from a well-designed control system that maintains near-optimum operating conditions, and to the very large financial penalty resulting from a control system that permits the compressor to surge. Further emphasis is given to the design of the system using data that is necessarily "fuzzy," and the need for judgment and "belt and suspenders" conservatism.

Mathematics up to calculus and operational calculus are used to illustrate the development of basic relationships. However, the basic tenets are not dependent on mathematics, and thus the development does not represent a barrier to those without a strong mathematical background.

The field of compressor operation and control is maturing. A wealth of technical literature exists, and most applicable writings are listed in the References. Practical control system configurations, proved in field applications, are correlated in this ILM. Advanced topics, e.g., multiple compressors and optimization, are included to show that concepts must change as new criteria are adapted. "Standard" configurations are described for widely applied operations, such as ethylene, ammonia, methanol, and so on.

This ILM, like all of the modules in the system, is designed for independent self-study. It is directed to the engineer, the first-line supervisor, and the senior technician. In addition, college, university, and technical students will find the material profitable.

It is intended that this ILM be both theoretical and practical—that it show the basic concepts of compressor control theory and the application of these concepts to daily practice.

Unit 1:
Introduction and Overview

UNIT 1

Introduction and Overview

Welcome to ISA's Independent Learning Module on compressor control. This unit provides you with an overview and the information you need to take the course.

Learning Objectives — When you have completed this unit, you should:

  A. Know the nature of material to be presented.
  B. Understand the general organization of the course.
  C. Know the course objectives.

1-1. Purpose

The purpose of this ILM is to convey to the student a comprehensive knowledge of the techniques for controlling centrifugal pumps and compressors. The documentation is logically categorized and designed for ready utilization in practical situations. Course structure is purposely divided into segments that represent certain aspects of compressor operation and control. Techniques for control are associated with appropriate types of compressors and process objectives. Using this information, the student should be able to determine the control configuration best suited to the process objectives. Once the configuration is chosen, a control system design must be developed to choose the instrument system to be used (from pneumatic to digital) and to provide calibration to meet process requirements.

1-2. Audience and Prerequisites

This ILM is designed for persons who wish to work on their own at their own pace and desire to learn the basics of the application of control to centrifugal compressors. The material will be useful to senior technicians, first-line supervisors, and engineers who are concerned with the application and operation of compressor controls as well as to students in technical schools, colleges and universities who wish to gain insight into the principles and practices of compressor control.

No particular prerequisites are required other than interest or motivation to complete the course. Some mathematical background is necessary for the development and utilization of equations. However, mathematics is used more in terms of an explanation of dynamics and thermodynamics and the development of generalized working equations than for specific solutions. Mathematics should not be a barrier to the understanding of compressor operation and the application of controls. In fact, this ILM is specifically designed to eliminate the mathematical barrier found in basic texts on mechanics, dynamics, and thermodynamics by stressing illustrations, graphical presentations, and applications.

1-3. Study Material

This textbook is the only study material required in the course. It is one of ISA's several Independent Learning Modules on fundamental instrumentation designed as an independent, stand-alone textbook. It is uniquely and specifically structured for self-study. A list of suggested readings in Appendix B provides additional references and study materials.

The student also may find it helpful to study other Independent Learning Modules available from ISA. In conjunction with compressor control, simultaneous study topics might be the measurement of flow, pressure, and temperature, or automatic control systems.

1-4. Organization and Sequence

This ILM is divided into ten separate units. Unit 2 describes centrifugal pumps, which add energy to liquids to force them to flow through pipelines and to a higher elevation. The remaining eight units describe the compression of gases and vapors to cause them to react, flow through pipelines, and eject through nozzles. Unit 3 discusses the basic concepts of the broad field of thermodynamics and the specific subject of compression. Unit 4 classifies compressor drivers, with the implications of their effect on the choice of a control system, and the ancillary compressor systems that must be monitored to guarantee reliability. The next three units describe the application of compressor controls to meet the process objectives of the compressor with safety.

Unit 8 discusses the coordinated control system developed and marketed by Compressor Controls, Inc., of Des Moines, Iowa. The next two units are concerned with large multistage, multiple-compressor systems.

Each unit is designed in a consistent format with a set of specific learning objectives stated in the very beginning of the unit. Note these learning objectives carefully; the material that follows the learning objectives is directed specifically toward them. The individual units often contain example problems to illustrate specific concepts. At the end of each unit you will find student exercises to test your understanding of the material. All student exercises have solutions contained in Appendix C.

1-5. Course Objectives

When you have completed this entire Independent Learning Module, you should be able to:

A. Understand the fundamentals of centrifugal machines, which add energy to liquids, vapors, and gases for the purpose of forcing them to flow.
B. Assess the advantages and disadvantages of each type of prime mover.
C. Apply capacity controls to various configurations of centrifugal compressors.
D. Appreciate surge, what it is and what it can do, and apply controls to protect against it.
E. Be familiar with the highly specialized services available from companies whose business is compressor control.
F. Calculate the huge amounts of energy (costs) required for compressor operation and the savings from even slight improvements in compressor performance.

In addition to these overall objectives, each unit contains a specific set of learning objectives—highlighted as in this first unit—that are intended to be very specific in order to help direct your study of that particular unit.

1-6. Course Length

The basic idea of the ISA System of Independent Learning Modules is that students should be able to learn best if they proceed at their own pace. As a result, there will be significant variation in the amount of time taken by individual students to complete this ILM.

You are now ready to begin your in-depth study of compressor control. Please proceed to Unit 2.

Unit 2:
Centrifugal Pumps

UNIT 2
Centrifugal Pumps

The centrifugal pump is universally used in the fluid processing industries to impart energy to a liquid to cause it to flow from one unit operation to the next. The pump imparts a centrifugal force to the liquid, which the casing converts to pressure. Thus it is a dynamic device, and its design determines the efficiency with which the conversion is made. Performance is described by the head-capacity curve, which varies in contour with the design of the pump.

Learning Objectives — When you have completed this unit, you should:

A. Have a general understanding of the underlying theory of centrifugal pumps.

B. Understand the significance of the head-capacity and system-load curves in determining control strategy.

C. Appreciate the results in performance of complex pump configurations—series and parallel.

D. Be aware of the usefulness of varying pump speed and the devices required to provide a variable speed pump.

2-1. Introduction

Pumping is the addition of energy to a fluid to move it from one point to another and is not, as frequently thought, the increasing of the pressure of the fluid. The increase in pressure is incidental to the forcing of the increased flow through the downstream resistance. Since energy is the capacity to do work, the addition of energy to a fluid causes it to do work, namely, flow through a pipe or rise to a higher level.[1]

Pumps are commonly designated by classifications that apply to the mechanics for moving the fluid. Three classes are in general usage—*centrifugal*, *rotary*, and *reciprocating*. These classifications can be lumped into two categories—*dynamic* and *displacement*. Rotary and reciprocating pumps are displacement-type pumps, wherein energy is periodically

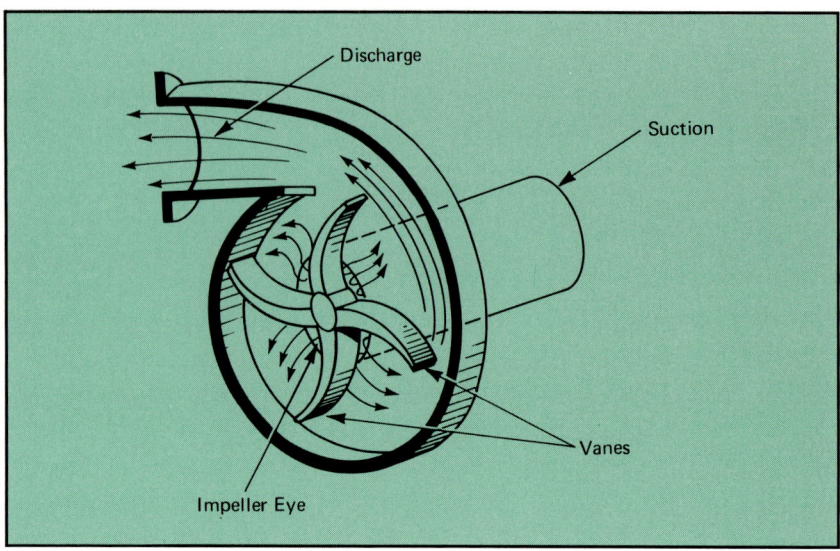

Fig. 2-1. Diagram of Internals of Centrifugal Pump[1]

added by the application of force to one or more movable boundaries of a number of enclosed, fluid-containing volumes. Centrifugal pumps are in the dynamic category, in which energy is continuously added to increase fluid velocities in excess to those occurring in the pipeline in such a way that excess velocity generates a pressure increase.[2] All classifications must be considered in selecting the most appropriate pump to meet process conditions.

A centrifugal pump consists of an impeller or rotor with an intake at the center, so arranged that when rotated it will discharge water or other liquids by centrifugal force into a casing that surrounds the impeller. This action is shown in simplified form in Fig. 2-1. The head or pressure developed by the pump is entirely the result of the velocity imparted to the liquid by the impeller and is not due to any impact or displacement.

2-2. Pump Casings

The pump impeller vanes and casing form are designed together to provide the smoothest transition from fluid velocity to downstream static pressure. Vane inlet and outlet slopes are determined from velocity vector diagrams, which consider impeller rotational velocity, fluid velocity through the impeller, and fluid velocity at the inlet or into the casing.

The resultant vectors are chosen to minimize turbulence (minimize energy loss) at impeller inlet and outlet. The design is modified as required to avoid such dangers as flow separation from the vane surface. Finally, a casing is chosen to provide an efficient transition into static pressure. Three casing configurations are in common use:

- *Volute Type*. Here the impeller discharges into a spiral-volute form of casing, proportioned to reduce the liquid velocity gradually. The volute casing provides a collector for the fluid discharged and converts the velocity energy into static pressure. Wall T, shown in Fig. 2-2, divides the beginning of the volute and the discharge nozzle and is called a "volute tongue". Double volutes produce near-radial symmetry, largely balancing radial loads on the pump shaft.
- *Diffuser Type*. The diffuser consists of a number of stationary vanes set around the impeller, as shown in Fig. 2-3. These gradually expanding passages change the direction of the liquid flow and convert the velocity energy imparted to the liquid by the impeller into static pressure. The flow from the vaned diffuser is then collected in a volute or circular casing and discharged into the outlet pipe.
- *Turbine Type*. This configuration is also known as the vortex, regenerative, or periphery pump. Liquid in this type is whirled for nearly one revolution in the annular channel

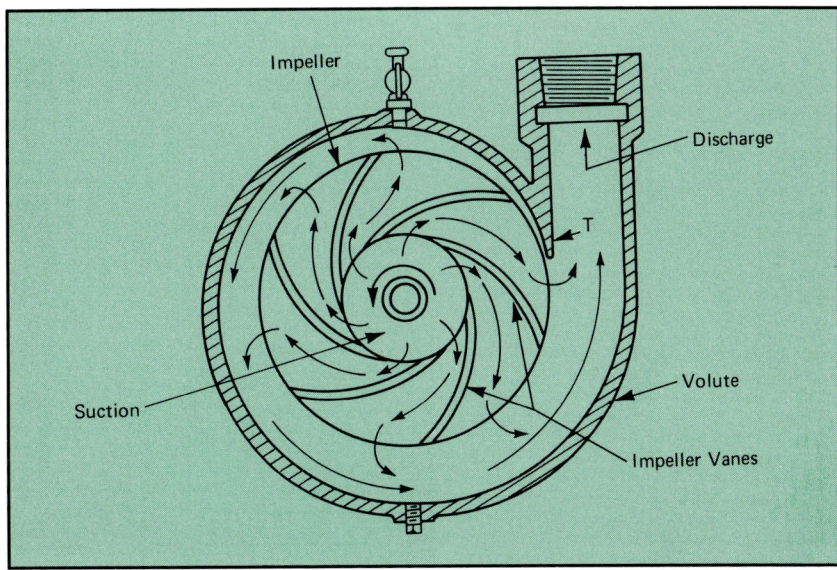

Fig. 2-2. Section through a Volute-Type Centrifugal Pump[1]

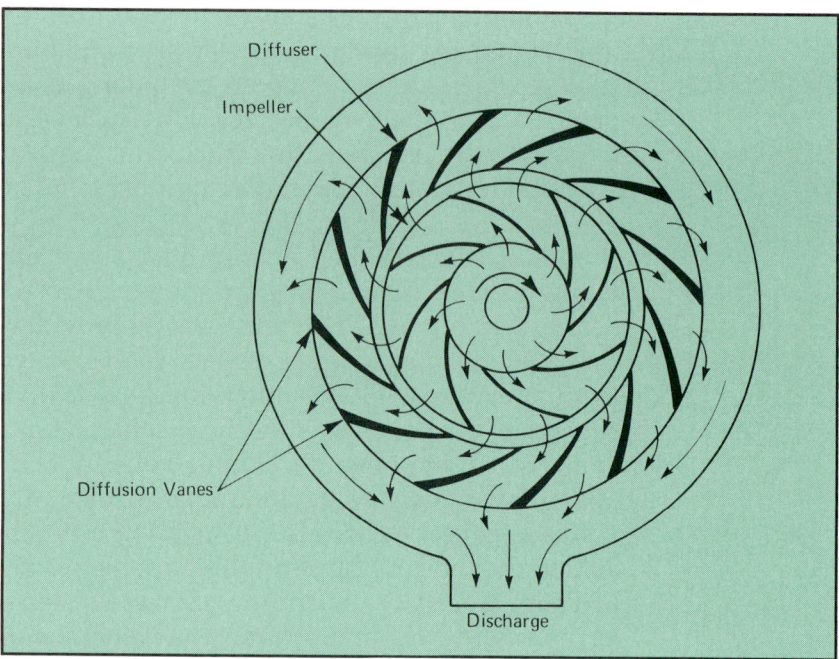

Fig. 2-3. Section through a Diffuser-Type Centrifugal Pump[1]

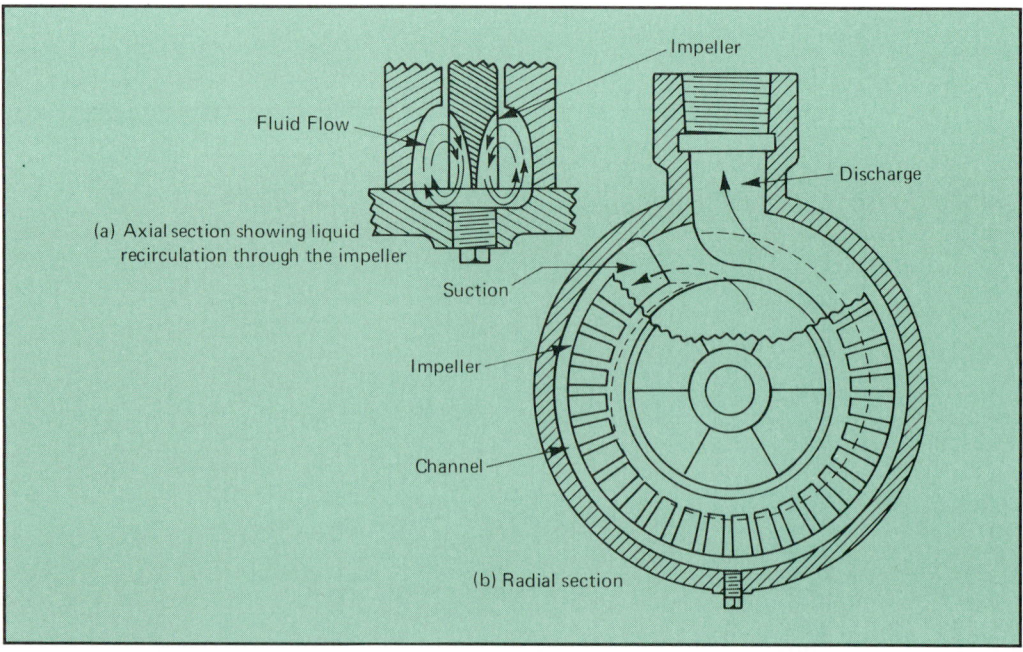

Fig. 2-4. Regenerative or Turbine-Type Pump[1]

in which the impeller turns. The liquid is circulated in and out of the impeller blades, as shown in Fig. 2-4. The actual path of the liquid is a circular spiral. These are sometimes classed as deep well turbine pumps rather than centrifugal pumps.

2-3. Pump Performance Curves

The performance characteristics of a pump rest partly on a sound theoretical base, such as the vector diagrams just discussed; partly on similitude;[3] and partly on many years of trial-and-error experience. Thus, for a given pump, it is desirable to see a graphical display of operating parameters, i.e., head, capacity, efficiency, and horsepower requirements. Such curves, called performance curves, are available for every commercially available pump and form the basis for the application of that pump. Figure 2-5 is an illustration of a performance curve for a centrifugal pump being operated at constant speed.

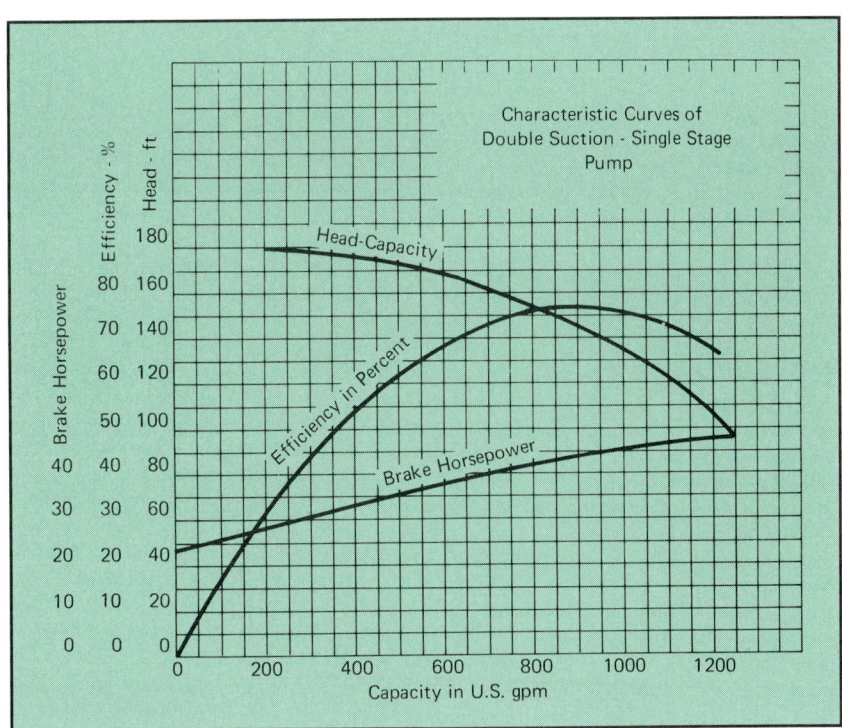

Fig. 2-5. Centrifugal Pump Performance Curve for Constant Speed Operation[9]

Performance curves, developed by the pump manufacturer, are based on theoretical considerations supported by extensive testing. The manufacturer will usually supply identical curves for identical (same model no.) pumps. Actually, no two pumps are identical but are slightly different due to manufacturing tolerances, casting differences, and so on.

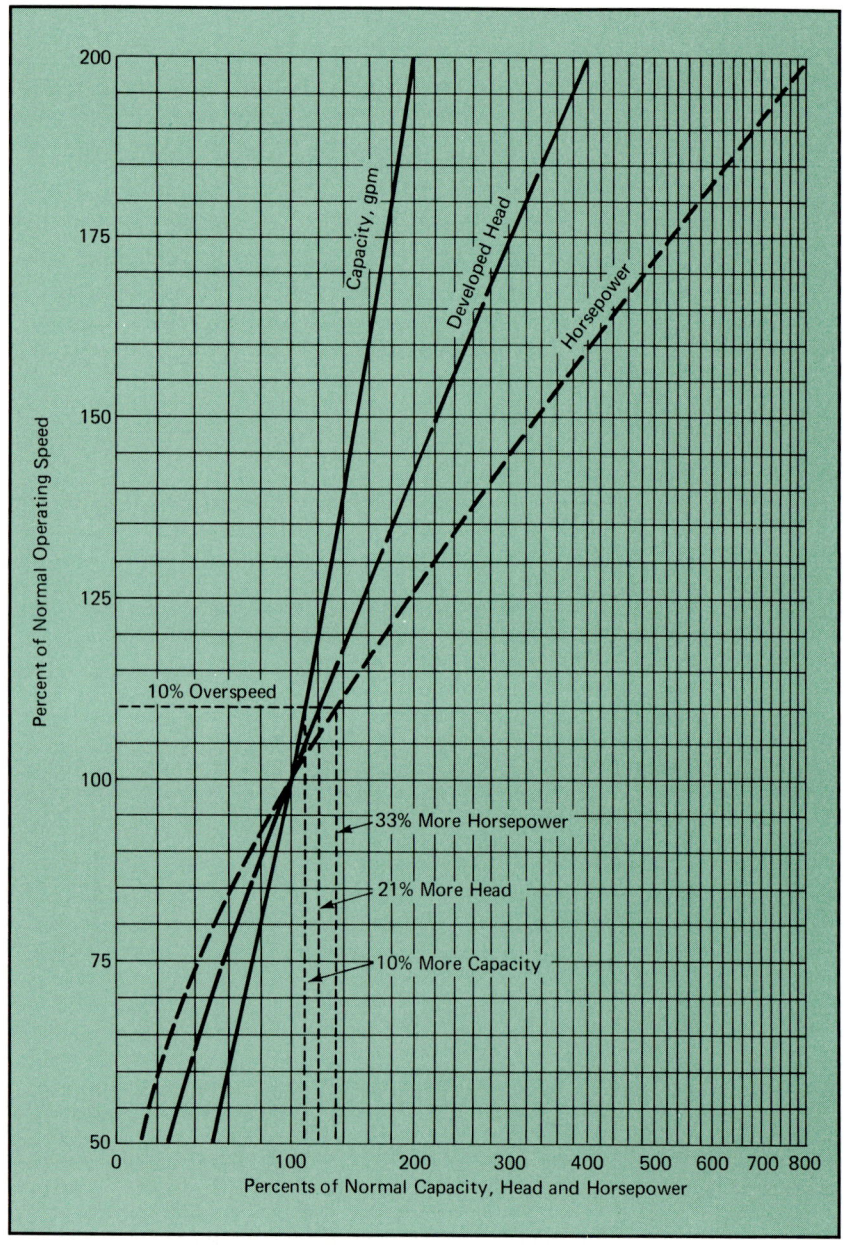

Fig. 2-6. Theoretical Effect of Changing the Speed of a Centrifugal Pump

Slight changes in operating characteristics due to manufacturing tolerances can be quickly estimated from charts such as Fig. 2-6.

Figure 2-6 is developed from three axiomatic relationships for rotating machinery. They are called the *Fan Laws* from their application to compressible fluids and can be stated as follows:

- Capacity is proportional to speed.

$$q = K_1 s$$

- Head is proportional to speed squared.

$$H = K_2 s^2$$

- Horsepower is proportional to speed cubed.

$$hp = K_3 s^3$$

The head-capacity curve, shown on Fig. 2-5, is used in the selection and application of a pump. It plots capacity in gpm versus increase in head of the flowing fluid. The pressure increase is, therefore,

$$p_2 - p_1 = \frac{H\rho}{144} \qquad (2\text{-}1)$$

The slope of the head-capacity curve can vary, as shown in Fig. 2-7, with the slope (angle) of the impeller vanes and the casing configuration. The various shapes can be identified:[2,4]

- Figure 2-7a—A rising head-capacity characteristic, which rises continuously as capacity decreases.
- Figure 2-7b—A drooping or unstable characteristic where the head rises to a maximum with decreasing capacity then decreases over the lower capacity range. This shape can cause instability at heads greater than the shutoff value, particularly if pumps are operated in parallel.
- Figure 2-7c—A steep head-capacity characteristic exhibits a large increase in head between that developed at design capacity and that developed at shutoff.
- Figure 2-7d—A flat head-capacity characteristic shows little variation in head throughout the capacity range.

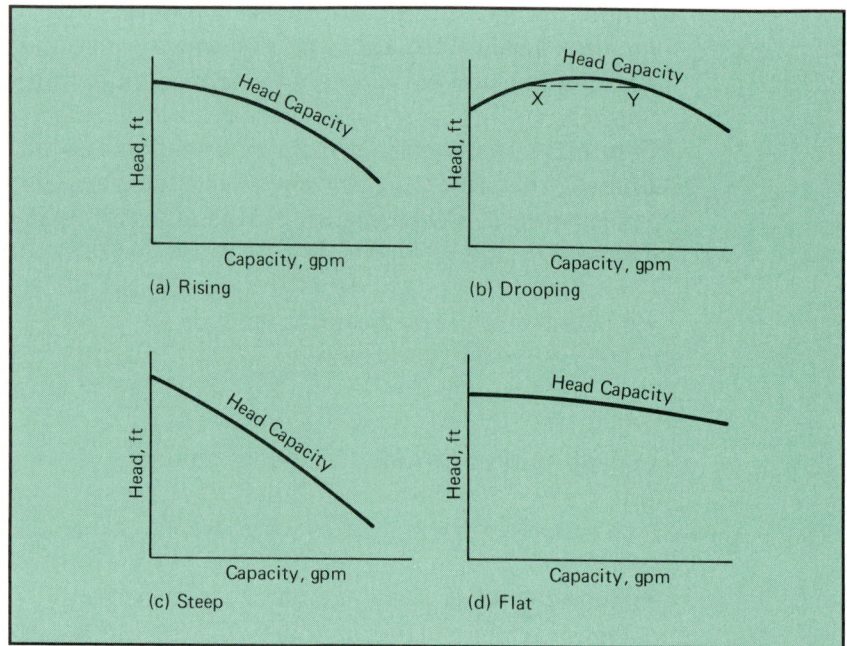

Fig. 2-7. Various Impeller Types and Designs Result in Different Shapes of Head-Capacity Curves[1]

Energy conservation is dictating the increasing use of variable speed operation of centrifugal machinery. The head-capacity curves describing operation of a variable speed pump are shown in Fig. 2-8. Energy optimization requires operation at minimum speed (minimum horsepower) and maximum efficiency. Obviously, more sophisticated controls are required to attain this goal.

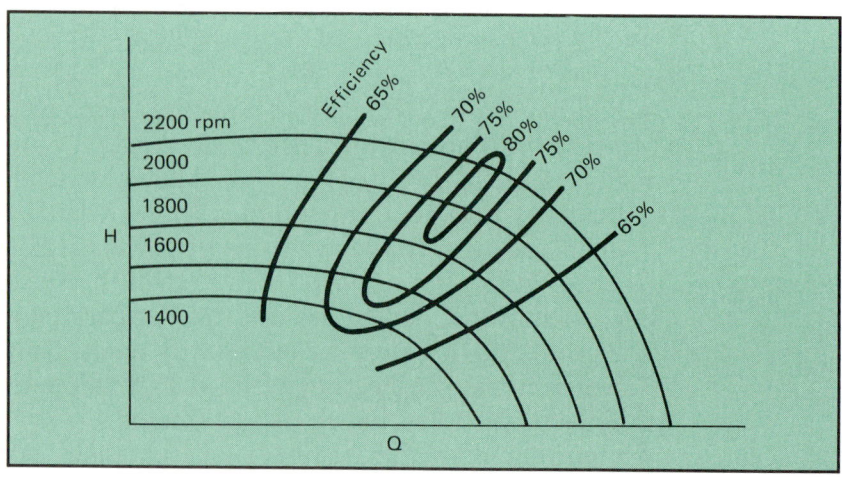

Fig. 2-8. Variable Speed Head-Capacity Curves[4]

2-4. Load Curves[4]

The load curve, or system-head curve, is obtained by combining the friction-head curve of the downstream system with the system static head and any pressure differences that exist. The friction-head curve is the kinetic energy lost due to pressure drop through piping, valves, and fittings in the suction and discharge lines. *Static head* is the difference in elevation between the liquid levels of the suction and discharge. The *kinetic pressure drop* is roughly a square law relationship between head and throughput, static head being a constant for all flow rates, as shown in Fig. 2-9.

Figure 2-9 is obtained by superimposing the load curve (for the system) on the performance curve (for the pump). The system will operate at the point where the two curves intersect. Thus, point A on Fig. 2-9 denotes the head and capacity for this particular system. Changing the resistance of a given piping system by partially closing a valve or the pluggage of a filter changes the slopes of the load curve. Thus, in Fig. 2-9, partially closing a valve in the discharge line

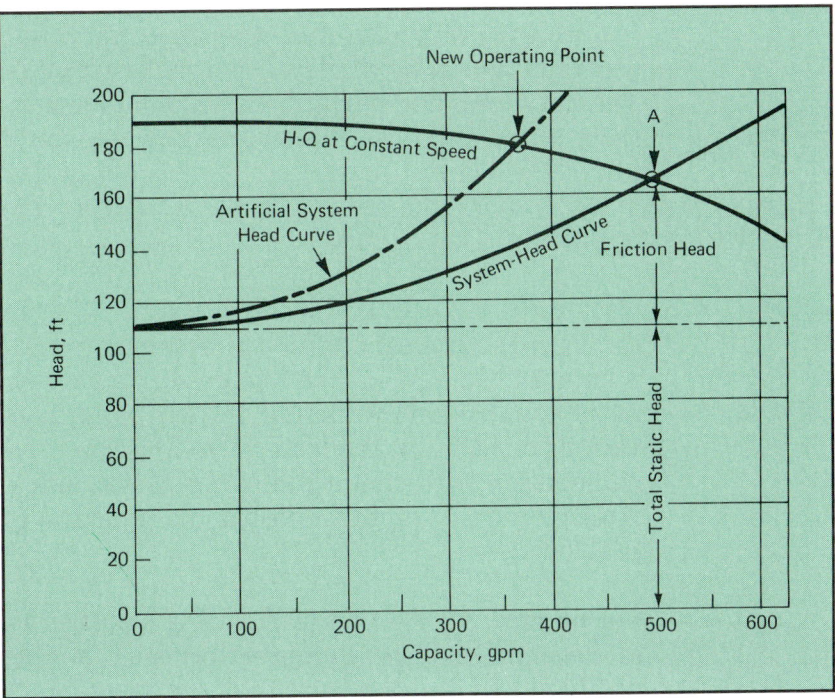

Fig. 2-9. System-Head Curve or Load Curve[4]

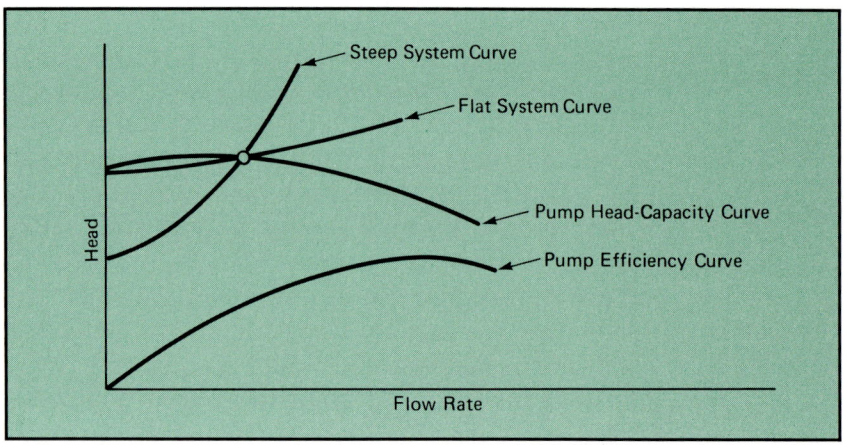

Fig. 2-10. Relationship of System-Head Curve and Pump Head-Capacity Curve[4]

produces the new load curve shown, shifting the operating point to a higher head but lower capacity. Opening the valve has the opposite effect.

Obviously, an infinite number of steep and shallow pump curves can be combined with an infinite number of load curves. Some combinations are to be avoided. Consider a shallow (all static head) load curve and a flat pump curve. Performance difficulty would be expected if the static head was represented by the level in a tank, which could change. The flat pump curve could not accommodate the change. A steep pump curve, as shown in Fig. 2-10, would provide trouble-free operation.

2-5. Series or Parallel Operation[4]

Should the rangeability of the demand, either in head or in throughput, be greater than can be accommodated by any one pump, multiple pumps can be used. For proper specification of the pumps and evaluation of their performance under various conditions, the load curve must be used in conjunction with composite pump-performance curves. Any multiple pump combination has the potential for causing problems.

For pumps in *series*, performance is obtained by adding heads at the same capacity. Pump performance in *parallel* is obtained by adding capacities at the same head. This technique is shown in Fig. 2-11.

- *Series Operation.* Pumps are frequently operated in series to supply heads greater than those of individual pumps. Identical pumps are normally used for the lowest cost installation, while non-similar pumps can be used if the pump curves are properly selected. A drooping pump curve can cause instability if throughput demand is near the peak of the curve.[1]
- *Parallel operation.* Parallel operation of pumps is often the best solution to a problem of capacity variation.

When two pumps with rising pump curves operate in parallel, system load should divide equally between them.[4] The combination of one rising pump curve and one drooping pump curve can result in the pump with the drooping curve being forced to the zero-flow condition, which can be detrimental to the pump. Two pumps with similar drooping pump curve characteristics will divide the load equally while operating on the rising portion of their characteristic curves, but, when throttled back to where the flow is double-valued, problems are introduced where (a) the starting head of one pump might not be adequate to get it on line with the other pump running, (b) instability (surging) can occur because of the double-valued nature of the curve, and (c) the pumps, even if identical, will not divide the flow equally. For these reasons, operating pumps with drooping characteristics in parallel is not good design practice. Beuter[5] has offered a valving technique to prevent surging when such a combination is unavoidable.

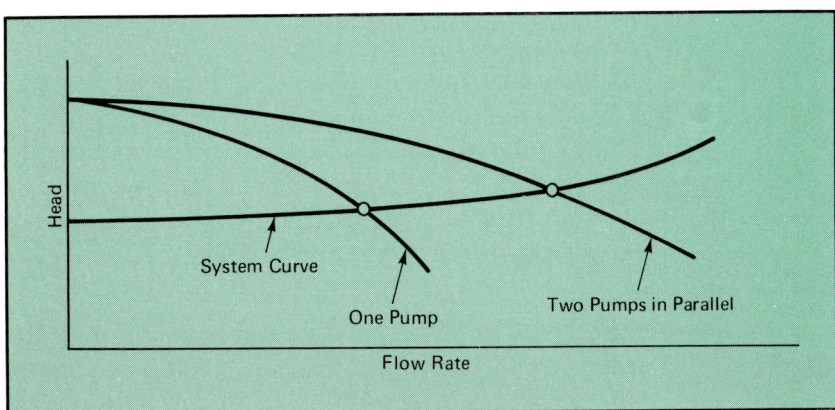

Fig. 2-11. System-Head Curve and Head-Capacity Curves for Pumps Operated in Parallel[4]

2-6. Net Positive Suction Head (NPSH)

Lack of net positive suction head (NPSH) is by far the most prevalent pump problem in operating plants. The condition reduces pump capacity and efficiency and can damage the pump.

A pump can pump only liquid; but every liquid will turn into vapor at some combination of pressure and temperature. If temperature is increased and pressure decreased, a point will be reached where the liquid vaporizes. This is its *vapor pressure*. Pump inlet pressure can be reduced by the piping and fitting losses in the inlet line and by trying to "pull" liquid from a tank below the elevation of the pump. Vaporization (*flashing*) will take place if a valve is closed in the inlet line, reducing inlet pressure below the NPSH, or closing a valve in the discharge line, causing temperature to rise by the same energy being added to less flowing fluid. The required NPSH must be obtained from the pump builder. Pump installation design must insure that the inlet pressure (head) can never fall below the NPSH.[4]

One result of flashing is to fill the pump with vapor, called gassing the pump or losing its prime, and the pump will cease pumping until it is "primed," where liquid is added to displace the vapor or air.

A more severe result of flashing is a phenomenon called *cavitation*. Here, the liquid being pumped is partially vaporized, resulting in bubbles in the liquid flow. The pump continues to pump, and the liquid flow carries the bubbles to a point of higher pressure where they are above their vapor pressure and will collapse. The collapse is generally near a metal surface of the pump. The collapse, called *imploding*, causes noise and vibration and damages the metal surface.[2]

The phenomenon called cavitation, also called eroding and pitting, is very detrimental to pump operation and pump life because it results in objectionable noise and in vibration that causes metal part fatigue, fastener release, and metal damage.

2-7. Capacity Regulation[2]

The most common method of capacity regulation in centrifugal pumps is *discharge throttling*. Partial closure of a valve in the discharge line will increase pump discharge pressure so that the load curve intersects the pump curve at a smaller capacity. This technique is by far the cheapest method of capacity regulation. It is expensive in operation. Economic penalties are incurred because of operating the pump at a point of low efficiency; pressure drop across the valve in the discharge line; overheating the pump; and wear on the pump shaft because of the imbalance on the impeller due to the pressure distribution.

Considerable power can be saved by diverting all or part of the pump capacity from the discharge line to the pump suction through a bypass line. The bypass line will typically contain a restricting orifice or a control valve.

Speed regulation can be used to minimize power requirements and eliminate overheating during capacity modulation. Steam turbine and combustion engine drivers are amenable to speed regulation at small cost. A number of variable speed techniques can be used with electrical motors. These are discussed below. They are generally so expensive that they are used only after justification by an economic analysis.

2-8. Variable Speed Drives

Polyphase ac motors are designed to run at or near a speed that corresponds to the rotating field impressed on the stator.[6] The synchronous speed of an ac motor is inversely proportional to the number of motor poles and directly proportional to the frequency applied to the stator. This inherent design criteria establishes one practical method for controlling ac motor speed: controlling the frequency of the supply power. A second method is to place an element between the electrical driver and the pump (or compressor), which causes pump rotational speed to change while motor speed remains constant.

Variable speed drives are widely used for factory automation and for energy savings in fluid processing plants.[7] In the factory, the motor drives are speed-paced according to the demands of the assembly logistics: car body presses use variable speed dc drives to synchronize with flexible production lines; paper coating machines use computer-based individual roll drives to provide torque differential; synthetic fiber manufacturing uses synchronized sequential and parallel drives; the upper and lower rolls in a hot-rolling reversing steel mill are independently driven; and a sectionalized paper machine uses variable speed to deal with intersectional speed-load interaction. For centrifugal loads, horsepower varies as the cube of the speed and torque as the square. Significant energy savings accrue from operation below the base speed. One pump installation,[7] assuming 50% flow, showed that recirculation would still cost 100%, throttling (a valve) energy would cost 70%, but adjusting pump speed would cost some 25%.

Older techniques for providing variable frequency include the motor-generator (MG) set, where dc motor speed is conveniently set by varying the voltage to the armature. The frequency of the ac generator output voltage is then proportional to speed. The ac motors wired to the generator will then follow the speed of the generator. The MG set requires extensive maintenance and has limited speed rangeability. Static inverters, such as the thyratron and the mercury arc rectifier, have also been used to produce adjustable frequency. The switching time of the thyration is too slow for optimum operation.[6] The mercury arc rectifier has been in use since World War I and suffers from low efficiency, low power factor, high maintenance, and large physical size. In recent years, these techniques have been largely superseded by solid-state circuits using power diodes, power transistors, and thyristors (silicon-controlled rectifiers).

Variable speed can also be provided by the coupling between the motor and the centrifugal pumping device. The eddy-current coupling is an electromechanical torque-transmitting device installed between a constant speed prime mover and a load to obtain adjustable speed operation. The input and output shafts of the motor are mechanically independent, with the output magnet member revolving freely within the input ring or drum member. The magnet member has a field winding, which is excited by a direct current, inducing eddy currents in the ring. The interaction of these currents and the magnetic flux develops a tangential force that tends to turn the magnet in the same direction as the rotating ring. An increase or decrease in field current will change the value of torque developed, thereby changing output speed. The efficiency of the coupling is essentially proportional to the ratio of output speed to input speed.[2] Eddy current couplings are available to transmit power from 3 to 5000 hp.

Adjustable speed belt drives also provide variable speed operation.[2] These drives are generally used for pump applications and are of the adjustable-pitch-sheave and rubber belt variety shown in Fig. 2-12. Horsepower ranges are from fractional to 100 hp.

In Fig. 2-12, the upper sheave assembly is the input element driven directly by the shaft of a constant speed motor. The input element has one stationary disk member on the left and a sliding disk member on the right. The sliding member is mechanically attached to a positive shifting linkage. The shifting linkage is actuated by a handwheel on which the desired speed ratio can be set. The lower sheave is the output element whose speed is adjustable. Here, the sliding disk is next to the spring carriage on the left, and the stationary member is on the right. This assembly is mounted on the adjustable speed output shaft of the drive. Moving the handwheel changes the position of the belt on the diameters of the two sheaves, thus changing the speed ratio.

The inherent design criterion of an ac motor causes it to follow the rotating stator field.[6] Therefore, the one truly practical method of controlling ac motor speed is to control the frequency of the supply power. The torque produced by an ac induction motor is proportional to the magnetic flux in the air gap. To meet the requirement of constant torque at the

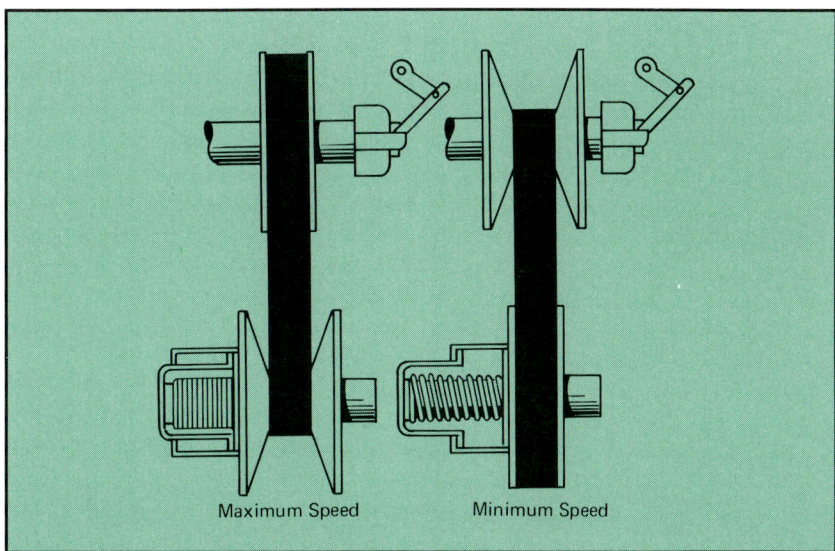

Fig. 2-12. Mechanical Adjustable Speed Belt Drive[2]

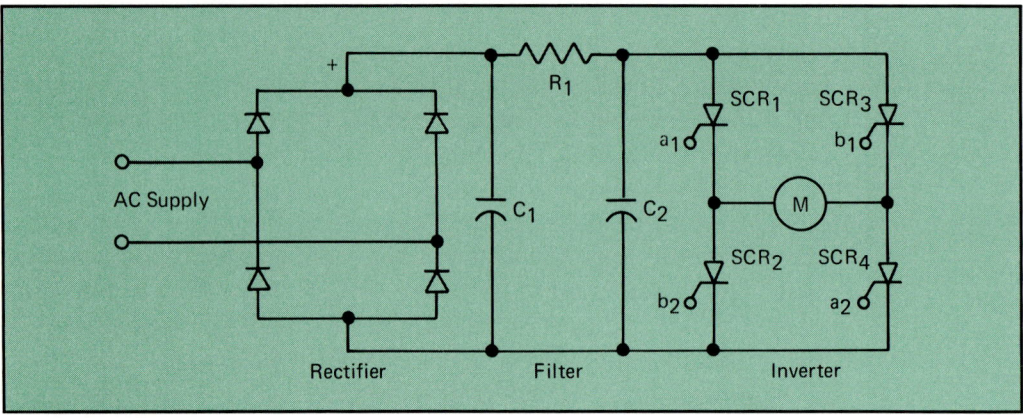

Fig. 2-13. Solid-State Single-Phase Rectifier-Inverter for Speed Control of AC Synchronous and Induction Motors[8]

output shaft with varying speed, the flux in the air gap must by maintained constant over the speed range. Thus, optimum adjustable frequency systems must consider voltage as well as frequency control to assure adequate torque-producing capability at all speeds.

One type of static rotating frequency, the dc link converter, is shown in Fig. 2-13. The ac supply power is rectified to dc with a standard rectifier, and the resulting "variable" dc is filtered to provide "clean" dc power to the static inverter. The inverter is a device that converts dc to ac power using thyristors, which are switched sequentially so that an alternating voltage waveform is delivered to the ac motor. A number of inverter circuits have been developed. The output frequency is determined by the rate at which the inverter thyristors are triggered into conduction. Triggering is by an oscillator circuit and logic elements.[3,8]

Exercises

1. What is the definition of pumping?

2. What is the difference between a pump and a compressor?

3. Draw a pump family line diagram.

4. What are the Fan Laws?

5. Prove the dimensions of eq. (2-1).

6. What is the operating point for the system described by

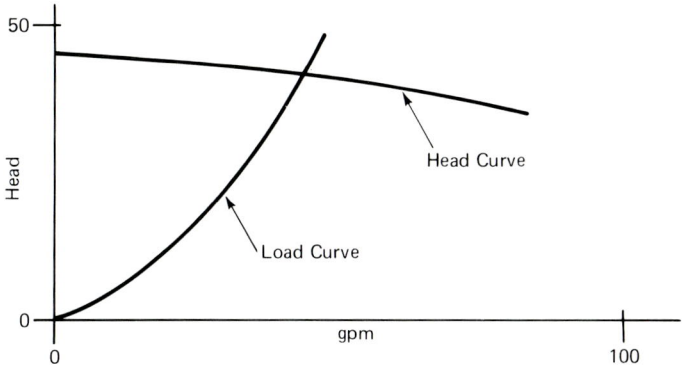

7. What other difficulty would be encountered in combining a shallow pump curve and a flat load curve?

8. Why not operate two pumps with drooping pump curves in parallel?

References

1. Carter, R., I. J. Karassik, and E. F. Wright, *Pump Questions and Answers*, McGraw Hill, New York, NY, 1949.
2. Karassik, I. J., W. C. Krutzsch, and W. H. Fraser, *Pump Handbook*, McGraw Hill, New York, NY, 1976.
3. Murphy, J. M. D., *Thyristor Control of A. C. Motors*, Pergamon Press, New York, 1973.
4. Hickes, T. G., and T. W. Edwards, *Pump Application Engineering*, McGraw-Hill, New York, NY, 1971.
5. Beuter, H. A. F., "Centrifugal Pump—Unstable Curve Correction by the Use of Control Valves," (personal communication), HAB Engineering, Three Rivers, MI, 1985.
6. Millermaster, R. A., *Harwood's Control of Electric Motors*, Fourth Ed., John Wiley & Sons, New York, NY, 1970.
7. Bailey, S. J., "Motor Drives 1985: Designed for Programmable Motor Control in the Factory Automation Era," *Control Engineering*, April 1985, p. 159.
8. Koslow, I. L., *Control of Electric Machines*, Prentice-Hall, Inc., Englewood Cliffs, NJ, 1973.
9. Anon., *Pump Engineering Data*, Economy Pumps, Inc., Sedgely Avenue at 19th and Lehigh, Philadelphia, PA, 1951.

Unit 3:
Thermodynamics of Compression

UNIT 3
Thermodynamics of Compression

Centrifugal compressors are adequately modeled by the fundamental laws of thermodynamics, hydraulics, continuity, and Newton's Laws. This unit reviews the applicable laws in a form that describes centrifugal compressors and develops those laws into relationships useful for design and for the estimation of performance of the compressors.

Learning Objectives — When you have completed this unit, you should:

A. Have a general understanding of the underlying theory of compression.

B. Be able to relate the theory to an operating compressor.

C. Understand the usefulness of the graphical compressor performance curves.

3-1. Basic Laws of Compression

The flow of compressible fluids, i.e., fluids in which the density varies materially with pressure, is modeled, in the case of steady-state flow or equilibrium, by the following hydraulic and thermodynamic relationships:

- *Continuity*, for steady unidirectional flow,

$$m_i = \rho_i A_i \nu_i \qquad (3\text{-}1)$$

which describes the mass flow (m) at point (i).
- *The First Law of Thermodynamics*, also shown as the General Energy Equation or the Law of the Conservation of Energy. For compression,

$$\Delta Q = \Delta h + W \qquad (3\text{-}2)$$

- *The Second Law of Thermodynamics*,

$$\Delta Q = T \Delta S \qquad (3\text{-}3)$$

- *The Equation of State* for a real gas,

$$\frac{p}{\rho} = ZRT \qquad (3\text{-}4)$$

The concept of entropy (S) in the Second Law of Thermodynamics involves the concept of reversibility. A process that can be made to retrace in the reverse order the various states of the original process and which yields exactly the same amount of energy in going in one direction as it consumed in going in the other direction is said to be reversible. Reversibility, therefore, introduces the idea of a perfect process and is useful in calculating efficiencies. The reversible process is called a conservative process, which has no losses. All real systems have losses, obviously, and are irreversible.

Equation (3-3) shows entropy to increase with the transfer of heat. However, the transfer of heat requires a "sink" of lower temperature to which heat will flow. *Entropy*, then, is a measure of the degradation or unavailability of heat energy and represents an inefficiency inherent in all energy that is converted into heat energy, e.g., friction.[1,2,3]

The First Law of Thermodynamics, eq. (3-2), is made up of the work energy (W), the heat energy (Q), the potential energy designated as enthalpy (h), and the kinetic energy, which can be assumed equal on the inlet and outlet of the compressor and cancels out of the general energy equation. The potential energy, h, is defined

$$h = u + pv \qquad (3\text{-}5)$$

Enthalpy describes the difference between compressible and incompressible flowing fluids. The displacement energy pv is the pressure head used to describe an incompressible fluid (Bernoulli's equation). However, each unit volume of a compressible fluid has the capacity to do work by expansion. This work can be done, for instance, by an expandable (compressible) fluid contained in a perfectly closed vessel. The capacity to do work by expansion is called the *internal energy* (u) of the fluid. It is of primary importance for positive displacement machinery, e.g., a steam engine, where the fluid does work by expansion behind a moving piston in an otherwise closed container (the cylinder).

A contrast of this situation with that of an incompressible fluid in a closed vessel is instructive. In this case, very little work is required to produce a very high pressure, but this pressure does not represent energy since, if the liquid were required to do work by moving a piston, the pressure would immediately drop to zero.

Enthalpy, then, is used to describe the continuous (steady) flow of a compressible fluid, since this fluid can do work by expansion as well as being displaced (pushed ahead) by the surrounding (following) fluid masses.[4] The steady flow displacement work is illustrated by Fig. 3-1 and is described by:

$$\text{Work} = \text{Pressure} \cdot \text{Volume Change}$$
$$W = pv$$

On the basis of one pound of flowing fluid,

$$W = pv = \frac{p}{\rho} \tag{3-6}$$

In differential form,

$$dW = pdv \tag{3-7}$$

Substitute in the First Law of Thermodynamics (eq. (3-1))

$$dQ = dh + dW$$
$$dQ = dh + pdv$$

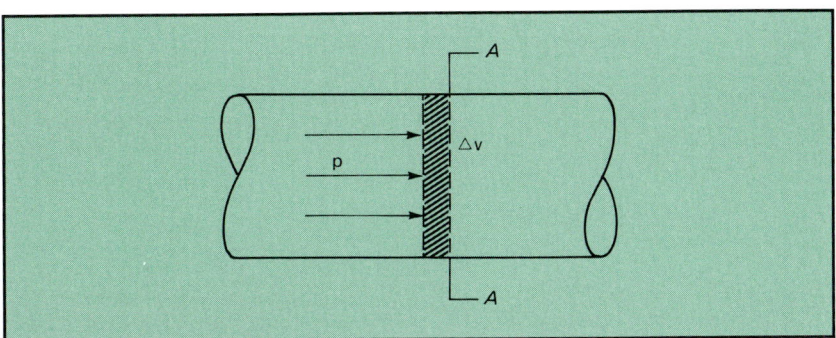

Fig. 3-1. Steady-Flow Work Pushing Each Finite Volume ΔV past Section A-A Is Equal to pΔV Where p Is the Mean Pressure[1]

and from eq. (3-5),

$$dh = du + d(pv) \qquad (3-8)$$

Specific heats can be defined as follows:

$$C_p = \left.\frac{\partial h}{\partial T}\right|_p \qquad C_v = \left.\frac{\partial u}{\partial T}\right|_v \qquad (3-9)$$

3-2. Thermodynamic Compression Processes

The heat transfer through the walls of a compressor is very small compared with the heat content of the gases involved, because a given volume of gas remains inside the machine for only a small fraction of a second. A process in which no heat is transferred is called an *adiabatic* process. This concept is of great practical importance in the analysis of turbomachinery, even if the heat insulation is far from perfect. Consider the general case of heating an ideal gas at variable volume and pressure. The First Law of Thermodynamics says that the thermal energy supplied to the gas must be equal to the increase in internal energy plus the work done by the gas:

$$dQ = du + pdv \qquad (3-10)$$

Since enthalpy (h) and internal energy (u) are independent of pressure and specific volume for a perfect gas, equations (3-9) become

$$dh = C_p dT \qquad du = C_v dt \qquad (3-11)$$

For an adiabatic process,

$$dQ = C_v dT + pdv = 0$$

From eq. (3-4), the Equation of State,

$$dT = \frac{d(pv)}{R} = \frac{1}{R}(pdv + vdp)$$

Substituting,

$$C_v pdv + C_v vdp + Rpdv = 0$$

then

$$R = C_p - C_v \qquad (3-12)$$

Substituting and integrating,

$$\frac{C_p}{C_v}\frac{dv}{v} + \frac{dp}{p} = 0$$

$$\frac{C_p}{C_v}\log_e v + \log_e p = \text{constant}$$

Since

$$k = \frac{C_p}{C_v}$$

then

$$pv^k = \text{constant} \qquad (3\text{-}13)$$

Equation (3-13) is the relationship between pressure (p) and specific volume (v) for the adiabatic compression process.

Following Stepanoff,[5] a more generalized thermodynamic process relationship can be developed. The object of a compressor is to raise the inlet pressure (p_1) to a discharge pressure (p_2). The work of compression of the gas can be thought of as the lifting of a given weight of gas at inlet pressure (p_1) and temperature (T) to a height (H) where the gas is discharged at the same pressure and temperature. This process is shown in Fig. 3-2.

Since gas density (ρ) and gas pressure (p) vary throughout the height of the column, a differential form of the equation relating head (H) and pressure (p) is required:

$$dp = \rho dH \qquad (3\text{-}14)$$

From eq. (3-4), the Equation of State,

$$\frac{p}{\rho} = ZRT$$

Substituting in eq. (3-14),

$$\frac{dp}{p} = \frac{dH}{ZRT} \qquad (3\text{-}15)$$

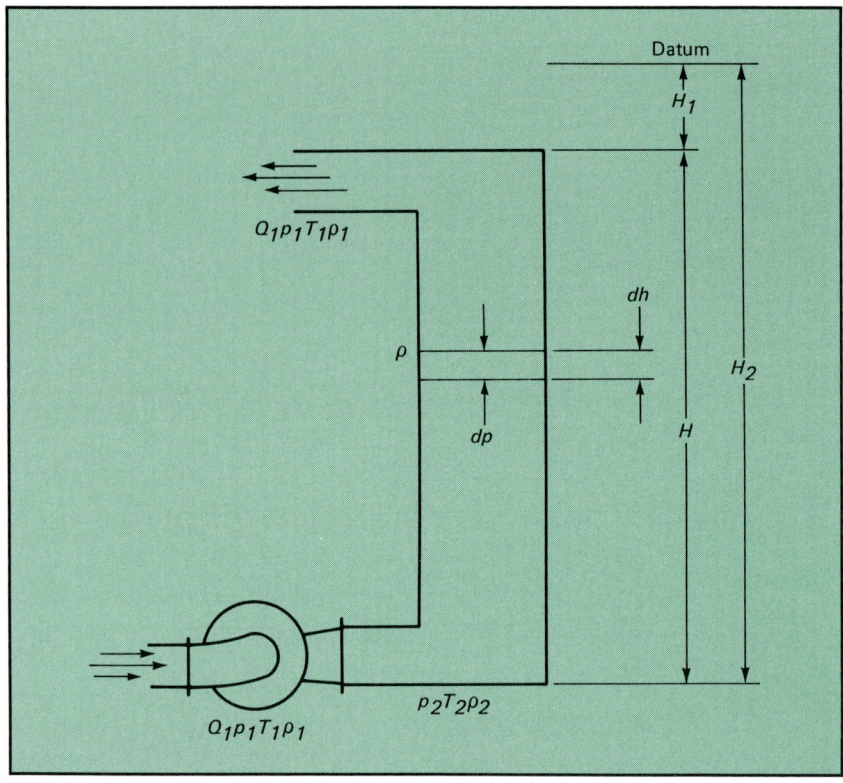

Fig. 3-2. Compressor Lifting a Given Weight of Gas at Inlet Pressure (p_1) and Temperature (T_1) to a Height (H) Total Head[5]

Equation (3-20) is a basic relationship in describing the thermodynamic flow process of compression. The process is described by the exponent "n" as follows:

As a first-order assumption, let temperature (T) vary directly with height (H):

$$\frac{dT}{dH} = C = \text{Constant} \qquad (3\text{-}16)$$

Substituting and separating variables,

$$\frac{dp}{p} = \frac{1}{CZR} \frac{dT}{T}$$

Integrating,

$$CZR \log p - \log T = \text{Constant}$$

or

$$\frac{p^{CZR}}{T} = \text{Constant} \qquad (3\text{-}17)$$

Unit 3: Thermodynamics of Compression

Substituting eq. (3-15),

$$\frac{p}{p^{1-CZR}} = \text{Constant} \tag{3-18}$$

Take the $1 - CZR$ root of both sides of eq. (3-18):

$$\frac{p}{p^{1/(1-CZR)}} = \text{Constant}$$

Then let

$$n = \frac{1}{1 - CZR} \tag{3-19}$$

Then,

$$pv^n = \text{Constant} \tag{3-20}$$

- *Isothermal.* Figure 3-2 shows that for a given compressor size and speed, but for different gas temperature distributions along the column, the height (H) of the gas column will remain the same. Thus, the pressure at the bottom of the column will depend on the rate of temperature change along the column. If temperature is constant along the column, the constant (C) in eq. (3-16) is equal to zero. Then $n = 1$ in eq. (3-19). This value of exponent "n" corresponds to isothermal compression.
- *Adiabatic.* A process where no heat is transferred and $n = k$ from eq. (3-13). The exponent "k" can be evaluated from its definition (C_p/C_v) and has a value of 1.4 for monotonic gases.
- *Isentropic.* An adiabatic process ($n = 1.4$), which is reversible or which has constant entropy.
- *Polytropic.* Any process that can be described by eq. (3-20). If an ideal gas process can be assumed to be polytropic, the exponent "n" can be evaluated from the definition of two state points.

The foregoing analytical development of thermodynamic processes is shown graphically on the pressure-specific volume (pv) and the temperature-entropy (TS) planes for ease in visualization. The effect of varying the exponent "n" is shown in Fig. 3-3. Since the work done (W) and the heat

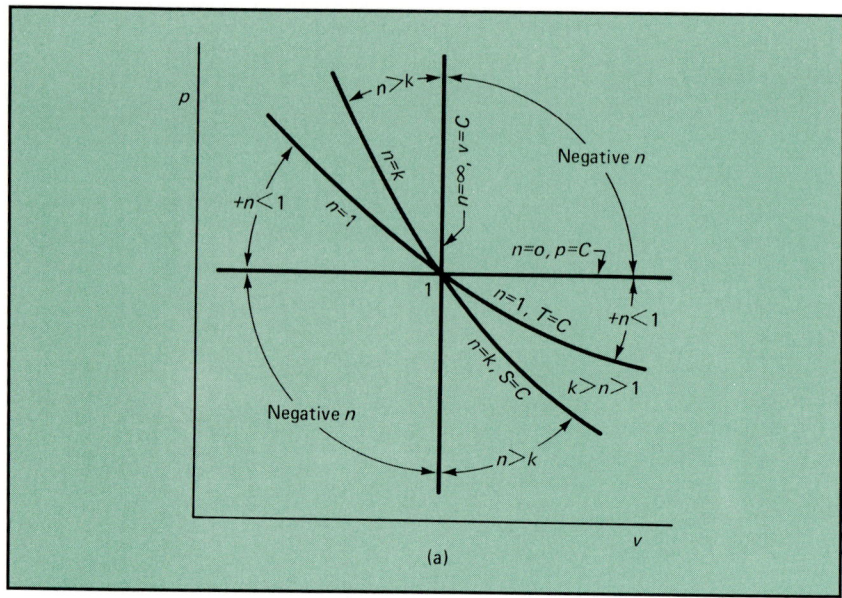

Fig. 3-3. (a) Effect of Varying n. Expansions or compressions are imagined to take place from some common point 1. All positive values of n give curves in the second and fourth quadrants on the pV plane; positive values of n may produce curves in all four quadrants on the TS plane

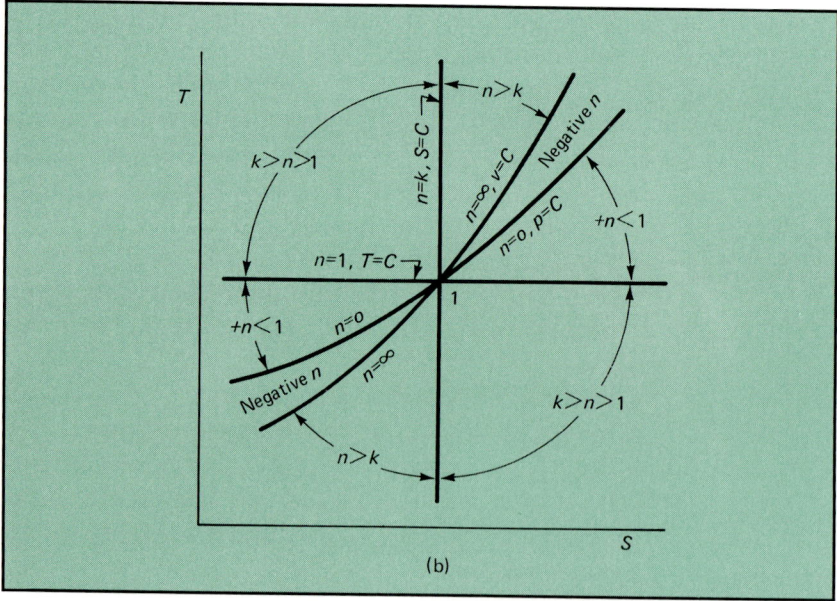

Fig. 3-3. (b) Curves with values of n between 1 and k will fall in the second and fourth quadrants on the TS plane and within a narrow region on the pV plane

transferred (Q) are defined by these graphs, the effect of "n" on these variables becomes apparent.

3-3. Compressor Cooling

Large compressors are constructed in sections, where a number of impellers are contained in a single casing and the casings are connected by a common rotating shaft. The word "stage" has become ambiguous in common usage, since it is commonly used in referring to both one impeller and a group of impellers in a single casing. To avoid misunderstanding, the casing (portion of a compressor between two flanges) is now formally called a "section." However, in operating plants, impellers are called "wheels" and casings are called "stages" and that terminology will appear in this book.

Multistage compressors use interstage cooling to lower gas temperatures. The work done on a gas in the compression process causes the temperature to rise. Cooling reduces temperatures within the compressor, which reduces fatigue on the metals of construction and reduces the outlet temperature to a reasonable value. Should condensate be formed in the interstage coolers, knockout pots are required to remove the condensate from the system. Liquid droplets will severely damage the downstream compressor wheels. However, condensation removes a portion of the flow to the next stage and has to be considered in the design of the compressor.[6]

Compression with intercooling approximates an isothermal thermodynamic process. Isothermal compressor head can be calculated from

$$H = RT_1 \ln (p_2/p_1) \qquad (3\text{-}21)$$

Cooling in high pressure compressors is economically justified because, for a given compression ratio, a lower head (H) is required. This means a lower speed or a smaller machine. The reduction in specific volume (with temperature) leads to a smaller size for a given weight flow. A saving in power results from the reduction in head necessary to produce the same pressure ratio.

Figure 3-4 shows the isothermal approximation of an isentropic process resulting from interstage cooling. The isentropic process is shown by AB on the pv and TS diagrams.

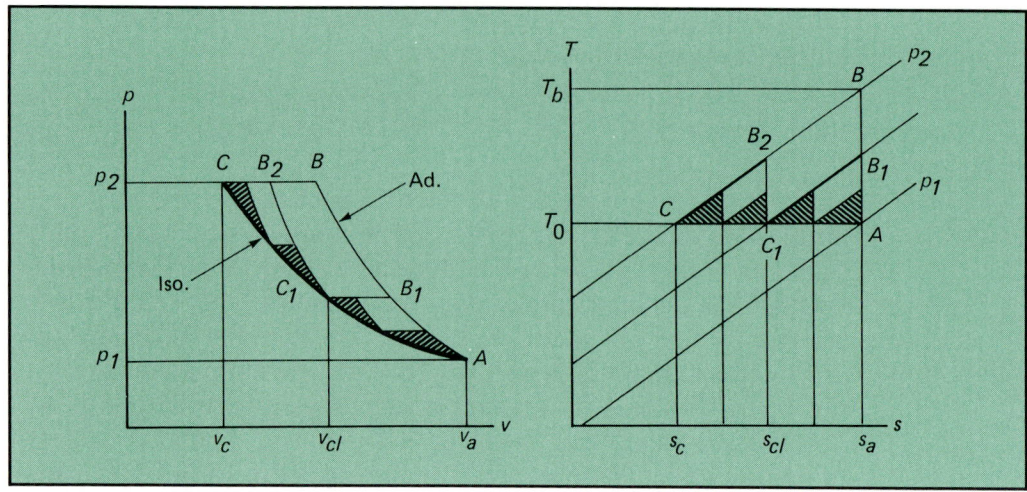

Fig. 3-4. Isothermal Compression Shown as an Adiabatic Compression Done in a Number of Small Steps with Cooling after Each Step[5]

Compression in two steps with intercooling is shown by triangles AB_1C_1 and C_1B_2C. The four cross-hatched triangles represent the process with four steps and three intercoolings. It is evident from Fig. 3-4 that the available energy being removed from the system decreases as the number of steps increases and the process approaches isothermal.[5]

The interstage cooling shown in Fig. 3-4 is seen to approach the Carnot cycle for a hypothetical reversible engine, which is shown in Fig. 3-5. While a real process cannot be reversible if it involves (1) friction, (2) transfer of heat through a finite temperature difference, (3) mixing of fluids at different temperatures, or (4) unrestrained expansion to lower

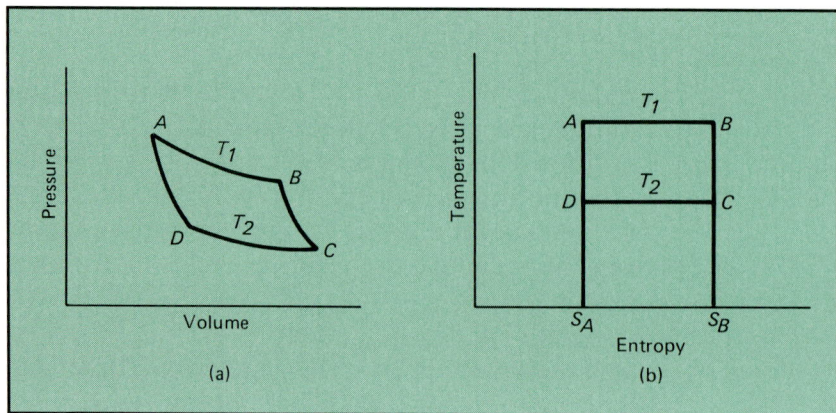

Fig. 3-5. Carnot Thermodynamic Work Cycle[7]

pressures, yet the concept of the reversible Carnot cycle is valuable as a process from which maximum work can be delivered. Reversibility is attained by the assumption of infinitesimal temperature differences and a frictionless mechanism.

It can be shown that no engine working between temperature limits T_1 and T_2 can have a thermal efficiency higher than that of an engine following the Carnot cycle. This development leads to the availability of a maximum efficiency basis with which to compare the efficiency of a real engine. Thus, a comparison of Figs. 3-4 and 3-5 shows intercooling to be improving compressor efficiency.[7]

3-4. Head

Figure 3-2 shows the concept of compressor head (H). It is a column of gas H feet high. It can be evaluated by integrating eq. (3-16),

$$T_2 - T_1 = C(H_2 - H_1) \tag{3-22}$$

But H_1 is an arbitrary datum:

$$H_2 - H_1 = H$$

Substituting for T_2 from eq. (3-17),

$$\frac{p_1^{CZR}}{T_1} = \frac{p_2^{CZR}}{T_2} \tag{3-23}$$

$$T_2 = T_1 \frac{p_2^{CZR}}{p_1} \tag{3-24}$$

Substituting,

$$H = \frac{T_1}{C}\left[\frac{p_2^{CZR}}{p_1} - 1\right] \tag{3-25}$$

and

$$H = \frac{ZRT_1}{(n-1)/n}\left[\left(\frac{p_2}{p_1}\right)^{(n-1)/n} - 1\right] \tag{3-26}$$

The head (H) is a fundamental property of the compressor. It is a function of the compressor design and of the speed and is not affected by the nature of the fluid, its thermodynamic properties, or the addition or subtraction of heat as it flows through the machine. This immediately suggests rating compressors in terms of head, since it is a property that does not change for an existing compressor.

The physical dimensions of a compressor are directly connected to its performance in terms of head; also, classification of compressors according to hydraulic performance is based upon the head.[5]

3-5. Compressor Map

The compressor map, also called the head-capacity curve or the performance curve, is a graph of the head and the throughput of the machine. It is required for the design of the capacity and the anti-surge control systems. The head-capacity curve is a straight line for an idealized compressor, as shown in Fig. 3-6.

The hypothetical straight line becomes curved at low capacities because of what are commonly known as "shock" losses. These losses are the head losses at the impeller entrance and exit. The nature of the hydraulic loss at the impeller entrance is due to the sudden expansion or diffusion after separation. The loss at the impeller discharge is mostly caused by the high rate of shear between the low velocity in the casing and the high velocity in the impeller.

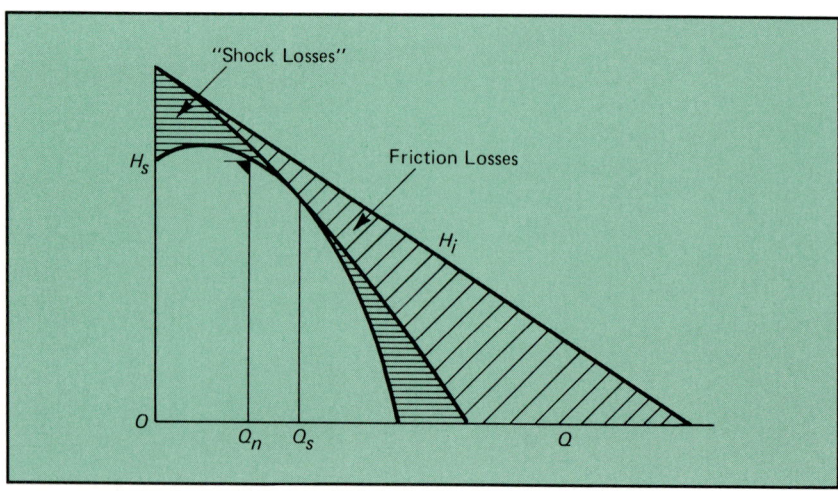

Fig. 3-6. Compressor Map Is Resultant of Subtracting Hydraulic Losses from Ideal Straight Line[5]

At high capacities, the straight line takes on curvature because of friction and diffusion losses. The general formula for friction loss is that the head loss is proportional to velocity squared. Using this concept, the head-capacity curve becomes a parabola with

$$H = Cq^2 \qquad (3\text{-}27)$$

at high throughputs.

An analytical relationship among the head (H), speed (s) and throughput (q) can be expressed:[5]

$$H = As^2 + Bsq + Cq^2 \qquad (3\text{-}28)$$

Here constants A, B, and C are chosen to fit the compressor characteristics. This equation has little practical usefulness, and the head-capacity curve is represented by regressions developed from actual data for computer analysis.[8]

Consider bringing the compressor shown in Fig. 3-7 on line. The compressor is brought up to operating speed by the variable speed electrical motor. The valve in the discharge line is wide open. Assuming no friction loss in either the inlet or outlet piping, both compressor inlet and outlet pressures are at atmospheric conditions. Equation (3-26) shows head to be

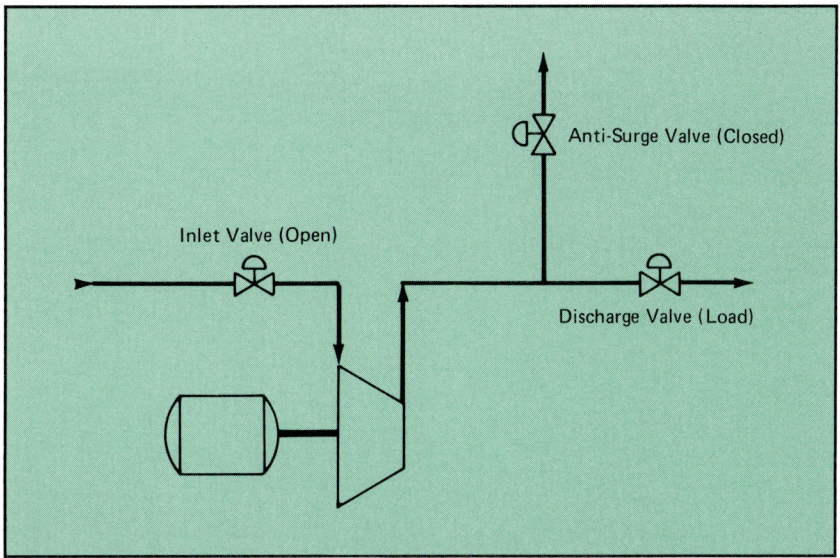

Fig. 3-7. Single-Stage Compressor with Typical Inlet and Anti-Surge Valves and a Discharge Valve Acting as the Variable Load

zero, and the compressor is operating at point 1 on the compressor map, Fig. 3-8. The discharge valve is then closed slightly. The discharge pressure increases to point 2, and the flow decreases only slightly. The vertical portion of the performance curve at high capacities is typical of all compressors. It is called the "stonewall" point, since reducing discharge pressure results in little increase in throughput.

Stonewalling occurs because the velocity approaches the sonic velocity of the gas somewhere in the compressor. This causes the gaseous fluid to change character. Shock waves result, which restricts the flow channels and causes an abnormal "falloff" in discharge pressure. The heavier the gas, the more severe the stonewalling problem. Improving the capacity limitation represented by the phenomenon depends on where in the compressor it occurs. A common location is at the impeller inlet. If so, a larger impeller will alleviate the problem.[9]

Closing the discharge valve even further causes discharge pressure to rise, while the greater resistance to flow causes throughput to decrease. This condition is shown by point 3, which is the point where the compressor would be designed to operate. As the valve is closed further, discharge pressure increases moderately, but throughput decreases significantly to where an aerodynamically unstable condition occurs in the compressor impellers. This instability can be extremely severe, with visible vibration and audible noise, and can significantly

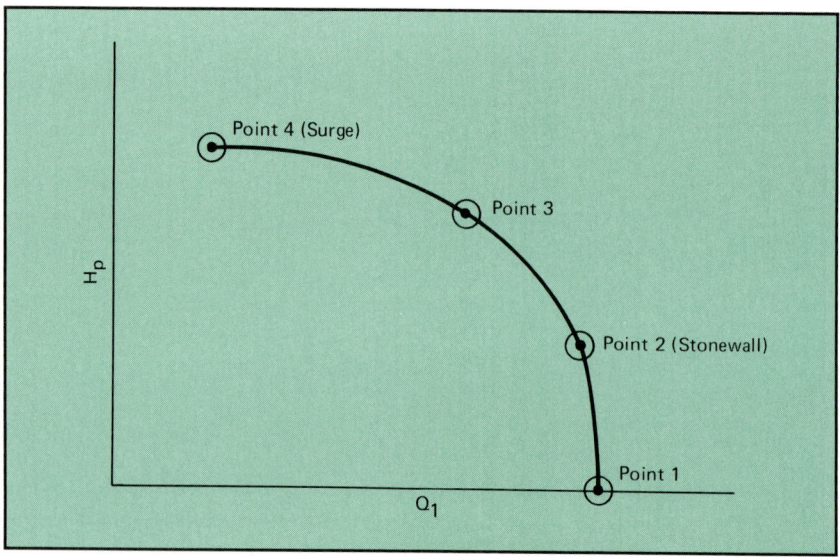

Fig. 3-8. Typical Performance Curve for a Centrifugal Compressor[10]

damage the compressor. This phenomenon is called "surge" and takes place at the point of highest head in the performance curve. Point 4 is the surge point.[10]

3-6. Surge[11,12]

Surge is an unstable operation that takes place at the peak of the performance curve, as shown in Fig. 3-8. It varies in intensity from an audible rattle to an ominous "womp" and from a minor vibration to a violent shock. Severe surge can cause the destruction of labyrinth seals, overload thrust bearings, cause the impellers to rub on the casings, and even bend the shaft. This damage, which can happen in seconds, will require the disassembly and rebuilding of the compressor.[13]

The phenomenon of surge can be qualitatively followed by reference to Fig. 3-9 with the extended operating curve HPDJGF.[11] Let the operating point of the compressor be moved from P to D. The load line OA is represented by a valve (as shown in Fig. 3-7). Closing the valve causes the load line to move from OA to OA' to OA". At point D, the compressor discharges more gas than the valve can pass, and pressure

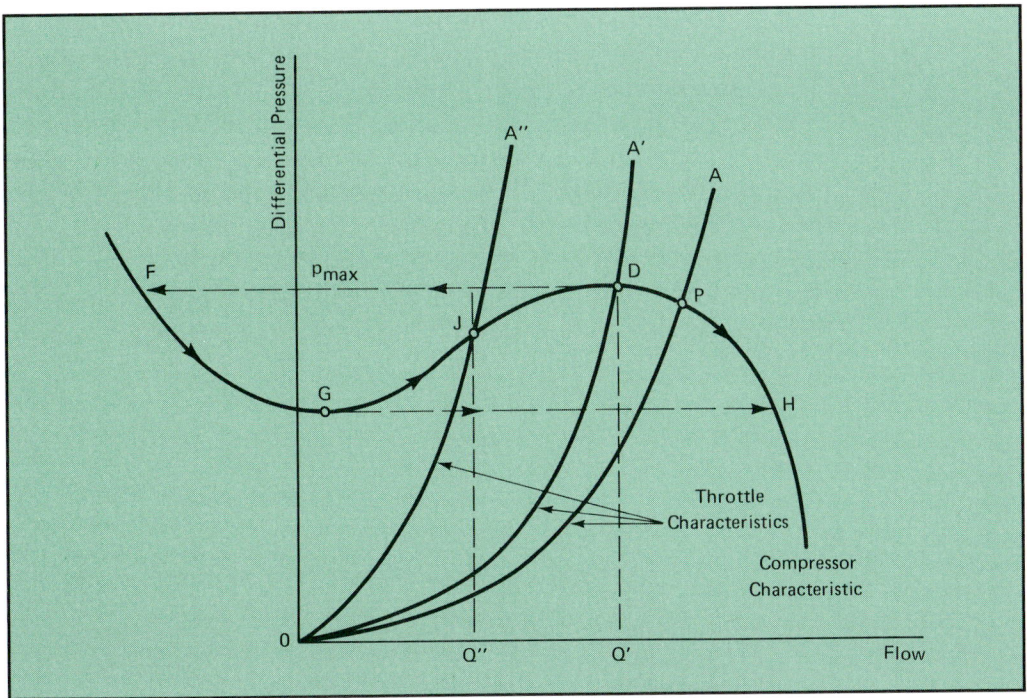

Fig. 3-9. Cyclic Surge Model for Centrifugal Compressor with Large Discharge Volume[11]

tends to rise. The pressure, however, cannot rise above D. The compressor cannot sustain this flow, and the operating point jumps to F (reverse flow). Operation will not go to point J because the capacitance of associated vessels and piping will prevent pressure from falling significantly during the short time period of the surge (as shown by Fig. 3-11). Further, point J in Fig. 3-9 is on a portion of the operating curve that has positive slope and is unstable (an increase in flow causes an increase in pressure, which further increases flow, and so on). The loud report, or "womp," is often associated with the "blowdown" as flow jumped from D to F.

The reverse flow at F causes the pressure to decrease along line FG. Forward flow at G is insufficient to make pressure rise, and flow again jumps to point H. The flow at operating point H is larger than the flow being passed by the load valve and pressure rises along HPD. Upon reaching point D, the process repeats itself.

Surge is seen to be a dynamic instability. It occurs in dynamic compressors. It can also occur in centrifugal pumps and blowers, but the occurrence is less frequent and the damage less severe. Blower surge will not damage the internal mechanism unless the pressure rise exceeds 2 psi and the blower size exceeds 150 bhp (brake horsepower). Small compressors or compressors with small outlet volumes will exhibit a compressor map cycle more like that of Fig. 3-10. The flow reversal between D and F (Fig. 3-9) involves a small inertia that will take a finite time. If the outlet volume is small, flow from the volume during that finite time will cause the pressure to drop. Thus, the pressure at point F will be lower than the pressure at point D, and the extent of the pressure loss will be a function of the magnitude of the volume and the impedance of the load. McMillan[12] shows the following relationship for defining the surge period:

$$T_s = \{2\pi \cdot \sqrt{(L_c \cdot V_p) A_c}\} a \qquad (3\text{-}29)$$

where:

a = speed of sound (ft/sec)
A_c = flow path cross-sectional area in compressor (sq ft)
L_c = flow path length in compressor (ft)
T_s = surge oscillation period (sec)
V_p = volume of the plenum (cu ft)

Unit 3: Thermodynamics of Compression

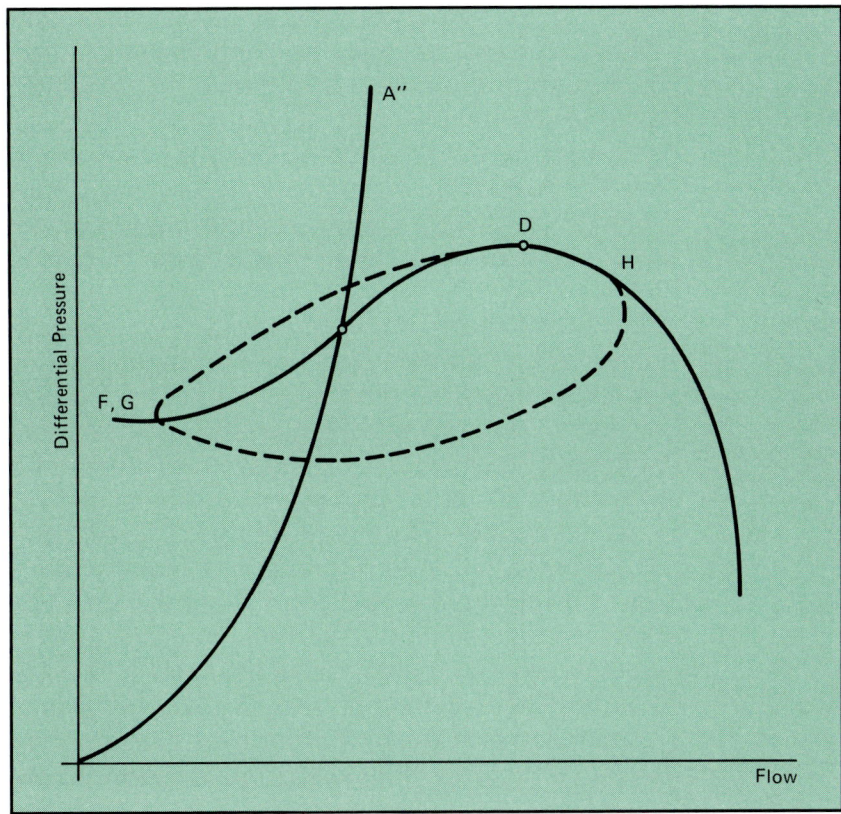

Fig. 3-10. Cyclic Surge Model for Centrifugal Compressor with Small Discharge Volume[11]

In addition to the phenomenon of surge, the positive slope of the performance curve results in a static instability in compressor operation. When operating on a portion of the curve with a positive slope, an increase in output pressure is seen as an increase in driving force across the load, which increases flow, which in turn again increases pressure and a positive feedback situation develops. A negative feedback causes pressure to decrease as flow increases, resulting in stability. This concept is of little practical interest, since surge prevents operation on any but the portion of the performance curve with negative slope.

The rapid flow reversals of surging can cause extensive radial vibration and axial thrust displacement. Further, the reheating of the same mass of gas during each surge cycle causes a large temperature increase. The thermal lags in sensors make this increase virtually impossible to measure. The rapid unloading of the impeller can cause overspeed of the compressor. Runaway speed can be attained so quickly that a speed controller (governor) cannot regulate it. The increase in vibration, thrust, temperature, and speed can cause extensive damage to the compressor in only a few seconds. Repair of the damage resulting from surge can cost hundreds of thousands of dollars, with additional comparable huge losses from the loss of production while the compressor is inoperative. Thus, controls for the prevention of surge are imperative on every compressor of large enough horsepower to damage itself.

Figure 3-11 shows a typical output pressure variation due to surge. Note the time scale. Flow through the compressor will typically drop from the design value to the minimum in 0.05 second. The detection of surge by flow measurement is preferred to detection by pressure measurement because pressure will not drop as precipitously as shown in Fig. 3-11 if the compressor discharge volume is large, and the oscillation amplitude in pressure is usually less than that in flow during surge. Typical process pressure and differential pressure transmitters, with time constants of 0.5 to 1.5 seconds, will attenuate variations such as that shown in Fig. 3-11 until they cannot be seen. McMillan[12] has made the comparison of flow transmitter time constants shown in Fig. 3-12. A diffused silicon electronic differential pressure

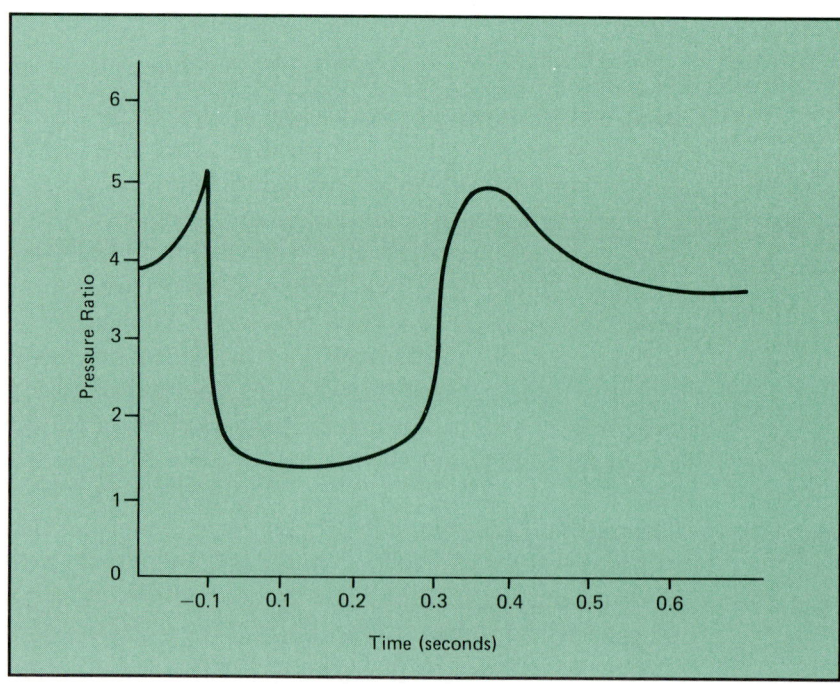

Fig. 3-11. Experimental Data Showing Period of Surge Cycle

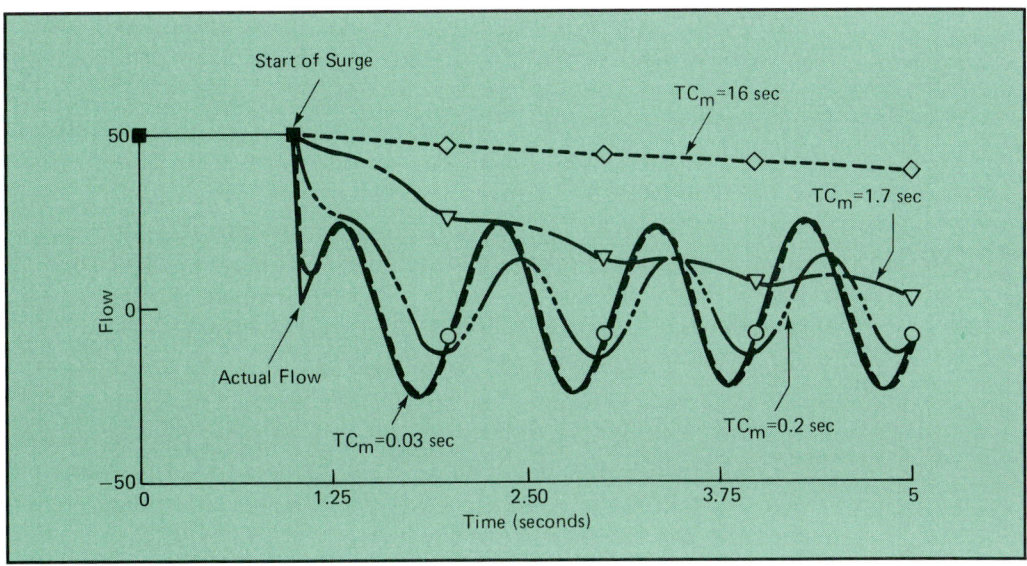

Fig. 3-12. Effect of Measurement Time Constant on the Recorded Surge Cycles[12]

transmitter is required to "track" the flow variations. Noise in the flow signal could be expected to mask the measurement, and material of construction of the transmitter is a limitation in a petrochemical plant.

The following conclusions summarize the discussion of the surge phenomena:

A. A negative compressor characteristic is necessary to insure stability.
B. Surge frequency is inversely proportional to the volume of the discharge system. The surge characteristic of a compressor with small associated volume is shown in Fig. 3-11.
C. An observer equipped with standard steady-state pressure and flow instrumentation would not detect surge but would observe only time averages.

3-7. System Load Curve[6,9,10]

The compressor is used to pump gas into a process. The characteristics of the process are very important in determining the choice of the compressor and in determining the control strategy. For example, centrifugal compressors are notorious for small rangeability, in which surge flow is typically two-thirds of stonewall flow. Thus, an analytical expression of the process characteristic must be added to the compressor map before the design of controls can proceed.

In general, there are two basic types of load curves, sometimes called demand loads, system resistance curves, or system curves. One type is a simple friction system, such as that found in mine ventilating, natural gas transmission, or a system of piping, exchangers, and vessels. This type of load curve is characterized by the typical square root curve, where the pressure drop varies as the square of volumetric flow. This load curve is shown as CD on Fig. 3-13.

A second type of load curve describes a process that operates at constant pressure or constant liquid head. Examples are liquid-full chemical reactors or aeration tanks for sewage, where the gas is blown into the bottom of the tank and bubbles up through the liquid. This process is characterized by curve AB on Fig. 3-13.

Actually, the most common type of load curve is a combination of the two foregoing basic types and takes the form AE on Fig. 3-13. For example, the liquid-full tank will always have piping (frictional resistance) in series with it, and

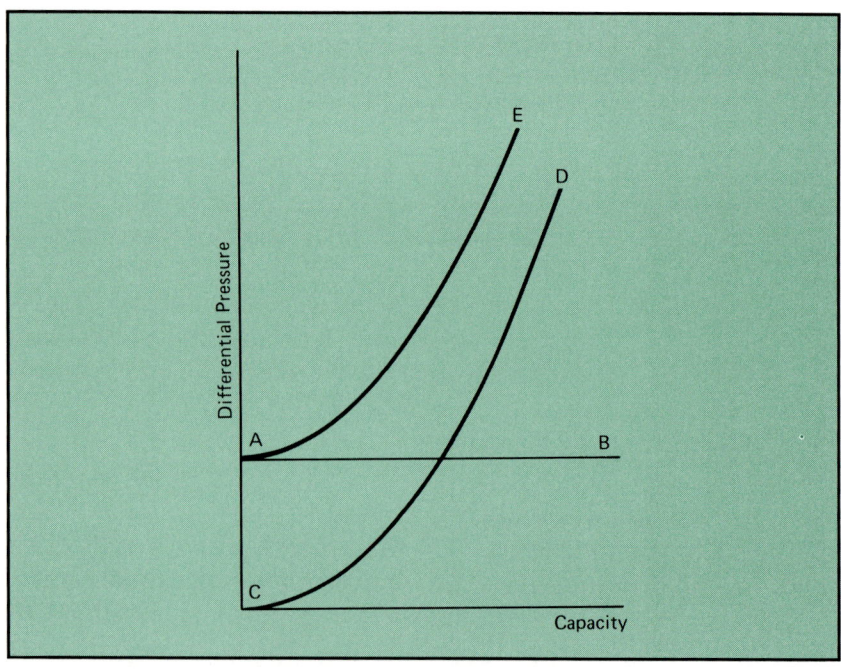

Fig. 3-13. Typical Curves Illustrating the Three Types of Compressor Loading[9]

all of the piping to a blast furnace will have a constant pressure in the furnace maintained by a pressure control system.

Figure 3-14 shows the load curve superimposed on a series of performance curves on the compressor map. Since the Affinity Laws, or the Fan Laws, state that both head and throughput are functions of compressor speed, each speed has its own performance curve, as shown. The load curve, a combination of both frictional and static resistance, transverses the family of performance curves. Since the compressor can operate only on the performance curve for a given speed at that speed, the operating point of the compressor is determined by the superposition of the two curves. Thus, for speed s_2, the system will operate at point A. Should speed fall to s_1, operation would be at point B.

The process characterized by the load curve is itself a dynamic element that has its own control loops manipulating control valves to bring variables to their set points. Changing valve positions can change the system resistance, and thus

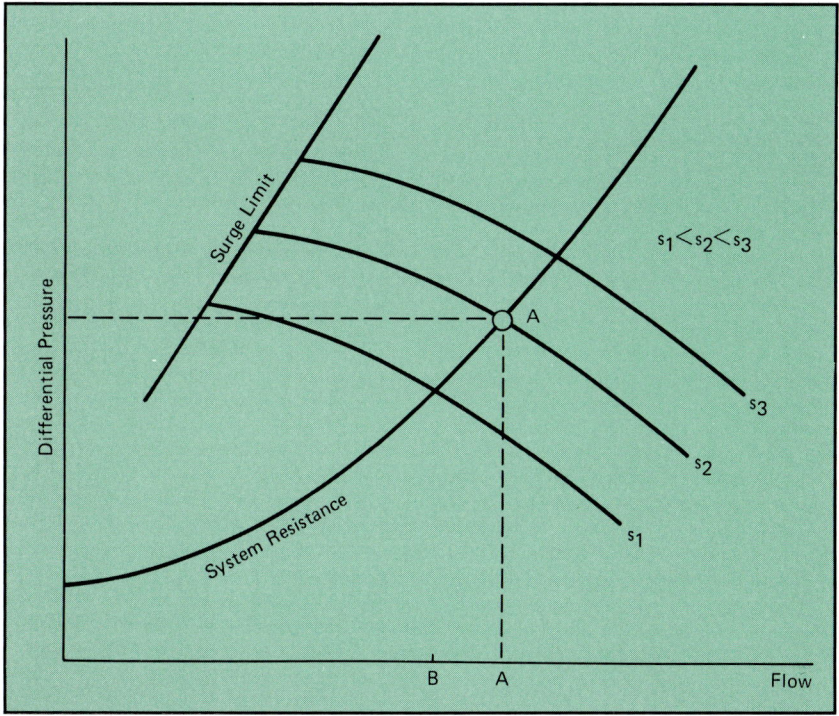

Fig. 3-14. Typical Performance Curves and System-Resistance (Load) Curve for Compressor Rated at A[10]

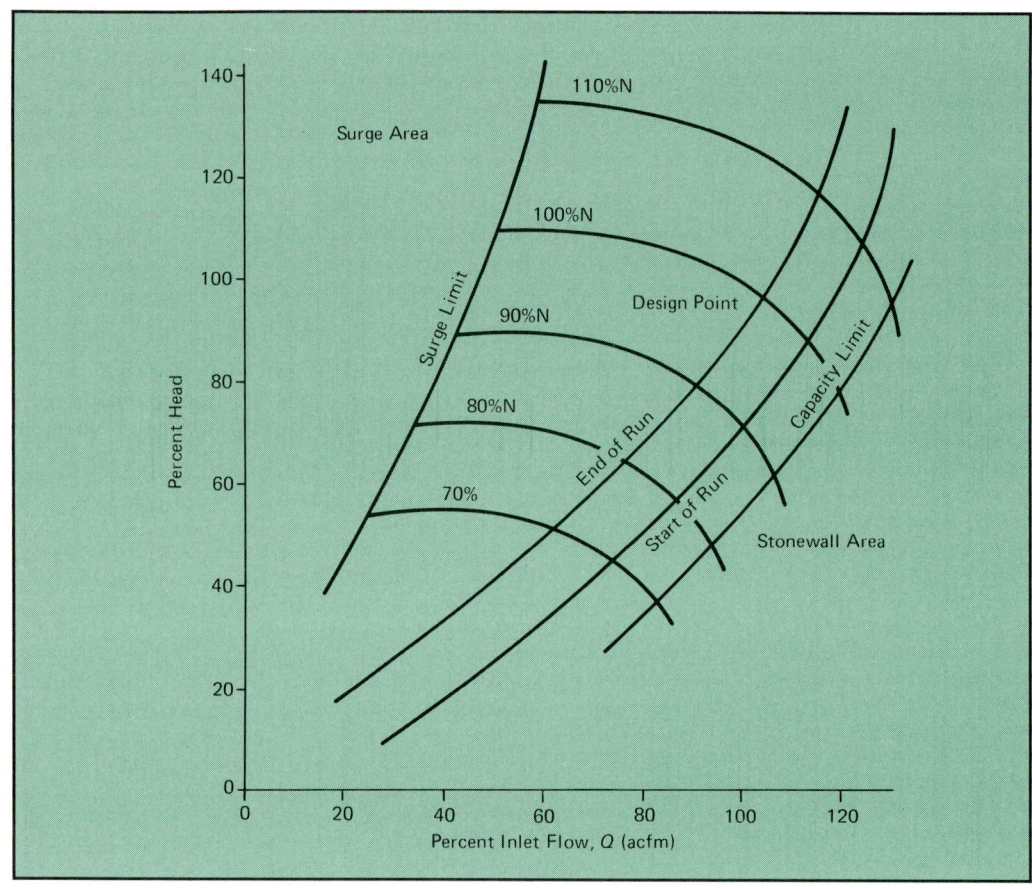

Fig. 3-15. Load Curves on a Compressor Map for a Reformer with Changing System-Resistance Characteristics from Start to End of Run[6]

change the load curve. Nisenfeld[6] has used the ammonia oxidation process (AOP) to illustrate this phenomenon. The result is shown in Fig. 3-15, where the load curve changes from the start of the run (SOR) to the end of the run (EOR).

3-8. Compressor Map Parameters

The compressor actually operates on polytropic head and actual inlet volumetric flow. The head is traditionally plotted as the y-axis of the compressor map and flow as the x-axis. For various reasons, these variables are replaced by associated variables in the actual plotting of the map. Polytropic head cannot be measured directly and is replaced by pressure ratio, pressure rise, and, for air compressors, discharge pressure. Inlet volumetric flow is replaced by discharge flow (because of the difficulty of measuring inlet flow) or by mass flow (because that variable is needed elsewhere in the process).

Polytropic head is a function of the mechanical design of the compressor and not of the thermodynamic relationships. Thus, it becomes a parameter for a given compressor, and eq. (3-26) can be solved for the compression ratio

$$R_c = \left[\frac{(M)(H)}{(1545)(Z)(T_i)\left(\frac{k}{k-1}\right)} + 1 \right]^{\frac{k}{k-1}} \quad (3\text{-}30)$$

And by definition

$$R_c = \frac{p_2}{p_1} \quad (3\text{-}31)$$

$$\Delta p = p_2 - p_1 \quad (3\text{-}32)$$

Combining eqs. (3-31) and (3-32)

$$\Delta p = p_1(R_c - 1) \quad (3\text{-}33)$$

Pressure rise (Δp) is a popular variable for the y-axis of the compressor map since it forms the basis for one of the widely applied anti-surge control systems and because it is easily measured. Substitution of eq. (3-30) into eq. (3-33) results in a messy equation that denotes the thermodynamic variables to which Δp is sensitive.

Compressor capacity, shown on the x-axis of the map, can be graduated in terms of inlet acfm, discharge acfm, or mass flow (lb/hr). Inlet volumetric flow is calculated from

$$q = K\sqrt{\frac{h\, T_i}{p_i\, M}} \quad (3\text{-}34)$$

Equation (3-34) represents a flow measurement by a head type flow element.[14] Measurement is accomplished by the use of a differential pressure transmitter, which measures the head (h) across an obstruction in the pipeline. The pipe is typically very large, and the straight piping runs required by an orifice installation are prohibitively expensive. An alternative is straightening vanes and a flow tube.[12] An averaging type of primary element has the advantages of being less sensitive to the skewed velocity distribution and the noise generated by

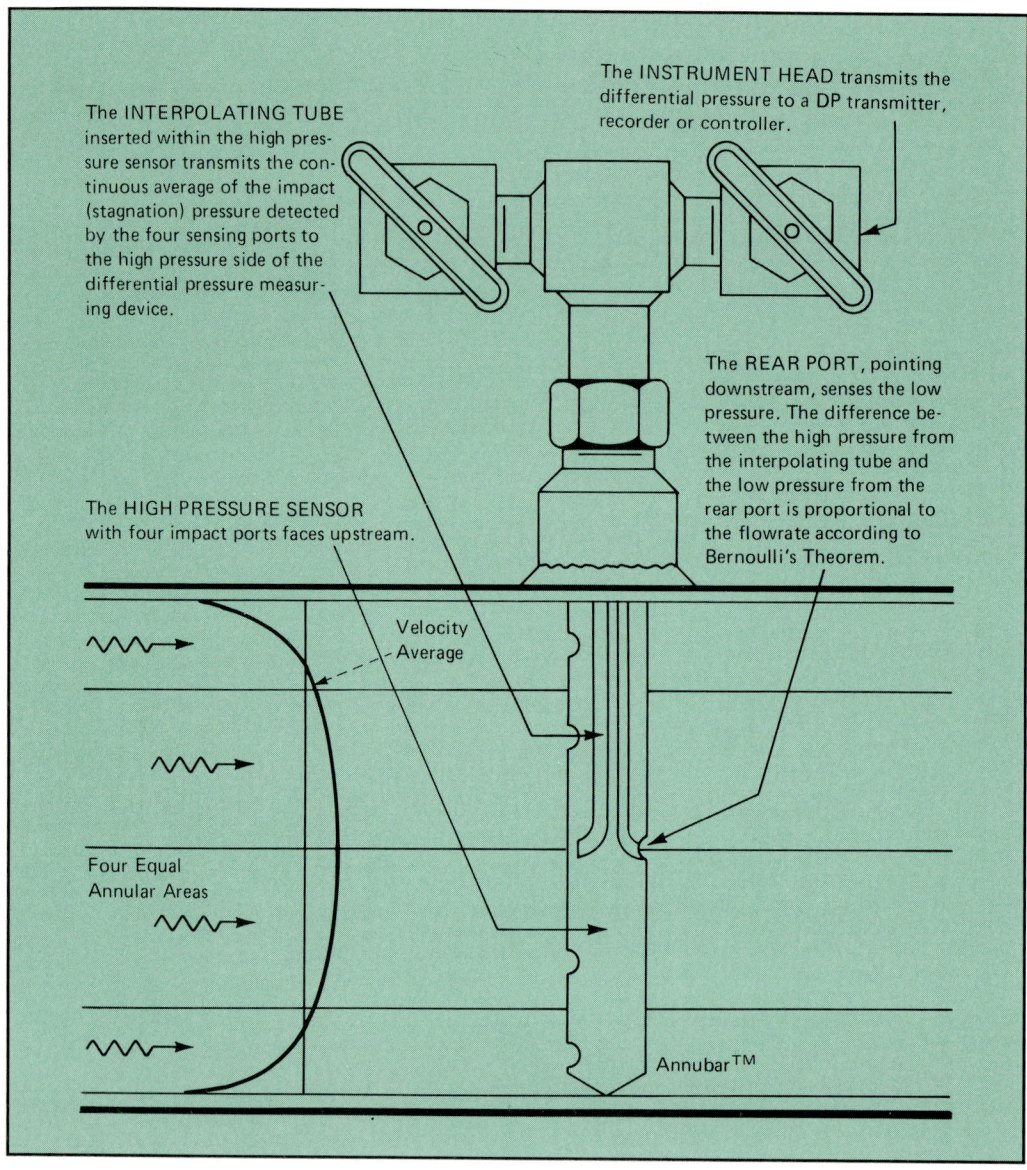

Fig. 3-16. Annubar™ Averaging Flow Sensor (Courtesy Dieterich Standard Corp.)

less than minimum required straight runs. One type of averaging element, the Annubar,* is shown in Fig. 3-16. Flow coefficients have been developed for the Annubar installed immediately downstream from a 90-degree elbow. Annubar

---

* Dieterich Standard Corporation, Subsidiary of Dover Industries, Box 9000, Boulder, Colorado, 80301.

Unit 3: Thermodynamics of Compression

size is chosen from the following

$$q = C' \sqrt{h/\rho} \qquad (3\text{-}35)$$

where:

$C' = (358.94)(K)(d^2)(F_{RA})(Y_A)(F_{AA})(F_e)$
K = Annubar flow coefficient
d = inside pipe diameter, in.
F_{RA} = Reynolds number correction factor
Y_A = Annubar expansion factor
F_{AA} = pipe and Annubar thermal expansion factor
F_e = elevation correction factor (corrects for local gravity)

Actual inlet volumetric flow is related to mass flow as follows:

$$m = \frac{(144)(60)(q\ acfm)(M)(p_f)}{(1545)(T_f)(Z_f)} \qquad (3\text{-}36)$$

The mass flow (m) is the same whether measured upstream or downstream. Since head-type flowmeters are inherently volumetric devices, the measured fluid velocity can be calibrated to mass flow for only a single value of pressure and temperature for a compressible fluid. The flowmeter is almost always compensated for pressure and temperature in order to provide an accurate measurement under varying conditions.[14]

Should the pressure rise (Δp) be plotted as the y-axis of the compressor map and volumetric flow rate (q) be plotted as the x-axis, then both variables are seen to be functions of the thermodynamic variables pressure, temperature, and molecular weight (per eqs. (3-30), (3-33), and (3-34)). Thus, the curves plotted on the compressor map will move with changes in these variables. Waggoner[15] and McMillan[12] have illustrated that movement as shown in Fig. 3-17.

Figure 3-17a shows the surge curve to move to the right on the compressor map with *decreasing* inlet pressure, as would be expected from eqs. (3-33) and (3-34). Equation (3-33) shows pressure rise (Δp) to decrease with decreasing inlet pressure, while eq. (3-34) shows actual volumetric flow to increase with decreasing inlet pressure. Thus, the curve moves to the right. Movement in this direction is not conservative and is dangerous, since surge occurs at a higher throughput flow rate.

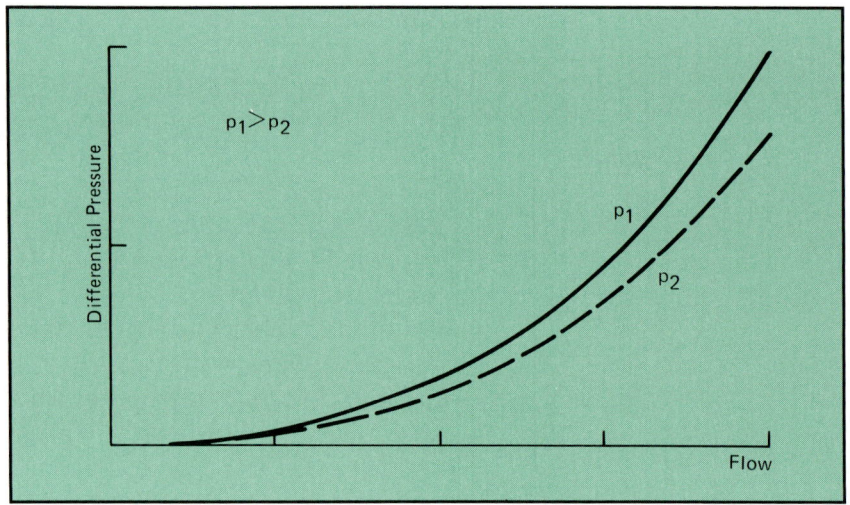

Fig. 3-17a. Change in Surge Curve due to Change in Inlet Pressure[12]

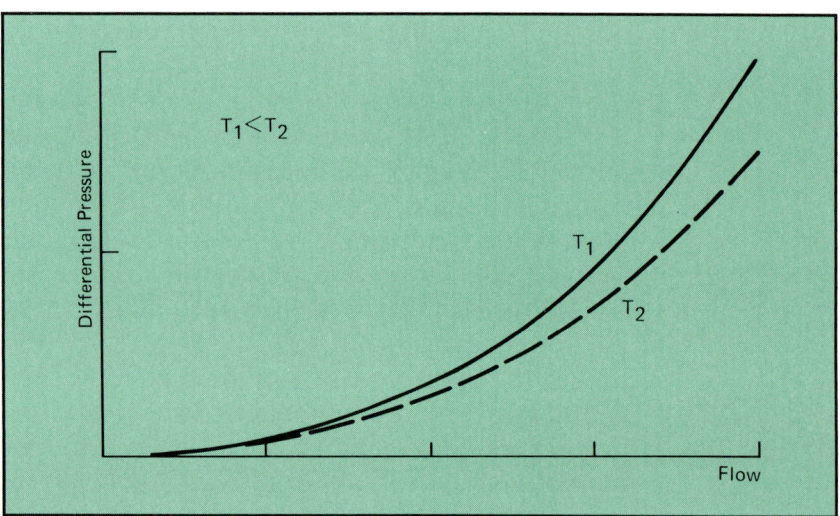

Fig. 3-17b. Change in Surge Curve due to Change in Inlet Temperature[12]

Figure 3-17b shows the surge curve to move to the right with an *increase* in inlet temperature. Again, the surge curve is now closer to the operating point and is less conservative. The movement is defined by combined eqs. (3-33) and (3-30), and eq. (3-34). By similar reasoning, Fig. 3-17c shows the surge curve to move to the right with decreasing molecular weight, and Fig. 3-17d shows it to move to the right with decreasing specific heat ratio.[12]

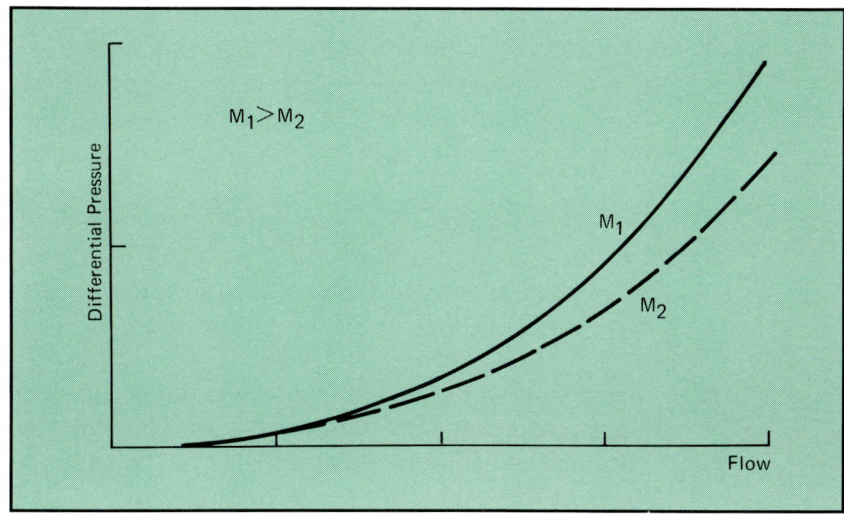

Fig. 3-17c. Change in Surge Curve due to Change in Gas Molecular Weight[12]

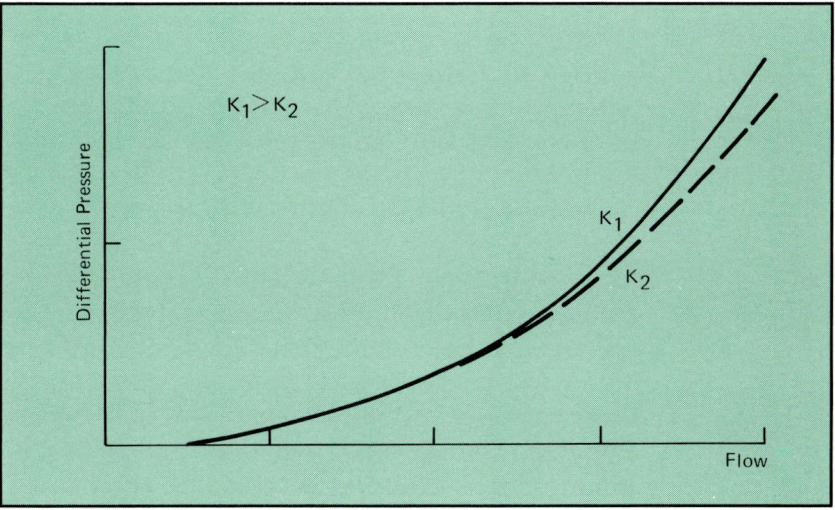

Fig. 3-17d. Change in Surge Curve due to Change in Gas Specific Heat Ratio[12]

3-9. Use of the Compressor Map

The single-speed compressor map, shown as Fig. 3-8, will typically have an operating range (between surge and stonewall) of about one-third of the stonewall capacity. Therefore, rangeability is severely limited. Great care must be exercised in selecting a compressor that is physically capable of meeting the most adverse operating conditions (load line) below the stonewall capacity, yet meets most of the capacity requirements without operating in the surge region. Similar care must be exercised in selecting the driver, since an underpowered driver will cause the stonewall capacity to move to the left on the map, further reducing the rangeability.

A control system, called an anti-surge system, is required to prevent the compressor from operating in the surge region. Surge is almost always prevented by opening a control valve in a recycle line around the compressor, even though surge can also be prevented by reducing the speed. The point at which this valve begins to open is chosen from the compressor map. Guidelines used for design are a 5% surge margin (the capacity at which the valve begins to open is 105% of the capacity at which surge takes place) for compressors that will not be substantially damaged by surge, an 8% margin for the typical compressor, and up to a 20% margin for compressors where surge must be prevented at all costs. Regrettably:

A. the optimum surge margin depends on the dynamics of the compressor and its associated piping, and
B. the only point "guaranteed" on the compressor map by the manufacturer is the design point, and the surge point can admittedly be substantially in error.[16]

As a result, the control system that protects a multimillion dollar machine from destruction is designed on shaky data and a generalized guideline. This leads to overconservatism. But, overconservatism is expensive, since opening the anti-surge valve when not necessary is wasting energy. The alternatives are to experimentally find the surge point on the compressor as it is commissioned in the field (a dangerous procedure in itself) and to simulate the compressor on a computer to determine the required surge margin.[6,17,18,19]

The surge point is one point on a surge curve generated by changing the speed of the compressor. The curve is assumed to be a parabola on the basis of the Affinity Laws:

$$q = K_1 s \qquad (3\text{-}37)$$

$$H = K_2 s^2 \qquad (3\text{-}38)$$

and

$$H = K_3 q^2 \qquad (3\text{-}39)$$

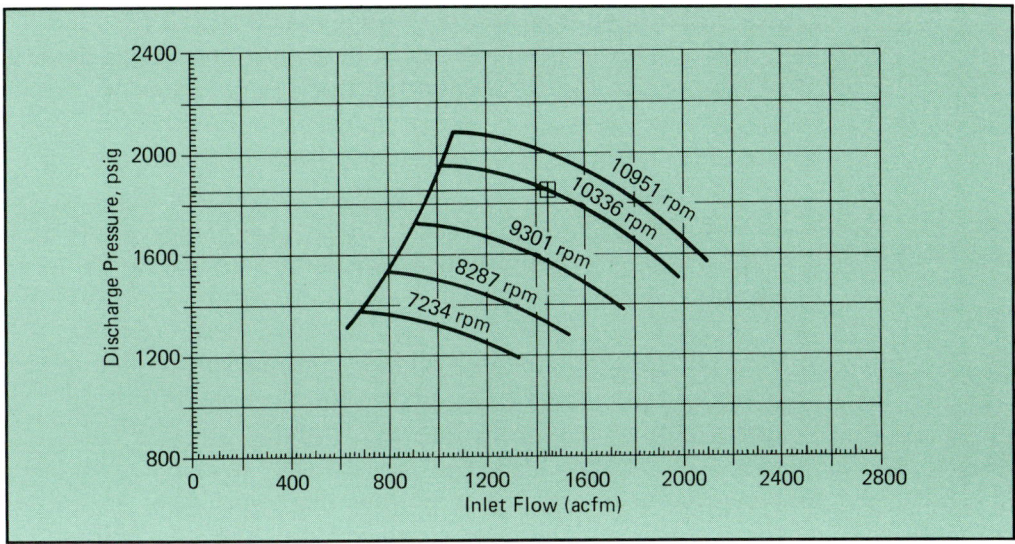

Fig. 3-18. Typical Compressor Map as Supplied by Manufacturer

Actually, this is a gross assumption because the head (H) is not the variable plotted, but it is replaced by the pressure rise (Δp), introducing all the nonlinearities of eqs. (3-30) and (3-33).

However, the complete surge curve, down to the origin of the compressor map, is needed to assess the possibility of surge on start-up and shutdown. The probability of serious surge on starting up and shutting down is increased with multistage compressors, where the stonewalling or underpressure of a given stage can drive its neighbors into surge. A compressor map for each stage of a multistage compressor is required for the design of the control system.

In summary, a compressor map showing the performance curve, the design operating point, the load line, and the stonewall point serves as basic data for the design of the control system. Unfortunately, necessity requires design on the basis of far more rudimentary information, such as that shown in Fig. 3-18, in many cases.

Exercises

1. *Hydrogen expands during a polytropic process from a state of 125 psia and 9.6 ft³ to atmospheric pressure and 48.25 ft³. Calculate n.*

2. The latent heat of water at 164°F is 1000 Btu/lb (778160 ft lb/lb). Calculate the increase in entropy in changing 1 lb of water from liquid to steam at this temperature.

3. A gas is compressed adiabatically from 14.7 psia and 100°F to 250 psia. Calculate to final temperature.

4. Does surge frequency increase or decrease with larger discharge system volume?

5. What is the sequence of events if compressor discharge pressure increases while operating on a portion of the performance curve with negative slope?

6. With regard to Fig. 3-8, why does the compressor go into the condition known as "stonewalling" between points 1 and 2?

7. Why doesn't the compressor characteristic follow line DG in Fig. 3-9?

8. How would a pressure transmitter with a time constant of 0.01 second track the oscillations in Fig. 3-12?

9. What are the limitations of the compressor map in Fig. 3-18 from the viewpoint of control system design?

References

1. Lichty, L. C., *Thermodynamics*, McGraw-Hill Book Co., New York, NY, 1948.
2. Faires, V. M., *Thermodynamics*, Fourth Edition, The MacMillan Co., New York, NY, 1962.
3. Kieffer, P. J., G. F. Kinney, and M. C. Stuart, *Principles of Engineering Thermodynamics*, John Wiley & Sons, Inc., New York, NY, 1930.
4. Wislicenus, G. F., *Fluid Mechanics of Turbomachinery*, McGraw-Hill Book Co., New York, 1947.
5. Stepanoff, A. J., *Turboblowers*, John Wiley & Sons, Inc., Publishers, New York, 1955.
6. Nisenfeld, A. E., *Centrifugal Compressors, Principles of Operation and Control*, Instrument Society of America, Research Triangle Park, N.C., 1982.
7. Jennings, B. H., and E. F. Obert, *Internal Combustion Engines*, International Textbook Co., Scranton, PA, 1943.
8. Anon., "Statistical Response Model," Applied Automation, Bartlesville, OK 74004.
9. Anon., *Compressed Air and Gas Handbook*, Compressed Air and Gas Institute, New York, NY, 1966.
10. Lapina, R. P., *Estimating Centrifugal Compressor Performance*, Gulf Publishing Co., Houston, Texas, 1982.

11. Dussourd, J. L., G. Pfannebecker, and S. K. Singhania, "Considerations for the Control of Surge in Dynamic Compressors Using Close Coupled Resistances," 1976, Joint Gas Turbine and Fluids Engineering Divisions Conference, American Society of Mechanical Engineers, United Engineering Center, New York, NY, 1976.
12. McMillan, G. K., *Centrifugal and Axial Compressor Control*, Instrument Society of America, Research Triangle Park, NC, 1983.
13. Scheel, L. F., *Gas and Air Compression Machinery*, McGraw-Hill Book Co., New York, NY, 1961.
14. Moore, R. L., "Flow Measurement," *Measurement Fundamentals*, Basic Instrumentation Lecture Notes and Study Guide, Third Edition, Instrument Society of America, Research Triangle Park, NC, 1982.
15. Waggoner, R. C., "Process Control for Compressors," *Centrifugal Compressor Operation and Control*, Instrument Society of America, Research Triangle Park, NC, 1976.
16. Boyd, D. M., "How Accurate are the Surge Curves?," Proceedings of the Fifteenth Annual ISA Chemical and Petroleum Instrumentation Symposium, San Francisco, CA, 1974, Instrument Society of America, Research Triangle Park, NC.
17. Franks, R. G. E., *Modeling and Simulation in Chemical Engineering*, John Wiley & Sons, New York, NY, 1972.
18. Davis, F. T., and A. B. Corripio, "Dynamic Simulation of Variable Speed Centrifugal Compressors," *Instrumentation in the Chemical and Petroleum Industries*, Volume 10, Instrument Society of America, Research Triangle Park, NC, 1974.
19. Schultz, H. M., R. K. Miyasaki, T. B. Leim, and R. A. Stanley, "Analog Simulation of Compressor Systems, A Straightforward Approach," *Instrumentation in the Chemical and Petroleum Industries*, Volume 10, Instrument Society of America, Research Triangle Park, NC, 1974.

Unit 4:
Prime Movers and Ancillary Systems

UNIT 4
Prime Movers and Ancillary Systems

While control system design is largely dependent on the compressor map and thermodynamic relationships, it also must accommodate the characteristics of the driving device. Further, information is available from the ancillary systems that are a part of the compressor, and the availability of that information must be considered in the design of the control system.

Learning Objectives — When you have completed this unit, you should:

A. Be familiar with the prime movers available to drive compressors.

B. Have some knowledge of the characteristics of the various drivers.

C. Understand the operation of speed governors.

D. Be aware of the ancillary systems that are part of the compressor.

4-1. Prime Movers

A prime mover has been defined as a mechanism that converts heat, hydraulic, fuel, or electrical energy into mechanical power. For the purposes of rotating machinery such as centrifugal compressors, this includes steam and gas turbines, gasoline and diesel engines, and electrical motors. These prime movers, varied as they are in construction and in principle of operation, have at least two things in common: first, they derive their power from the flow of some energy medium; and, second, they are rotating machines.[1]

The purchaser of a prime mover is primarily interested in its economy, efficiency, and service life. Control system design requires knowledge of both steady-state and dynamic operating characteristics. This sort of information is not readily available. Considerable effort will be required in securing such data and in enlisting the cooperation of prime mover manufacturers in providing it.

Centrifugal compressor drives are primarily concerned with rotating machines in which masses are forced to follow circular paths, rather than straight lines. Rotating machinery is described by the following analytical parameters and variables for the purpose of control system design.

- *Torque (A)*—The measure of turning effort exerted; proportional to the product of the tangential force and the length of the lever or crank arm on which it acts.
- *Radius of gyration (K)*—The distance from the axis at which all the area could be located and have the moment of inertia remain the same.
- *Mass (m)*—The mass of the rotating members.
- *Speed (s)*—The speed of rotation.

The *moment of inertia* is illustrated by the disk of moment of inertia mK^2 shown in Figure 4-1. The disk is attached to a shaft with a known torsional stiffness. Consider the twisting motion of the disk under the influence of the force (F) and the moment arm (r) as it moves through the angle (ϕ). Then,

$$\text{Torque (A)} = \text{Force (F)} \times \text{Moment arm (r)}$$

and by Newton's Law for a rotating body:[2]

$$\text{Torque} = \text{Moment of inertia} \times \text{Angular acceleration}$$

$$A = (mK^2)\left(\frac{d^2\phi}{dt^2}\right) = (mK^2)\left(\frac{ds}{dt}\right) \qquad (4\text{-}1)$$

This relationship among torque, moment of inertia, and acceleration is of vital importance in predicting control performance, since it permits calculation of acceleration, or rate of change of speed, of a prime mover-compressor combination for which the moment of inertia is known. The value of the moment of inertia for each mechanical device should be available from its manufacturer.

The moment of inertia is sometimes known as the "flywheel effect". Any difference between torque developed by the prime mover and the torque required by the compressor must go into acceleration or deceleration of the machine. For a given torque difference, positive or negative, the acceleration is inversely proportional to the moment of inertia. Thus, speed will change half as fast if the inertia is twice as large.

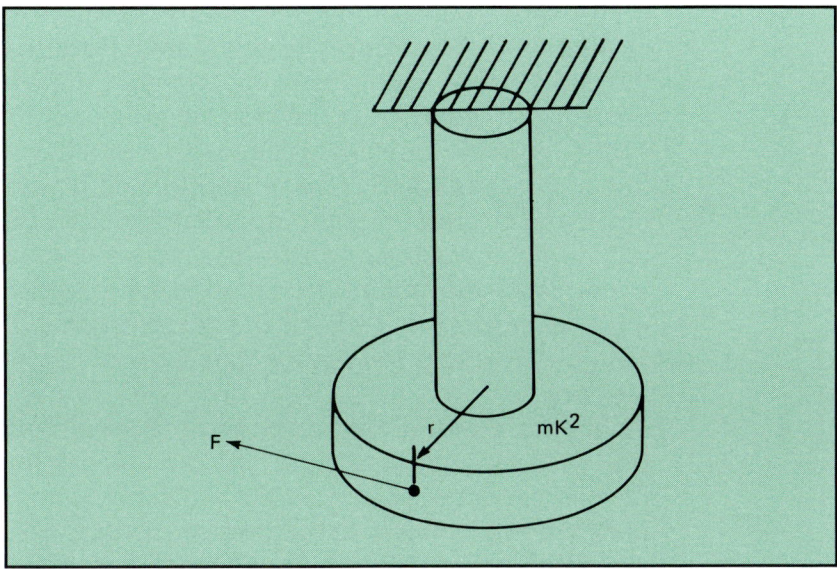

Fig. 4-1. Disk with Moment of Inertia (I) Attached to Shaft with Torsional Stiffness (k) Twisted with Torque (Fr)

Therefore, a greater moment of inertia allows more time for the throttle to move in the corrective direction and permits maintaining speed with less deviation. A flywheel, obviously, purposely increases the moment of inertia and is sometimes added to provide speed stability. Conversely, additional inertia can slow the speed response of the machine to where it cannot accommodate the load requirements of the compressor, making it uncontrollable.

The moment of inertia has a profound effect on the time required to start up and shut down (coasting) the compressor.

When starting up a compressor powered by a single-speed synchronous electrical motor, the motor windings represent a short circuit until rotation approaches design speed and a "back emf" (electro-magnetic force) is developed. The short circuit causes all of the electrical power to turn into heat, rapidly raising the temperature on the windings. The manufacturer will invariably provide a high temperature switch on the windings, which will disconnect power from the motor when actuated by high temperature. Should the moment of inertia be too high, the motor will not be able to get the compressor up to speed before it loses power because of high temperature. Thus, the compressor cannot be started.

Conversely, suppose that the compressor has a small moment of inertia but feeds into a large volume. Beyond the volume is the check valve that protects the system from backflow. The anti-surge (recycle) valve, which relieves pressure in the volume, opens (too) slowly. Since the impeller wheels have a low moment of inertia and remain loaded by the persistent downstream pressure, they stop very quickly. But, with remaining downstream pressure, flow reverses, and the wheels are driven with rotation in the opposite direction. Reverse rotation causes the loads on the thrust bearings to reverse, with the potential for bearing damage.

Figure 4-2 shows a multistage compressor driven by a turbine that has large volumes (pipe and vessels) before the downstream check valve, which prevents reverse flow. This installation is subject to the reverse rotation described in the preceding paragraph.

Possible mismatches between driver and compressor show the necessity for a coordinated design. Computer simulations* are available to assure compatability as purchase specifications are written.

Kinetic energy is energy stored in the rotating mechanism by means of its rotating motion. It is proportional to the moment of inertia and the speed of rotation squared:

$$KE = (mK^2)(\omega^2) \qquad (4\text{-}2)$$

A large moment of inertia results in a smaller speed change to absorb a given amount of excess energy. Similarly, additional load will cause less speed change with a large moment of inertia.

A final useful concept is that of energy expended per unit time, or power. *Horsepower* can be defined in terms of torque (A) and speed (s).

$$bhp = \frac{As}{33000} \qquad (4\text{-}3)$$

---

*CRC Bethany International, Inc., P.O. Box 3227, Houston, Texas 77001

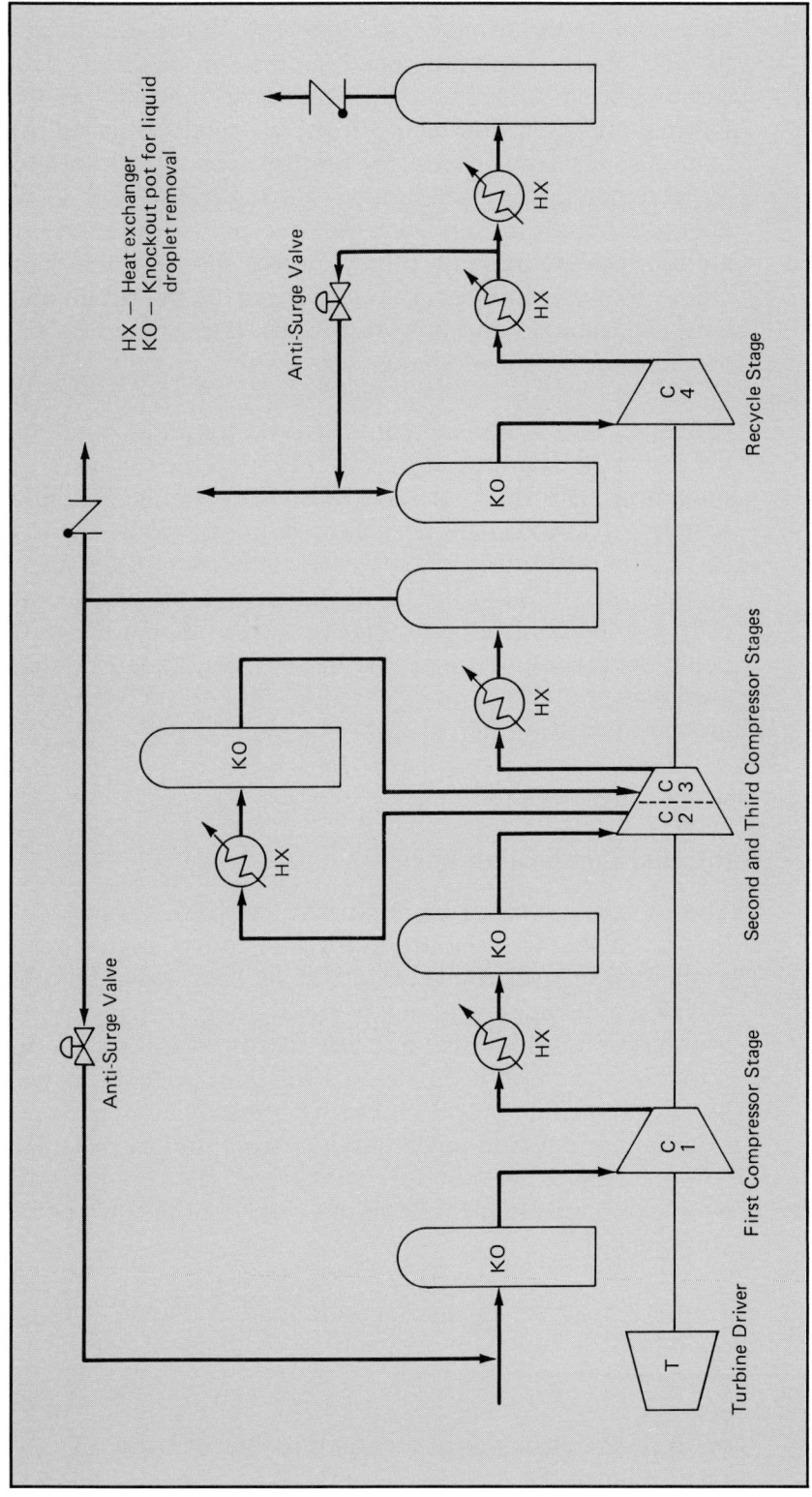

Fig. 4-2. Multistage Compressor with Large Volumes between Third and Fourth Stages and their Respective Check Valves

The brake horsepower (bhp) is the mechanical shaft power delivered by the driver. Equation (4-3) shows that torque is proportional to power if speed is constant. When a driver is running at equilibrium at a given speed, the required torque (corresponding to the load power) is exactly equal to the compressor output power (or torque). The output torque is equal to the developed torque resulting from the flow of energy medium to the driver less the power required by driver auxiliaries, friction, and other internal losses of the machine. Under these conditions, the net torque, or the difference between the load and output torques, is zero, and no acceleration or speed change can result.

However, when the load (the required torque) is suddenly decreased, the output torque will remain momentarily unchanged, resulting in an excess torque being available, which will produce an acceleration in accordance with eq. (4-1). The magnitude of the speed increase will depend on the moment of inertia of the rotating components and the ability of the speed control system (governor) to match the power being developed to match the new demand. During the transient, the excess energy will go into an increased kinetic energy of rotation (eq. (4-2)).

4-2. Prime Mover Characteristics

Internal Combustion Engines[3]

The internal combustion engine differs from steam-actuated drivers in that fuel is burned directly in the engine and not in a separate furnace. Because of this feature and a high thermal efficiency, it represents one of the lightest (in weight) power-generating units known. For this reason, it finds wide usage in portable or mobile vehicles—all the way from the power lawn mower to airplanes and including portable compressors. The internal combustion engine also powers remote or isolated compressors, where the fuel supply can be stored nearby or even extracted from the pipeline carrying the fluid being compressed.

Figure 4-3 shows an internal combustion engine. The cycle of operations is typical of all reciprocating spark-ignition engines:

- An intake stroke draws a combustible mixture into the cylinder of the engine.

Unit 4: Prime Movers and Ancillary Systems

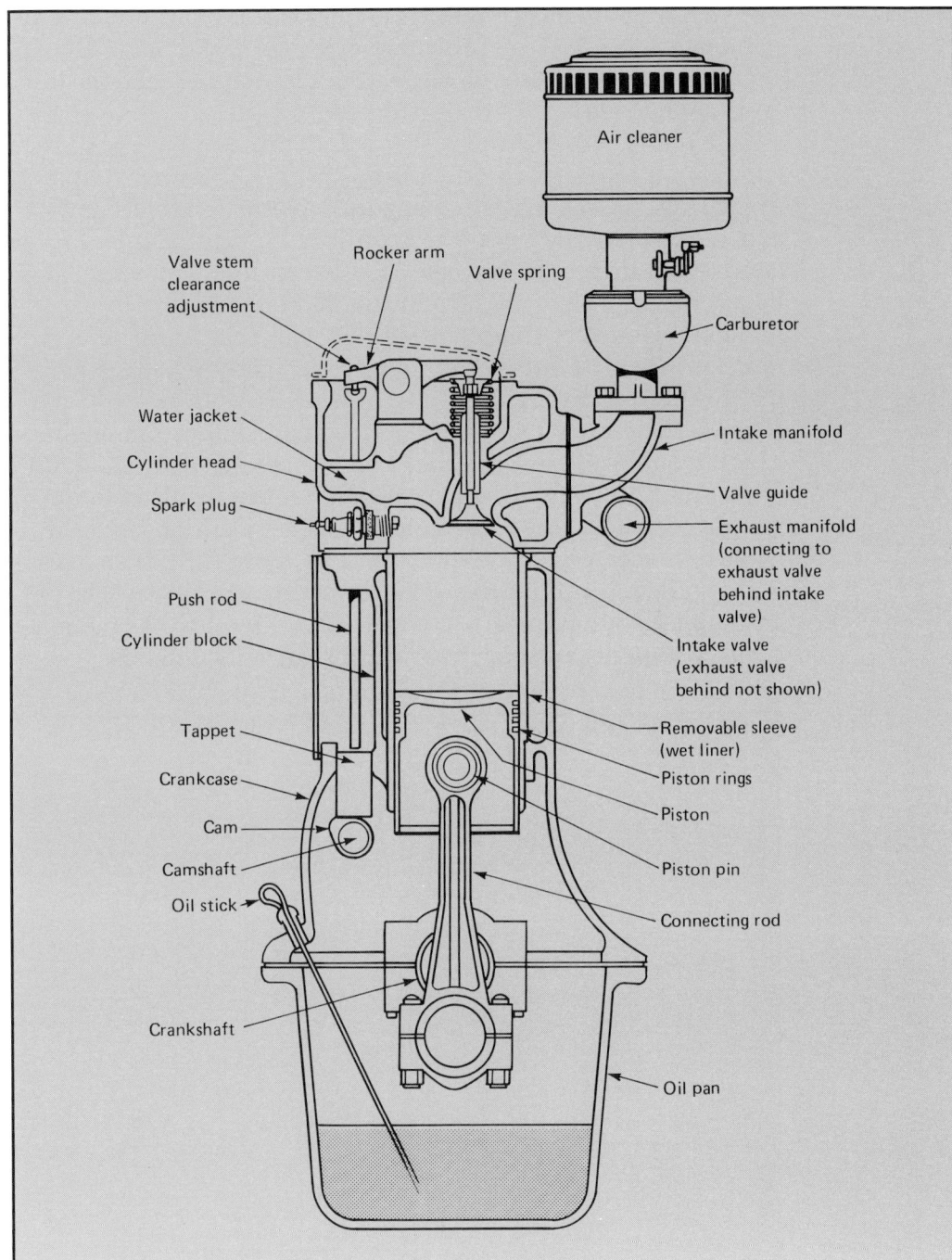

Fig. 4-3. Cut-Away View of Spark Ignition Internal Combustion Engine[4]

- A compression stroke raises the temperature and pressure of the mixture.
- A power stroke, through which the ignition and combustion of the mixture drives the piston downward.
- An exhaust stroke sweeps the cylinder free of burned gases.

The speed of a reciprocating engine driver is relatively low, and speed-increasing gears are required for centrifugal compressor drives. However, speed is continuously variable by a throttle, which can be adjusted either manually or remotely. One very important steady-state characteristic is the manner in which output torque varies as speed changes with the throttle remaining fixed. Figure 4-4 shows this characteristic for a spark-ignition internal combustion engine. The torque decreases with increasing speed, providing a kind of self-regulation to the driver. That is, if the required torque remains constant, the increment of torque resulting from the change in speed is in the direction to oppose further change. Self-regulation in the driver makes stable control easier to accomplish, but automatic control will be required to maintain speed within acceptable limits during large load changes.

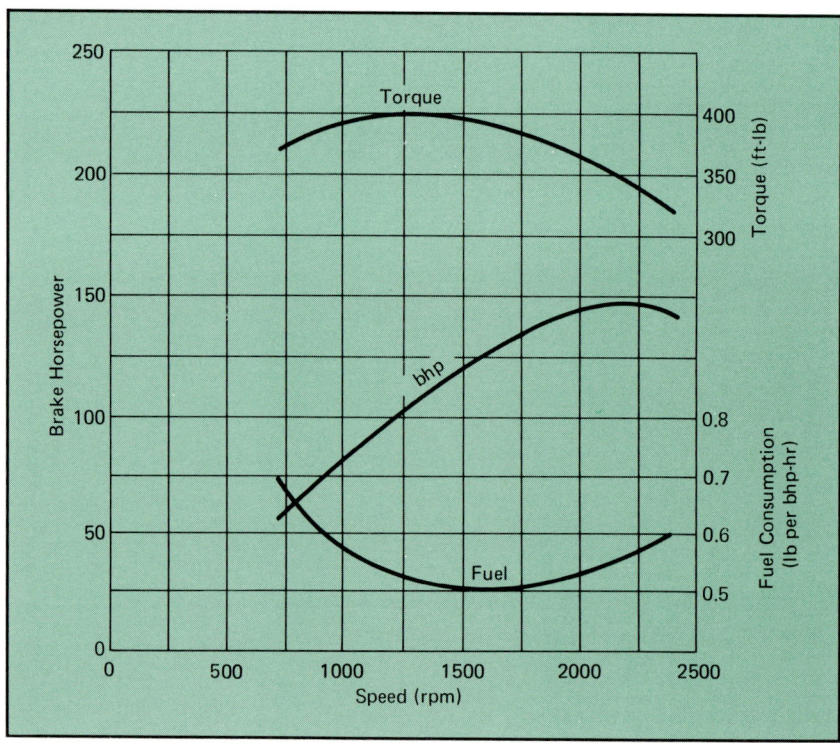

Fig. 4-4. Variation of Torque and Horsepower with Speed for an Internal Combustion Engine with Throttle Wide Open[4]

An alternative to the spark-ignition engine shown in Fig. 4-3 is the compression-ignition *diesel* engine. The diesel engine draws air alone into the cylinder, raises its temperature to the combustion point of the diesel fuel by a very high compression, then injects the fuel into the combustion chamber. The fuel then ignites spontaneously. Figure 4-3 can be used to visualize a diesel engine: imagine the spark plug replaced by a fuel injection valve and the carburetor removed. Then the firing cycle is:

- Intake stroke (air only), with intake valve open.
- Compression stroke (both valves closed).
- Injection of fuel, spontaneous burning, power stroke.
- Exhaust stroke with exhaust valve open.

The diesel engine, for a given fuel rack setting, will produce a torque that is essentially independent of speed, as shown in Fig. 4-5. Therefore, it will not have the inherent self-limiting characteristics of the gasoline engine shown in Fig. 4-4. The speed will be less stable, and a feedback device, a governor, becomes a necessity to avoid runaway.

Diesel engines are known for economy, low maintenance, and reliability, but also for their difficulty in starting, noise, and evil smell.

The *gas turbine*, shown in Fig. 4-6, is a form of internal combustion engine that eliminates the pistons, which are

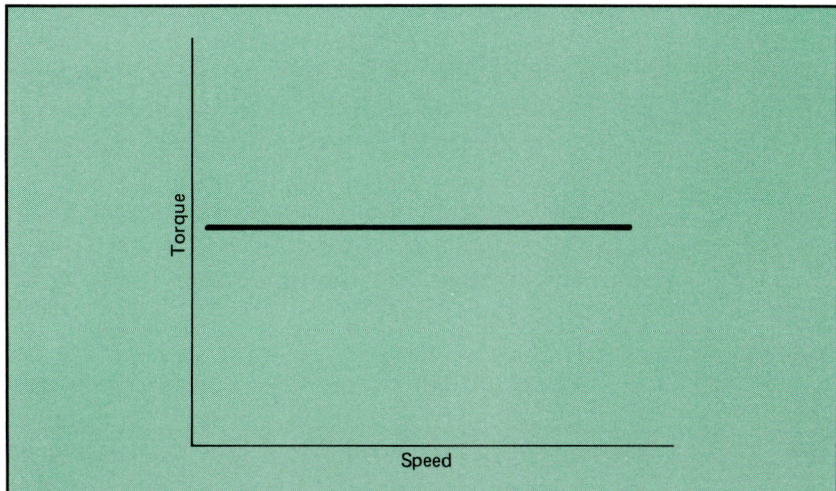

Fig. 4-5. Variation of Torque with Speed for a Diesel Engine

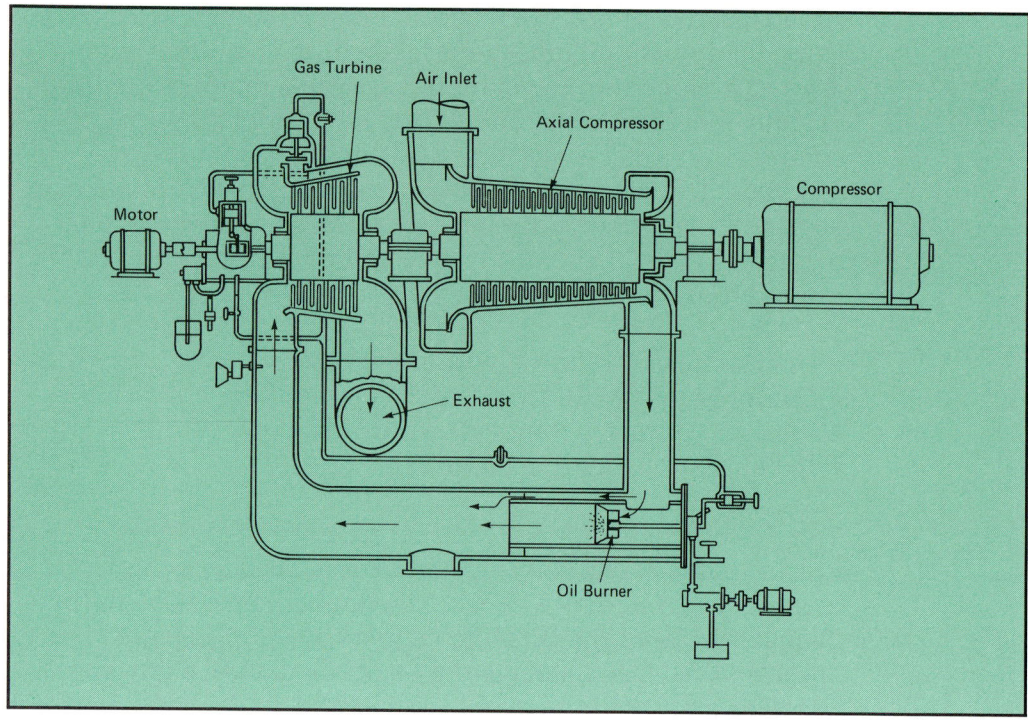

Fig. 4-6. Cut-Away View of a Continuous Combustion Gas Turbine Driving a Compressor[4]

inefficient because of the necessity of reversing their direction of movement. The inherent high-speed characteristic of gas turbines makes them well suited for driving compressors. These turbines are normally selected for applications where electricity or steam is not readily available, such as in isolated pumping stations.

The continuous combustion gas turbine shown in Fig. 4-6 compresses air to 20 to 30 psig and forces the air into and around a combustion chamber. A part of the air is used in the combustion of the fuel. The remainder is used for cooling the gaseous products of combustion. The very high-temperature mixture enters a reaction turbine, which drives the air compressor and also provides useful shaft work in driving the centrifugal compressor.[4]

An important consideration of compressor prime movers is their response dynamics, or the time that elapses between the movement of the throttle to a new position and the development of torque at the corresponding new level. This time response has a major effect on the control of compressor speed but varies significantly among prime mover types.[1]

The dynamic response of a reciprocating internal combustion engine has three components: the lag in the manifold, the dead time between the charging of the cylinder with fuel and its conversion to torque, and the time required for all the cylinders to be firing at the new level. The first time response listed applies only to gasoline engines and is proportional to the volume of the intake manifold. It has the greatest magnitude of all three effects. The other two effects are inversely proportional to the rotational speed. The time delay due to the manifold volume also applies to the gas turbines, as shown in Fig. 4-6, but, obviously, the latter two delays do not apply.

Steam Turbines[5]

The steam turbine is well suited for driving centrifugal compressors because it has a speed range of some 2500 to 12,000 rpm and can be coupled directly to the compressor. Further, speed can be varied by adjustment of the steam throttle valve, thereby accommodating any requirement for variable speed compressor operation.

The steam turbine can be designed for compatibility with almost any existing plant steam system. A short, single-stage turbine will be utilized for available low pressure exhaust steam; a multistage condensing turbine can be used for steam economy; and an automatic extraction turbine can be designed to provide low pressure steam required elsewhere.

Steam turbines are always equipped with speed governors. For power ratings up to 2000 hp, turbines are generally equipped with a single, governor-controlled steam inlet valve, while larger turbines will be equipped with multiple governor-controlled inlet steam valves for tighter speed control and greater economy.

Because turbine governors are proportional feedback devices, they will exhibit "droop" (a change in controlled speed with a change in load). The typical technique for eliminating offset or "droop" is to add the reset function to the feedback control system. This addition cannot be made on a turbine governor since the reset mode (a function of time) slows the response to the point where it cannot regulate the turbine speed.[6]

Gas Expansion Turbines[5]

The gas expansion turbine, sometimes called an expander, is a supplementary driver that uses energy from high-pressure gas

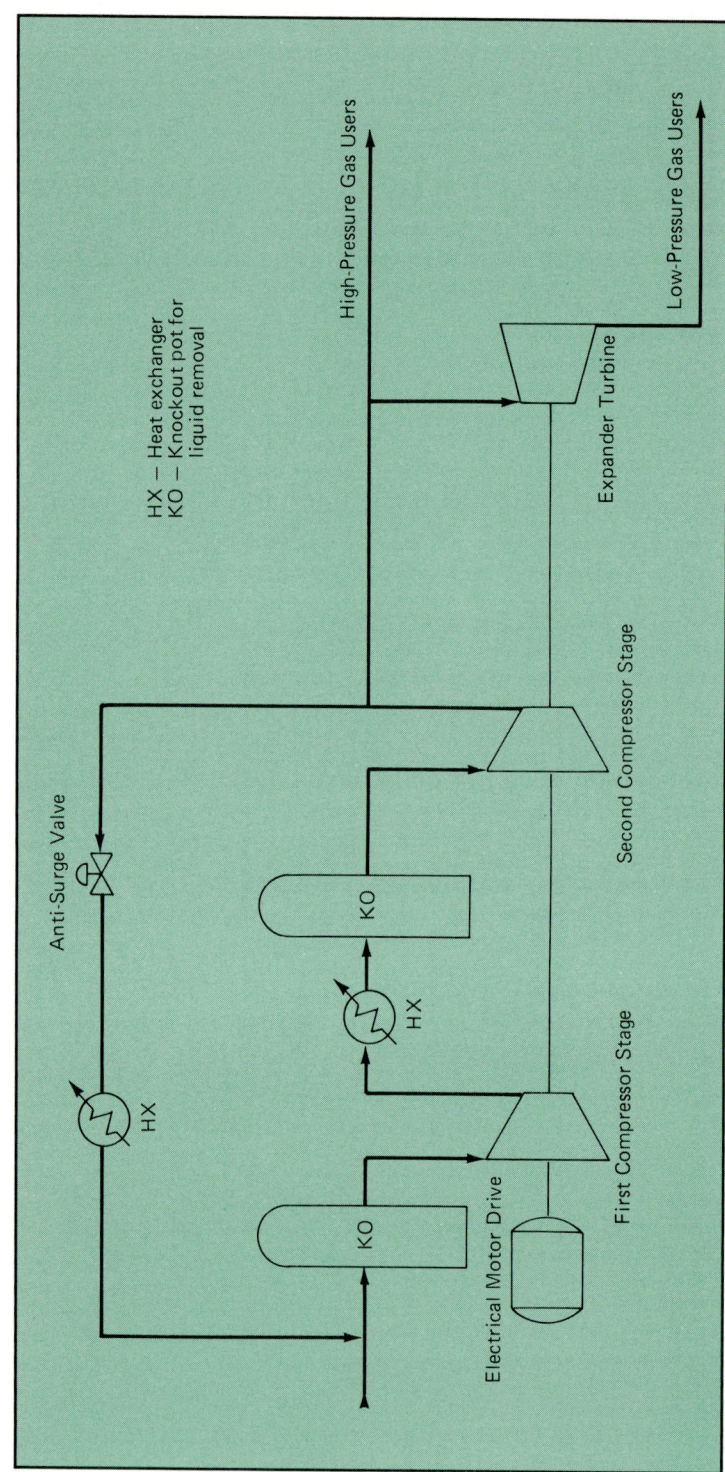

Fig. 4-7. Multi-stage Compressor with Expander Turbine Driven by an Electric Motor

that would otherwise be discarded by throttling across a pressure reduction valve. Figure 4-7 shows a driver used in conjunction with a synchronous electrical motor to drive a multistage compressor. An induction-type electrical motor will have a speed-torque relationship such as is shown in Fig. 4-8. As more high-pressure gas becomes available to the expander, it will accept more of the load. Less torque will be required of the electrical motor, and it will increase speed slightly, accommodating the varying contribution of the expander very gracefully.

Electrical Motors

Standard electric motors are limited to a maximum speed of 3600 rpm and will usually require speed-increasing gearing between the motor and the compressor. An economical driver for the standard compressor installation is the three-phase, alternating current, squirrel-cage induction or synchronous motor. This motor is inherently a constant speed device, so compressor capacity must be varied by the use of a throttling valve in the inlet, inlet guide vanes, or diffuser guide vanes as shown in Fig. 5-8. The simplest mechanism is the inlet

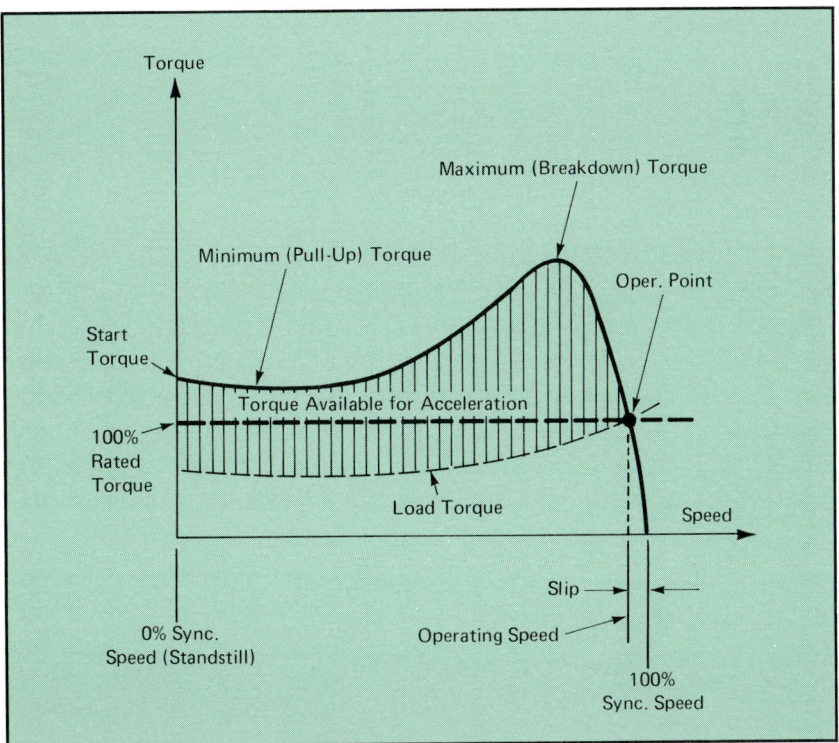

Fig. 4-8. Speed-Torque Characteristics for a Conductive-Type Electrical Motor

throttling valve, which merely reduces the inlet pressure until the discharge pressure, as dictated by the pressure ratio, attains the value required by the load curve. Conversely, the inlet valve can be viewed as a flow-regulating device that forces flow to the value specified by the intersection of the load and performance curves.

The start-up of a compressor driven by a constant speed induction motor requires special consideration. Figure 4-8 shows the start-up torque to be less than the maximum torque capability of the motor. Further, eq. (4-1) shows torque to be increased by the acceleration of the mechanical parts. Therefore, it is necessary to unload the compressor while starting.

The compressor is unloaded by partially closing the inlet valve, which lowers the pressure, and thus the density, of the gas at the compressor inlet. The less dense gas requires less power to pump and effectively unloads the compressor. The question then arises as to how far to close the inlet valve. Obviously, if the valve stops flow completely, the compressor will surge. On the other hand, opening the valve too far will demand excessive torque, overloading the motor.

Equation (4-1) shows a relationship among torque, speed, and time. Torque and speed are plotted on Fig. 4-8, where time, during the start-up transient, is the time for speed to increase from the start-up torque to the breakdown torque. This portion of the curve is known as the pull-up torque. Should the torque load be too high, the motor will take too long to cross the pull-up torque region of the curve. The motor will pass a high current during the prolonged starting period because it represents a low resistance (low back emf). The high current will cause overheating and a rise in temperature. A high-temperature interlock switch is supplied to cut off power from the motor should temperature rise high enough to damage the motor. This switch essentially prevents starting the motor with a load so high that the acceleration is more than a few seconds.

The position of the inlet throttling valve can be quantitatively determined in one of two ways. One is to simulate[3,7,8] the driver-compressor. This is an expensive step, but the ancillary information gained will invariably prove to be serendipitous. Cost and complexity can be reduced by using one of the available programs to calculate the starting time. An alternative procedure that generally proves adequate is to utilize the compressor map near the origin, say, at 5% or 10% of throughput flow. Assume that the anti-surge valve is wide open and, thus, that throttling valve inlet and compressor discharge are at the same pressure. Under these conditions, the pressure rise across the compressor is the pressure drop across the control valve. Since the flow is known from the compressor map and the valve size is known, the percent opening can be calculated as follows.

Example 4-1. Consider an air compressor operating under the following conditions:

$$G = 1$$

$$T_i = 60°F$$

$$C_v = 2500 \text{ for 10-in. butterfly valve}$$

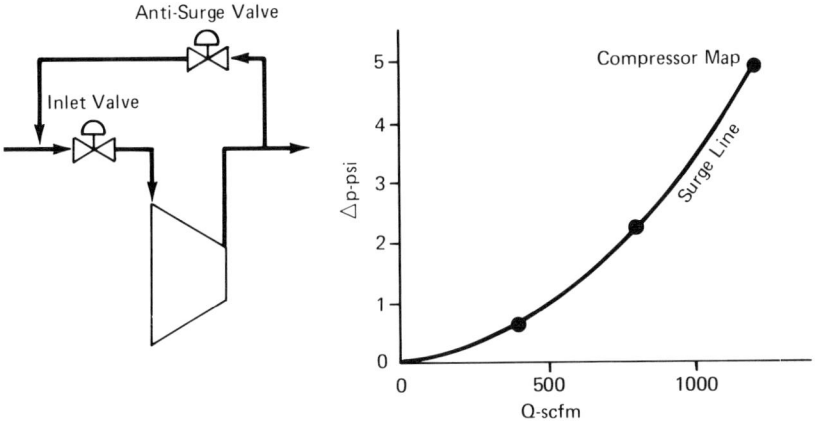

Choose a point to the right of the surge curve, say,

$$Q = 1000 \text{ scfm}$$

$$\Delta p = 2 \text{ psi}$$

At this point, the valve C_v must be:

$$C_v = \frac{60Q}{963} \sqrt{\frac{GT}{\Delta p(p_1 + p_2)}}$$

$$C_v = \frac{(60)(1000)}{963} \sqrt{\frac{(1)(520)}{(2)(14.7 + 12.7)}}$$

$$= 191.9$$

For the typical equal percentage flow characteristic of the butterfly valve,

$$\frac{C_v}{C_{v/max}} = \alpha^{x-1}$$

$$\frac{191.9}{2500} = 20^{x-1}$$

$$x = 0.142$$

where:

α = valve rangeability $\approx$ 20
x = valve stem position, normalized

The foregoing indicates that this particular valve should have a limiter (a mechanical, pneumatic, or electrical stop) that prevents the valve from closing to less than 15% of stem movement.

To re-emphasize the foregoing discussion, it is possible to purchase and install a motor-valve-compressor combination that cannot be started up because of the high-temperature interlock on the motor. Some care is necessary to avoid this embarrassment.

The inlet throttling valve or the guide vanes incur pressure drop, thereby introducing a power penalty in their operation. Energy costs are increasingly dictating the use of variable speed to provide rangeability in compressor capacity. Variable speed devices for electrical motors, described in Unit 2, represent a major component in the cost of the driver.

4-3. Speed Governors[1]

The speed governor is a speed control system. It is a feedback device designed as a servomechanism, where the servomechanism responds well to set point changes. The speed is sensed, and a proportional signal is transmitted to a control device, which compares the measured speed with the set point. The output of the controller then moves a manipulated device in such a way as to bring speed closer to the set point. The advantage of the use of a servomechanism is that it will apply sufficient power to bring the speed to the set point more promptly than if the manipulated device is actuated directly.

Governors vary in design all the way from the extremely simple device found on a power lawn mower to the complex mechanical, hydraulic, and electronic devices used to regulate the speed of prime movers. The flyball governor is probably the oldest and certainly one of the simplest of speed sensing devices. A schematic of the flyball configuration is shown as Fig. 4-9. It consists of a pair of weights, usually spherical, at the ends of two arms pivoted near the axis of rotation in such a way that the flyweights can move radially in a plane through the axis. The weights move away from the axis of rotation because of centrifugal force, which is proportional to both the speed of rotation squared and to the first power of the distance of the mass from the axis of rotation. Figure 4-10 shows the forces involved and the resulting torque balance attained at a given speed. Additional links are attached to the arms and collar about the axis to form a parallelogram configuration. Thus, when the weights move outwardly, the collar moves up, providing a linear motion that is a function of speed. The collar is then linked to the throttling mechanism on the engine.

The governor, whether mechanical, hydraulic, or electrical, lends itself to mathematical analysis in the complex and the

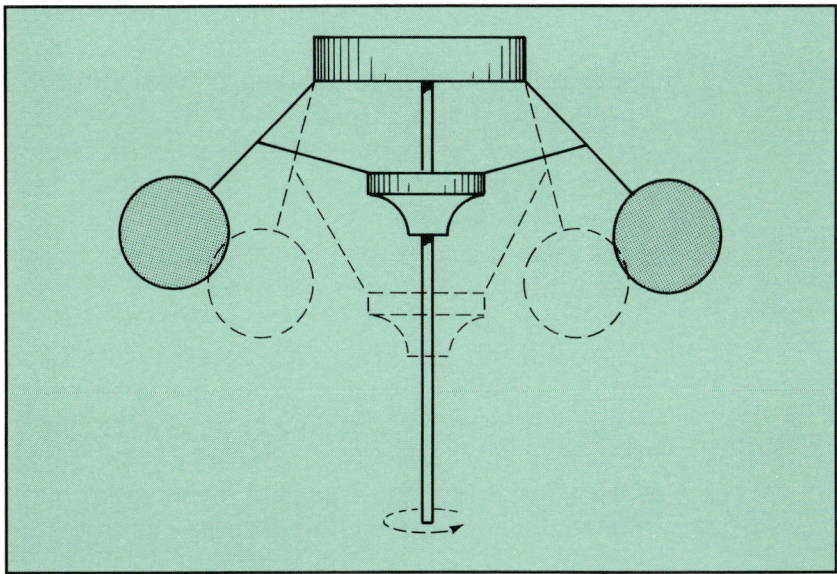

Fig. 4-9. Schematic of a Fly-Ball Governor[1]

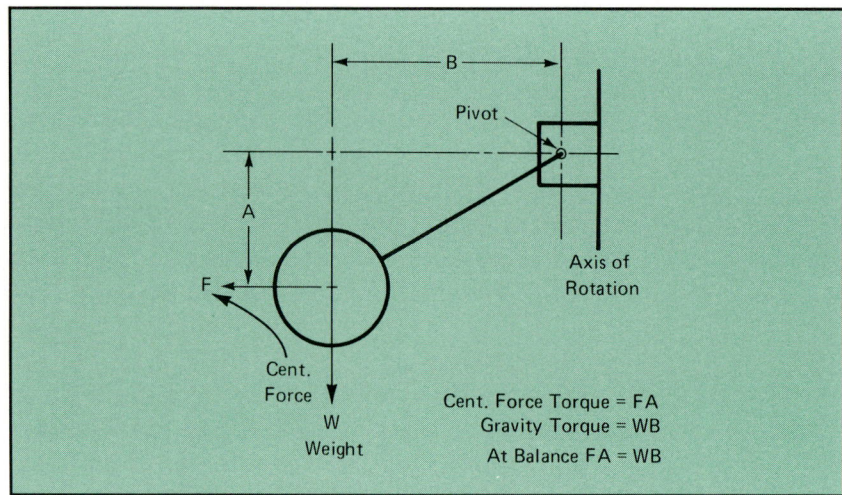

Fig. 4-10. Torque Balance on a Ball of a Fly-Ball Governor[1]

frequency domains.[1,6,9] The analysis, when complete, permits the development of design parameters (weights, spring rates, valve part dimensions) for a desired governor stability, response, and offset (or "droop").

Figure 4-11 shows the modern configuration of a governor used on internal combustion engines. It consists of two weights with their centers of mass about the same distance from the axis of rotation as the pivots about which they swing. The "toes" that transmit the force are arranged at right angles to the body of the flyweight (it is not a "ball"). As the weight moves toward and away from the axis of rotation, the toes convert this motion to an axial movement of the speeder rod through a suitable thrust bearing. The centrifugal force is balanced by the force exerted by a compressed speeder spring instead of gravity. The spring is of a trumpet-shaped or conical configuration to provide a linear relationship between speed and speed setting, since the flyweight position is proportional to the speed squared.

The ballhead type of speed-sensing device has been developed into a highly reliable sensor and has been widely used in conjunction with a servomotor device for speed regulation. However, it is invariably connected to the servomotor by a series of mechanical links and levers, all of which require maintenance and which add hysteresis to the servomechanism. For this reason, it is being strongly challenged by electronic-hydraulic and electronic-pneumatic

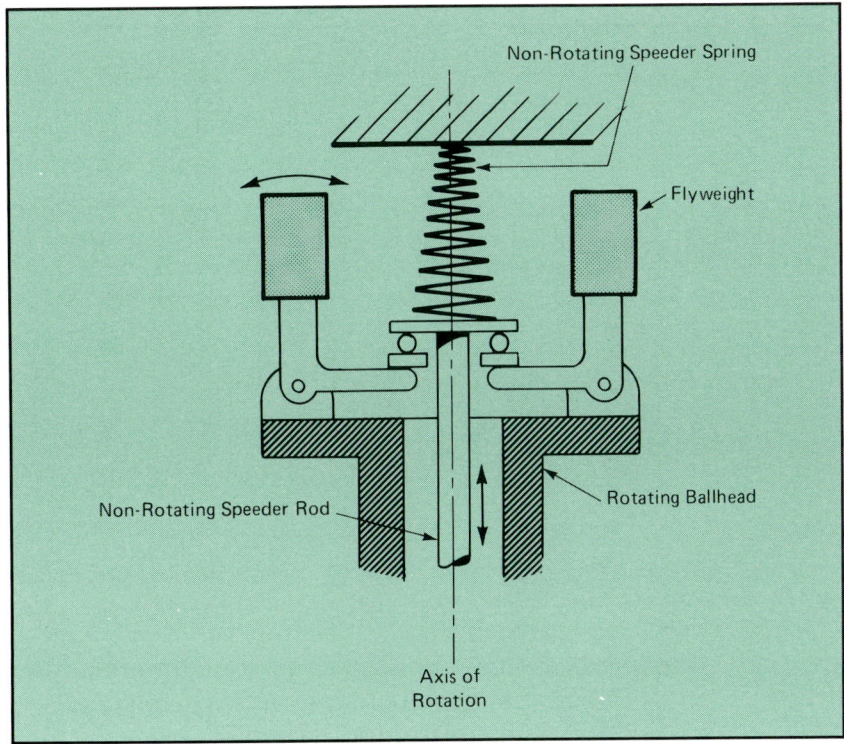

Fig. 4-11. Diagram of a Flyweight Governor for an Internal Combustion Engine[1]

speed control. Electronically, speed is sensed by an inductive-type sensor, which transmits a frequency proportional to the passing of a magnet attached to the rotating shaft past the sensor. The electronic-pneumatic system is used on small turbines. Figure 4-12 is a schematic of an electronic-hydraulic system.[10]

The energy required by most speed-regulating devices demands a servomotor for high power level output. Hydraulic servomechanisms are widely applied because of their ability to amplify simply and their speed of response. The necessity of maintaining clean oil and their tendency to leak represent disadvantages. The hydraulic system shown in Fig. 4-13 shows a four-way spool valve being positioned by the speed sensor or controller (Fig. 4-12). The four-way valve is known as the pilot valve, and it directs high-pressure oil to one side or the other of the power piston. The opposite side of the piston is drained by the four-way valve. The power piston operates the control valve or throttling device that modulates fuel or energy flow to the prime mover.

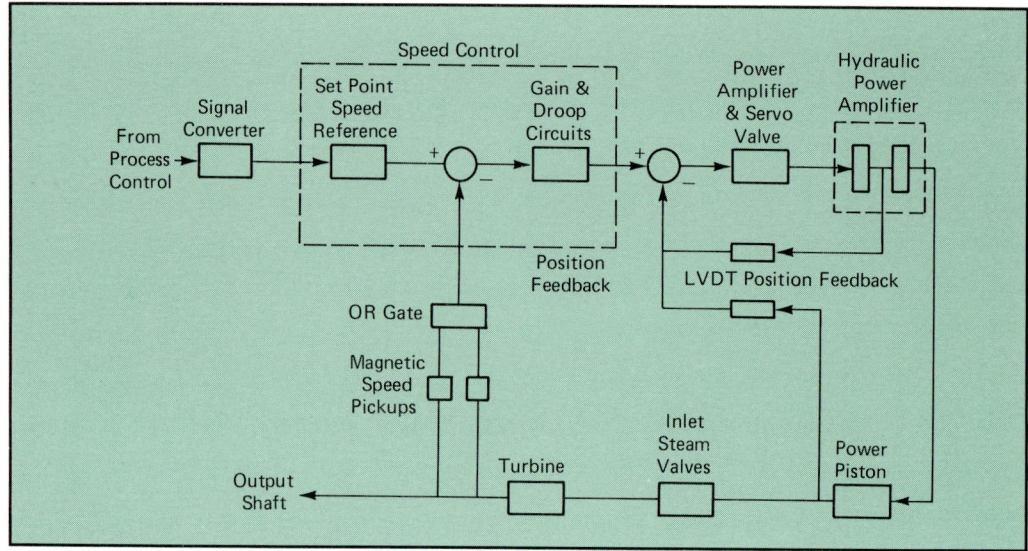

Fig. 4-12. Electrohydraulic Speed Control for Steam Turbine Driving Compressor[10]

Electric motors are used as servomotors in applications where power requirements are low and where speed of response is not essential. Their use is desirable from the viewpoint of their wide availability, but control of the motor, which is continually started, stopped, and reversed, is a serious problem.

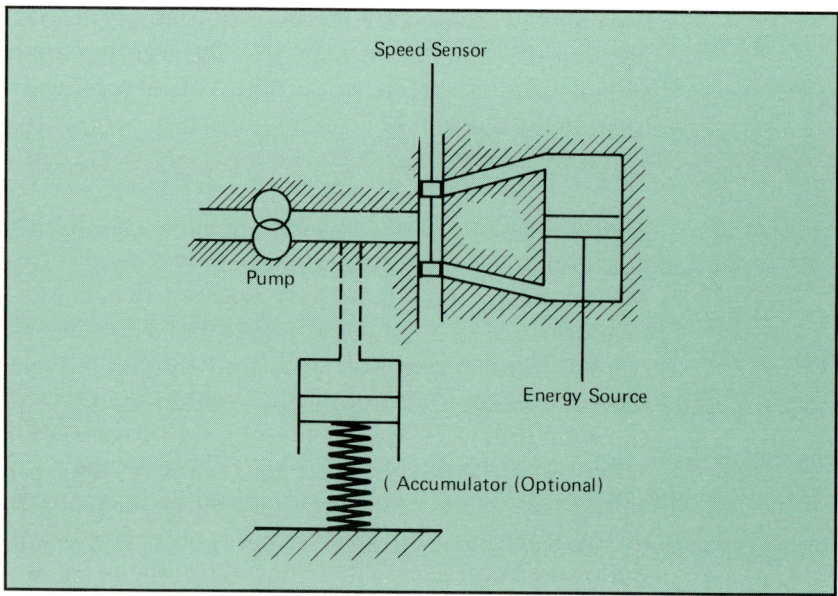

Fig. 4-13. Hydraulic Servomechanism Consisting of Reciprocating Power Piston and a Spool Valve Actuated by the Speed Sensor[1]

4-4. Turbine Overspeed Protection[7,10]

The rapidity with which compressor speed can increase with a sudden unloading has already been discussed. Overspeed with a modern governor is rare, but should it occur, serious internal damage (e.g., throwing rotor blades) to the compressor can result. As a result, overspeed trip interlock devices are mandatory. Except for very small compressors, redundant overspeed trip devices should be considered. A mechanical delatching trip can be provided which interlocks oil pressure down on the hydraulic servomotor, shutting off the source of fuel or energy. An independent electrical-electronic magnetic overspeed sensor is also provided. This type of sensor has the advantage of being compatible with the interlock-alarm system serving the compressor. Good practice dictates that mutliple magnetic speed sensors be installed and that they be connected to an "auctioneer" circuit as insurance against spurious shutdowns.

4-5. Oil Supply Systems[10,11]

The centrifugal compressor is usually the most sophisticated and complex mechanical device to be found on a petrochemical plant site. Its close tolerances and precision machining compare to an old-fashioned mechanical Swiss watch. The lubrication system has been called the heart of the compressor, and the circulating oil compares to the lifeblood of the machine. It provides lubrication for the bearings and muscle for the control. The results of failure of the system can range from increased maintenance costs to a catastrophic failure of the compressor. Thus, the lubricating oil (lube oil) system is a process within itself, with its own sensors, controls, interlocks, and alarms.

Actually, up to four oil systems can be required on a given compressor: compressor lube oil, compressor seal oil, turbine lube oil, and turbine hydraulic system control oil. Reservoirs can be combined or dispersed as required by the compressor. Lubricating system components must be within the compressor's specifications, of course, and can be supplied by the manufacturer. However, competitive pressures can require that the manufacturer provide marginal or even inferior ancillary system components, leading to operating difficulties after commissioning. The vendor's package should be examined carefully to assure that it meets the purchaser's operating philosophy.

Figure 4-14 shows a typical *lube oil system*. Oil from a reservoir is pumped, cooled, filtered, distributed, and returned to the reservoir. The reservoir is sized to provide 5- to 8-minute residence time for degassing and sufficient volume to contain all the oil upon shutdown. A temperature-controlled heater provides heat at start-up. The oil becomes hot with running, and the control system turns the heater off.

Two, and possibly three, redundant lube oil pumps are provided. Steam turbine-driven pumps are economical if the

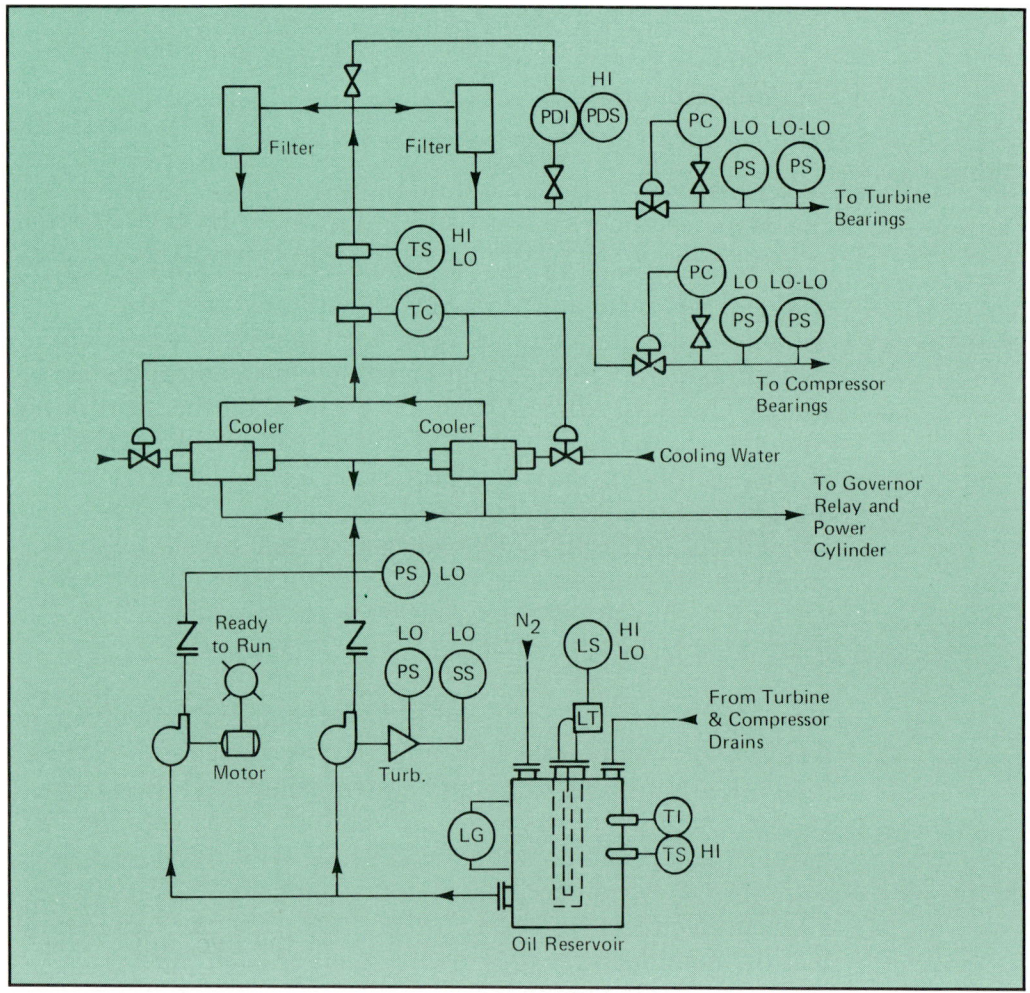

Fig. 4-14. Diagram of a Typical Lube Oil System on a Centrifugal Compressor[10]

compressor is steam turbine-driven. However, steam turbines do not start up well or reliably and cannot be used for emergency service unless they are kept running slowly on a "slow roll". Electrical motor drives can be started immediately, so a common configuration is a steam turbine drive on the pump in continuous use with an electrical drive on standby. The standby pump will be started automatically on low oil header pressure or low turbine pump speed.

Figure 4-14 shows temperature and level indication and alarm switches. The switches will typically actuate alarms in the control room on high temperature and high or low oil level and will interlock the compressor down on dangerously high (hi-hi) temperature and hi-hi or lo-lo level. The oil pump discharge pressure switch will start an auxiliary pump and alarm on low pressure and will interlock the compressor down on lo-lo pressure. Low steam chest pressure or low turbine speed will start an auxiliary pump.

Cooler temperature is regulated by a temperature control system, alarmed at high and low values and used to interlock the pressure down on hi-hi and lo-lo values. High differential pressure across the oil filters actuates an alarm in the control room.

A similar oil supply system, the *seal oil system*, insures that no leakage takes place from or within the compressor. The seal oil must balance pressures within the compressor, so pressures up to 2500 psi must be provided. Positive displacement pumps are commonly used to provide the required pressure. Two systems are used to insure a flow of oil to each shaft seal. In one, level-controlled head tanks are located above the seals; in the other, seal oil is provided on the basis of the differential pressure between the gas in the compressor and the seal oil itself. Both systems are widely used, and both have intensive alarms and interlocks to protect against failure.

The third oil system supplies the *turbine governor hydraulic servomotor*. Figure 4-14 shows this system to be supplied from the lube oil pumps. Since the pressure required by the power cylinder is greater than that required by the shaft bearings, the pumps supply the pressure required by the hydraulic system, and pressure is reduced by pressure control systems for the oil supplied to the bearings.

4-6. Bearing Temperatures

The lube oil system is very carefully designed and operated to supply clean oil to the bearings at the proper temperature and pressure. However, since the compressor is expected to run for months without shutdown for service, conservative practice is to monitor the bearing metal temperature and/or the temperature of the oil leaving the bearing to determine whether a bearing is overheating. High temperature is a warning of potential failure. Temperatures will be measured as follows:

- Bearing metal temperatures are measured by a thermocouple installed in the Babbitt metal of the bearing.
- Radial bearings have thermocouples in the loaded area.
- Thrust bearings have thermocouples in each shoe.
- Oil drains will have a thermocouple as close to the bearing as possible.

Typically, some 15 to 20 thermocouples will be monitored and will actuate high-temperature alarms. The multipoint temperature indicator, on which temperatures could be called up one by one, is being largely replaced by high-speed scanners. The scanner will alarm on high temperature, identify the alarmed temperature point, and accumulate historical data for trend evaluation.

4-7. Vibration Monitoring

Vibration measurements are an important indication of the proper operation of high-speed rotating machinery. At least five vibration sensors are used. Two radial movement detection probes, each 90° apart, are located at each end of the shaft, and one additional probe senses axial movement of the shaft. The use of the signals transmitted by the probes varies. At a minimum, they will actuate high-vibration alarms. Generally, continuous digital indicators on each probe will show vibration amplitude. Orbital shapes and real-time frequencies can be displayed on an oscilloscope or high-performance recorder. Photographs of machine "signatures" can be taken and compared with other pictures taken at other times as evidence of a change in compressor operating condition, thus serving as a troubleshooting tool.

Exercises

1. Consider a 22,000-hp compressor with a moment of inertia of 250,000 lb-in.-sec^2. Estimate the time required to change speed from 6000 to 7000 rpm.

2. Redo exercise 1 with an inertia of 125,000 lb-in.-sec^2.

3. What inherent advantage do turbines and reciprocating engines have over electrical motors as turbine drivers?

4. What are the advantages and disadvantages of hydraulic governors?

5. What is the danger of starting a compressor represented by a large inertial load with a synchronous electrical motor?

6. What is the danger in stopping a compressor representing a small inertial load and a large discharge volume?

7. What advantage do turbines and internal combustion engines have over electrical motors when used as compressor prime movers?

8. What is the function of the compressor seal oil system?

References

1. Anon., "The Control of Prime Mover Speed," Bulletin 25031, Woodward Governor Co., Rockford, Illinois.
2. DenHartog, J. P., *Mechanical Vibrations*, McGraw-Hill Book Co., New York, NY, 1948.
3. Franks, R. G. E., *Modeling and Simulations in Chemical Engineering*, John Wiley & Sons, New York, NY, 1972.
4. Jennings, B. H., and E. F. Obert, *Internal Combustion Engines*, International Textbook Co., Scranton, PA, 1943.
5. Anon., *Compressed Air and Gas Handbook*, Compressed Air and Gas Institute, New York, NY, 1966.
6. Moore, R. L., *The Dynamic Analysis of Automatic Process Control*, Instrument Society of America, Research Triangle Park, NC, 1985.
7. Nisenfeld, A. E., *Centrifugal Compressors, Principles of Operation and Control*, Instrument Society of America, Research Triangle Park, N.C., 1982.
8. Davis, F. T., and A. B. Corripio, "Dynamic Simulation of Variable Speed Centrifugal Compressors," *Instrumentation in the Chemical and Petroleum Industries*, Volume 10, Instrument Society of America, Research Triangle Park, NC, 1974.

9. Thoma, J., "Inverting Functional Diagrams Eases Control Analysis," *Control Engineering*, November, 1968, p. 75.
10. Arant, J. B., "Centrifugal Compressor Auxiliary Instrumentation," *Centrifugal Compressor Operation and Control*, Instrument Society of America, Research Triangle Park, NC, 1976.
11. Wislicenus, G. F., *Fluid Mechanics of Turbomachinery*, McGraw-Hill Book Co., New York, 1947.

Unit 5:
Capacity Control

UNIT 5
Capacity Control

Control of a centrifugal compressor must meet two fundamental requirements. First, the capacity or throughput of the compressor must be controlled to meet process requirements. Second, the compressor must be kept out of surge. The control systems for both objectives manipulate the mass balance around the compressor, so strong interaction is to be expected. A strong secondary requirement is the minimization of energy usage for the operation.

Learning Objectives — When you have completed this unit, you should:

A. Understand the objectives of capacity control.

B. Be aware of the various techniques for accomplishing capacity control.

C. Appreciate the complexity introduced by secondary systems installed to minimize energy or increase rangeability.

5-1. Typical Control of an Air Compressor

The typical operating plant will use large quantities of air for a multitude of purposes, including the operation of pneumatic instruments and the actuation of control valves. The air is supplied by one or more compressors located in a compressor room or engine room, which is a part of the power house. Air is supplied at a constant supply pressure, say, 100 psig, and is reduced in pressure as required by the various users. Therefore, the compressor is required to supply air at a constant "header" pressure. Air, obviously, is a nice gas to handle because unused amounts can be exhausted directly to the atmosphere.

Figure 5-1 shows a compressor supplying air to a pressurized header system. The compressor is driven by a single-speed electrical motor. The users take whatever air they need at the header pressure. The control system must regulate the compressor to supply just the amount of air needed by the users at the header pressure. Thus, the pressure control system

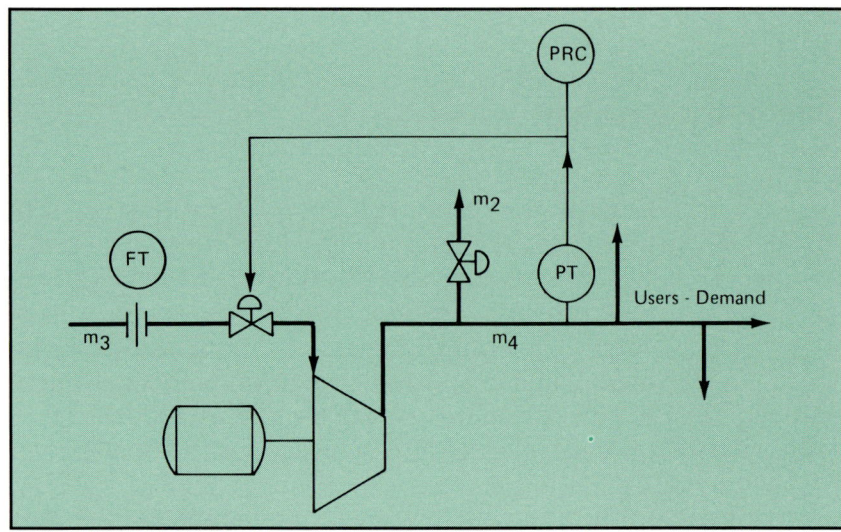

Fig. 5-1. Typical Constant Speed Air Compressor with Pressure Control to Regulate Discharge Mass Balance

actually regulates the mass balance:

$$C \frac{dp}{dt} = m_3 - m_2 - m_4 \qquad (5\text{-}1)$$

where C = capacitance, cu in./in. (see Ref. 6).

Equation (5-1) shows header pressure (p) to increase if the compressor throughput (m_3) is larger than the demand flows (m_2, m_4) and to decrease if smaller. Therefore, maintaining header pressure (p) constant insures that the compressor supplies a flow that equals the demand.

Figure 5-2 is a compressor map for the single-speed compressor shown in Fig. 5-1. As previously discussed, operation will always be on the single performance curve. Discharge pressure is attained by reducing the inlet pressure until the product of inlet pressure and pressure ratio provides the desired value. The inlet pressure is reduced by closing the inlet throttling valve, thus increasing its pressure drop. Since air is being drawn from the atmosphere, the inlet pressure of the control valve remains at essentially atmospheric pressure, resulting in vacuum conditions at the compressor inlet.

Since inlet conditions are essentially atmospheric and constant, an air compressor map will plot discharge pressure

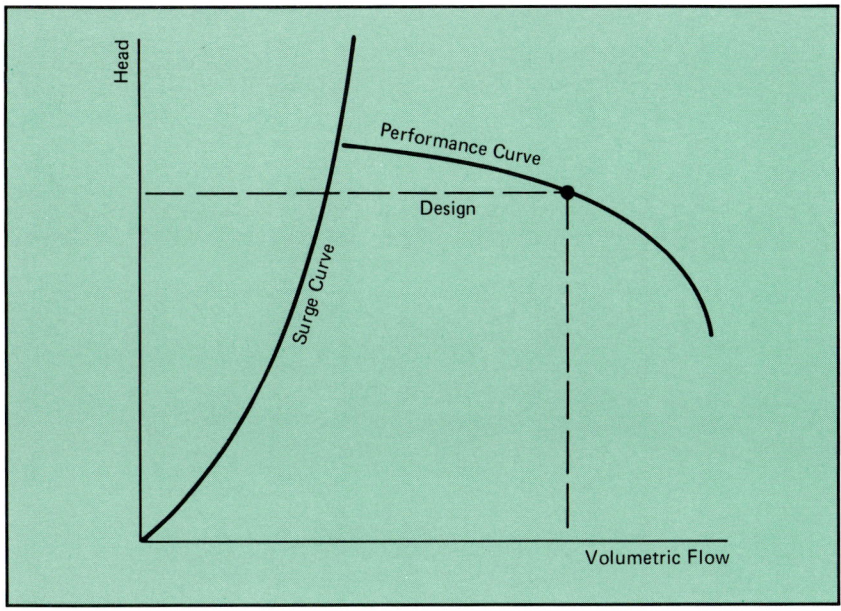

Fig. 5-2. Compressor Map for a Constant Speed Compressor Showing Surge Curve and Design Point (dotted line)

against the volumetric throughput, as shown on Fig. 5-3. However, now the performance curves and surge curves begin to move, as predicted by eqs. (3-30), (3-33), and (3-34), with the changing compressor inlet pressure. The movement is shown, in terms of the opening and closing of the inlet throttling valve, on Fig. 5-3. Each performance curve has an associated surge curve. However, with constant speed operation, the surge curve can be reduced to a surge point. The surge curve under varying inlet conditions becomes the locus of those surge points. Note that the locus surge curve is concave instead of convex as seen from the operating area of the map.[1]

Discharge throttling can also be used for capacity regulation, where the control valve is placed in the discharge of the compressor. Valve actuation then results in moving the load line (see Fig. 3-15). This incurs many disadvantages. First, the flow cannot be decreased or the pressure increased too much without driving the compressor into surge. Limits are placed on either the controller output or the valve stem position to prevent the valve from closing too far. Second, at reduced flow the compressor develops more head (pressure) than the process requires. This extra pressure is throttled before passing to the process equipment, and the horsepower

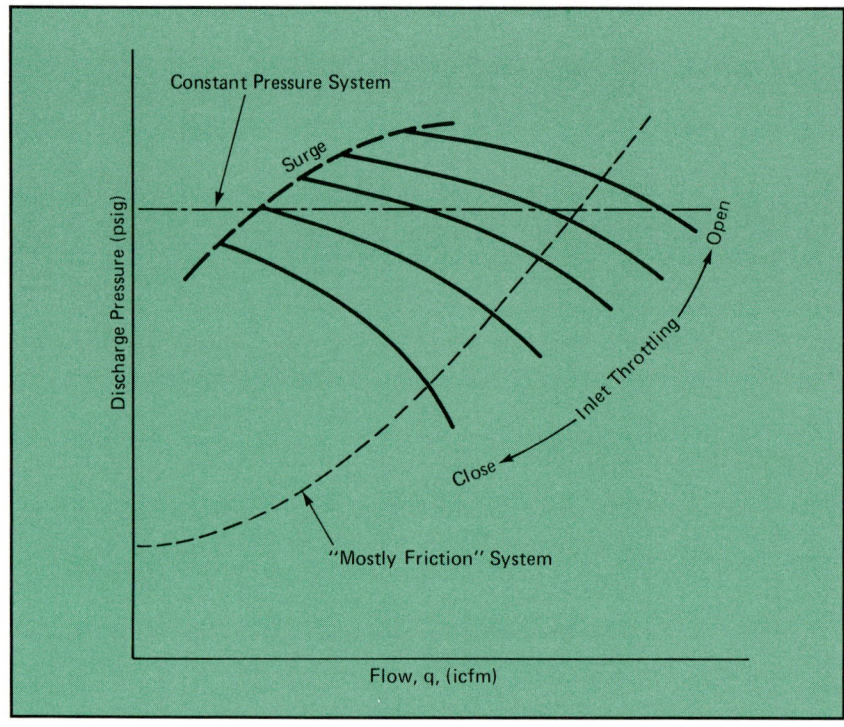

Fig. 5-3. Compressor Map for a Constant Speed Compressor with Inlet Throttling Valve[1]

consumed in compressing it is wasted, resulting in relative inefficiency.[2,3,4] Discharge throttling provides less turndown, results in lower efficiency, and is applied only on small horsepower machines where the inefficiency can be neglected.

Small horsepower compressors are classified in several ways according to their application and operating conditions. They are designated as follows:

- *Fans* are simple compressors ordinarily used in unconfined spaces. If confined in a duct or piping system, a fan will create less than 2-psi differential pressure.
- *Blowers* are compressors that operate in confined systems of piping and create differential pressures of between 2 and 10 psi. Blowers make up a large proportion of industrially installed compressors.
- *Compressors* operate over a wide range of inlet and outlet conditions and can provide 100,000 acfm and pressures of 6000 psi.

Thus, the inefficiencies of discharge throttling would not be costly in absolute terms when used with a fan or blower.

Likewise, so little energy is involved in fan or blower operation that the danger of damage from surge is very minimal.

The simplest capacity control on an air compressor is *total recycle control*. Figure 5-4 shows this arrangement to be not only simple but the least costly in installation, since it can be combined with the anti-surge system through an override.[27] Only one control valve is required. As long as the flow to the process, as measured by transmitter FT1, is greater than the surge set point flow on controller FIC2, the process flow controller FIC1 will modulate the control valve. Should the process flow demand be less than the compressor throughput flow, the controller FIC1 will open the control valve and vent the excess. The vent flow plus the process flow will equal compressor throughput flow, and throughput flow will remain constant while less process demand will increase vent flow. However, should inlet flow transmitter FT2 measure a flow less than the surge flow set point on controller FIC2, the output of that controller will decrease until it takes over the manipulation of the control valve through the low selector relay and opens the valve until transmitter FT2 indicates the required flow to avoid surge.

The low selector relay shown in Fig. 5-4 will always open either the capacity control loop around controller FIC1 or the

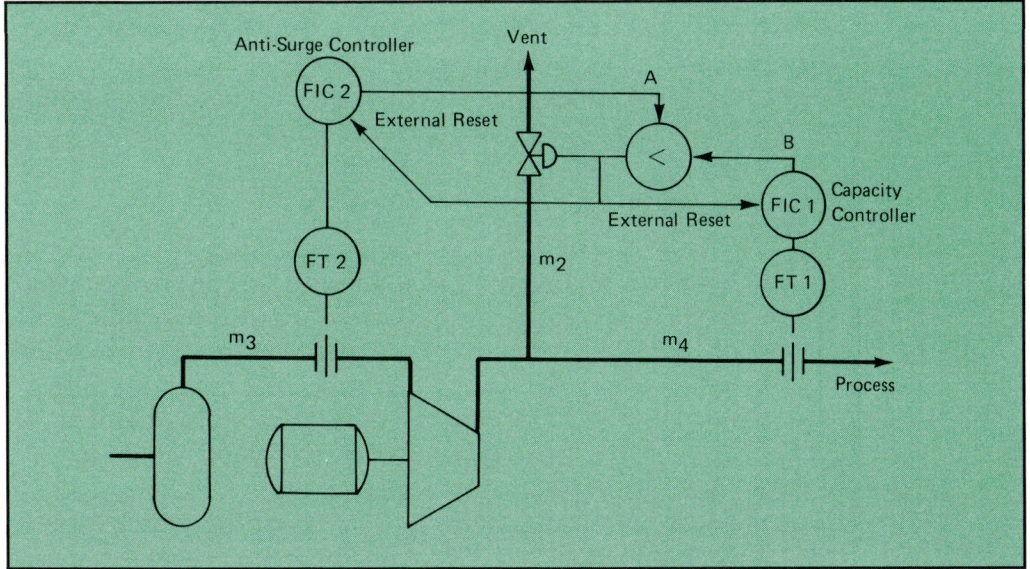

Fig. 5-4. Constant Speed Compressor with Total Recycle Control of Capacity and Surge[5]

anti-surge control loop around controller FIC2. Under open-loop conditions, the controller cannot manipulate its final control element (the valve) to bring the measured variable (flow) to the controller set point value. The "open-loop" controller will have a continuous internal error, and the reset mode will drive the controller output to saturation in trying to eliminate the error. This condition is known as "reset windup". The controller, with its output and reset mechanism saturated, will require significant time to again reach equilibrium upon being reconnected to the valve by the low selector relay. The external reset connection, shown in Fig. 5-4, forces the reset mode to attain a value that will cause the controller output to equilibrate within the active signal span. Therefore, it is ready to assume control quickly and gracefully.[1,5,6]

The total recycle control system for the compression of air is actually a total vent system. Total recycle refers to gases or vapors that cannot be vented but are recycled from the discharge to the inlet of the compressor. Such a recycle must include a heat exchanger to remove the heat accumulated from the repeated compressions.

Total recycle control is seldom used because of the high energy costs. Since compressors are large energy users, considerable attention is given to maximizing efficiency and minimizing energy usage. Figure 5-5 shows guide vanes and variable speed to be manipulated variables that provide greater efficiency than either total recycle or a throttling valve. This figure shows total recycle to require 100% of compressor horsepower, but less horsepower as capacity is reduced with other compressor-manipulating techniques.

Guide Vanes

Guide vanes can be installed at the inlet of a compressor stage, where they are called *inlet guide vanes*, or at the discharge of a compressor stage, where they are called *diffuser guide vanes*. The vanes are designed for the control of the compressor so that its output will be changed to meet the varying demand of the process with as little loss in efficiency as possible.[7].

Inlet guide vanes are based on Euler's Equation, which expresses the theoretical interchange of energy between a rotor

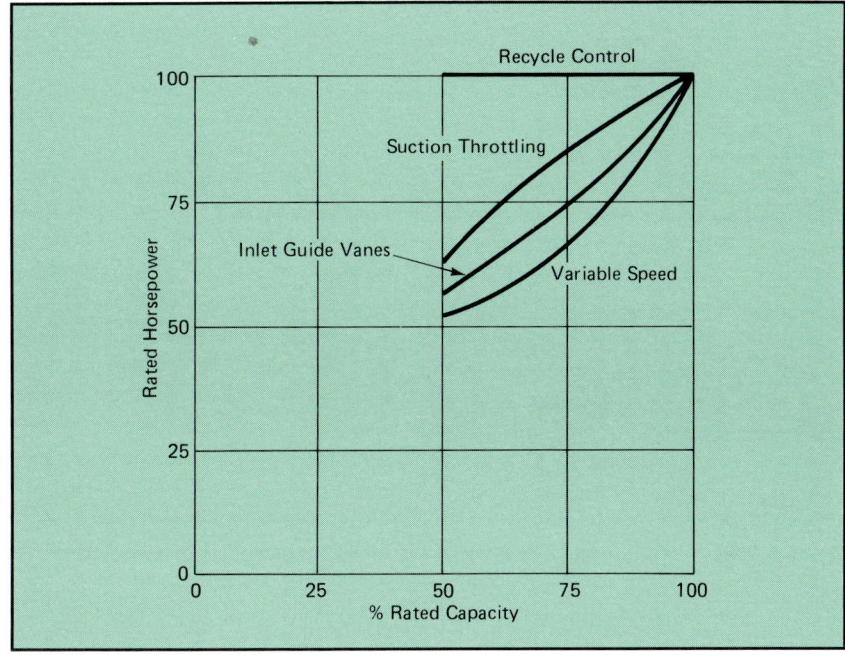

Fig. 5-5. Relative Horsepower Demand of Different Capacity Control Mechanisms[16]

and a "perfect" or frictionless fluid:

$$H = \frac{1}{g}(U_2 C_2 - U_1 C_1) \qquad (5\text{-}2)$$

where:

    H = head, ft-lb/lb
    U = linear velocity of blades, ft/sec
    C = tangential velocity of fluid, ft/sec
    R = relative velocity of fluid, ft/sec
    V = absolute velocity of fluid, ft/sec
    1 = subscript, inlet
    2 = subscript, discharge

The velocity components of the compressor are shown in Fig. 5-6. Normally, it is assumed that fluid approaches the inlet of the rotor in a generally axial direction; then, after passing into the inlet of the rotor, the fluid turns and flows radially away from the axis. As a result, no rotational (or spin) flow exists, and term C of Euler's Equation is zero.

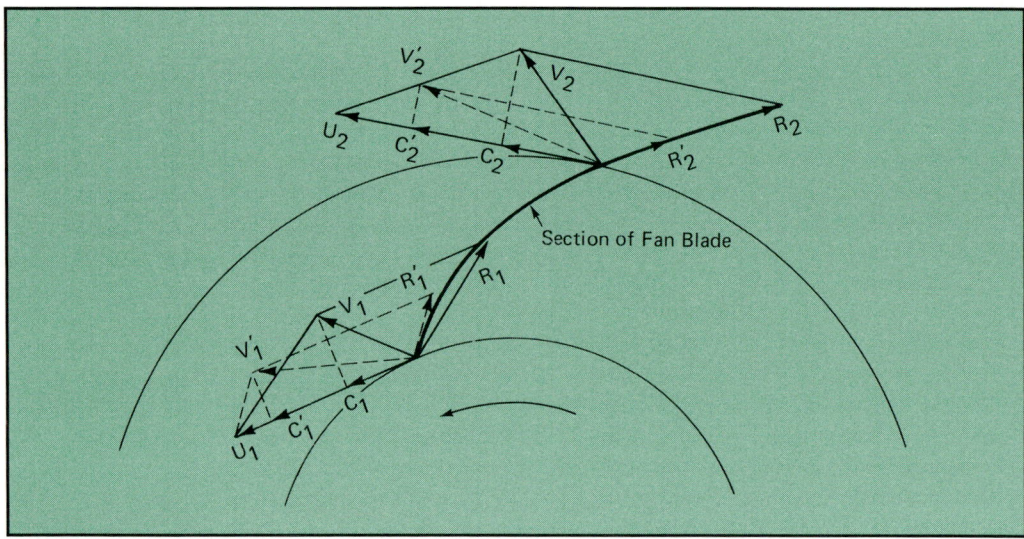

Fig. 5-6. Velocity Vector Diagram for the Inlet and Outlet Edges of the Backward Curved Rotor Blade[7]

Guide vanes control the flow of entering fluid. The fluid is given a spin in the direction of rotation. The entrance spin is controlled by the position of the guide vanes and thus sets the value of C_1 in Euler's Equation. For a given compressor operating at constant speed, velocities U_1 and U_2 are constant, so increasing C_1 by positioning the guide vanes will reduce the head (H) produced by the compressor.[7]

While the Euler concept is quite simple, the practical aspects of the problem remain difficult. A real gas moving over real blade surfaces introduces surface friction and turbulence. The problem becomes one of designing guide vanes that can be moved by an external device and can change the velocity and direction of the flow of gas without introducing throttling. A number of designs have evolved. In one position of adjustment they must enable the full capacity of the compressor to be produced. Other adjusted positions must effect a reduction in the discharge pressure and in the volume of gas moved concurrent with a marked saving in power input. Whatever the design, the guide vanes represent a complex mechanism inside the compressor that cannot be maintained without shutting down the compressor. Should the gas being handled be dirty, there will be a tendency for solids or deposits to build up on the vanes and on the moving links, causing them to stick. Further, variable guide vanes are a relatively expensive compressor feature, so additional purchase costs and additional maintenance costs must be considered in providing more efficient operation at partial loads.

As the vanes are moved to a more nearly closed position, the spin velocity is increased, causing the gas to approach more closely the velocity of the rotor blades. An increase in spin velocity is shown graphically by the dotted positions of the velocity vectors in Fig. 5-6. Vectors C_1^1, R_1^1, and V_1^1 show the entering spin to be rapidly approaching the peripheral velocity U_1 of the inlet edge of the compressor blade. This results in less opportunity for the blades to do work on the gas or impart pressure to it. Since less pressure is produced in the outgoing gas, less volume flows and less input power is required.

A new performance curve and a new surge curve are produced for each position of the vanes. Figure 5-7 shows several performance curves produced by different settings of the control vanes. The performance curve noted "max. open" is the full-capacity characteristic of the compressor. Besides influencing the compressor output, moving the guide vanes also changes the slope of the surge curve. In other words,

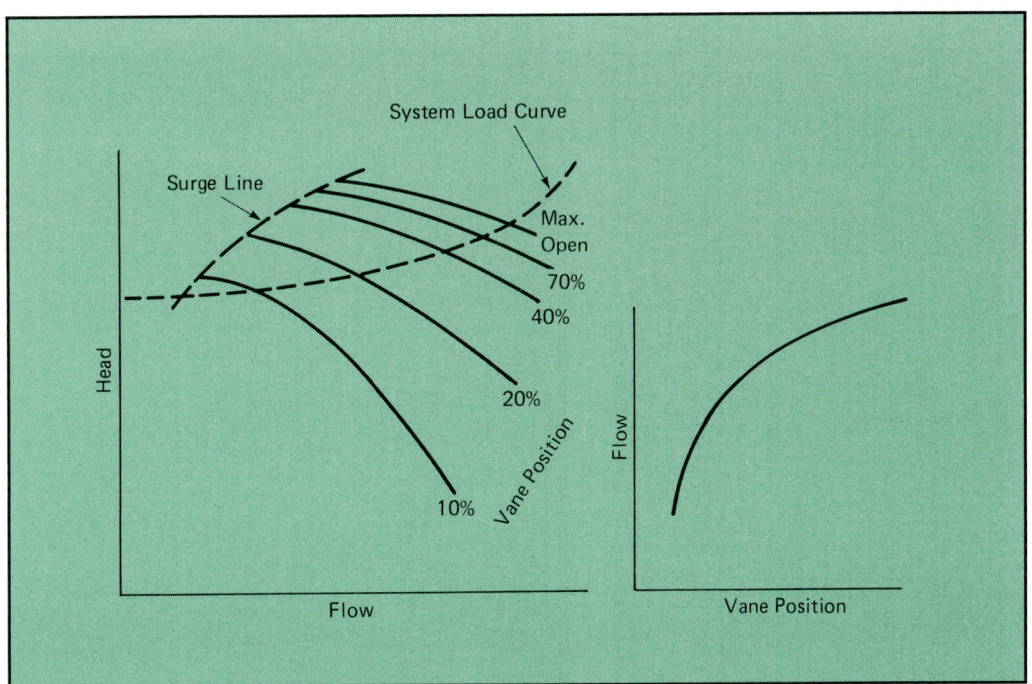

Fig. 5-7. Performance Curves for Constant Speed Compressor with Adjustable Inlet Guide Vanes[24]

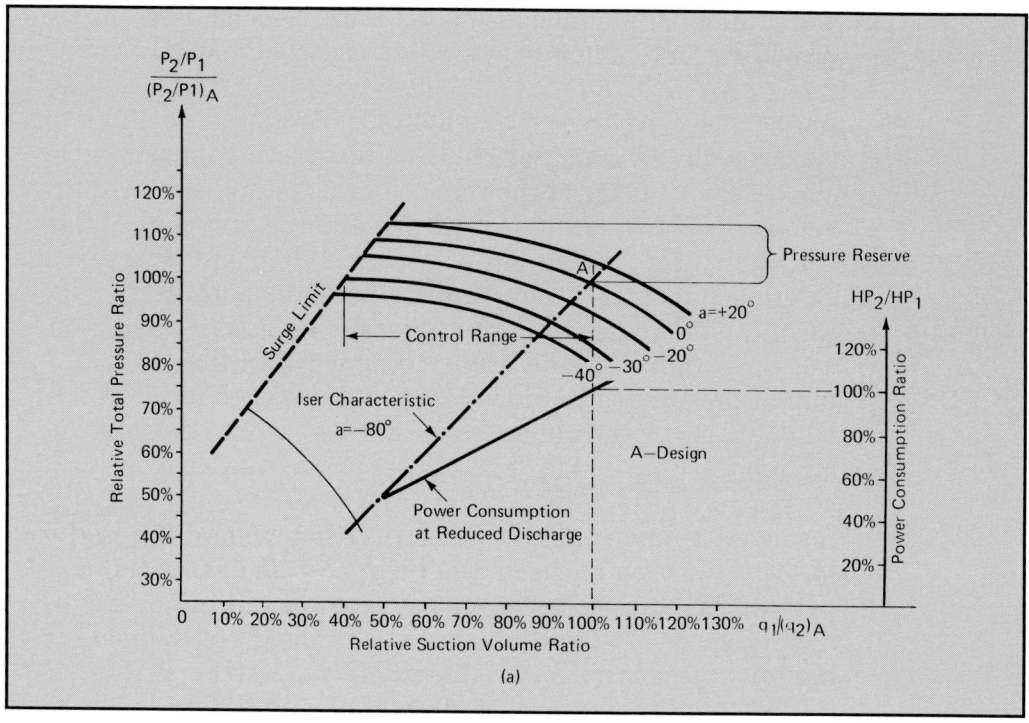

Fig. 5-8. (a) Typical Characteristic of a Constant Speed Compressor with Adjustable Inlet Guide Vanes[9]

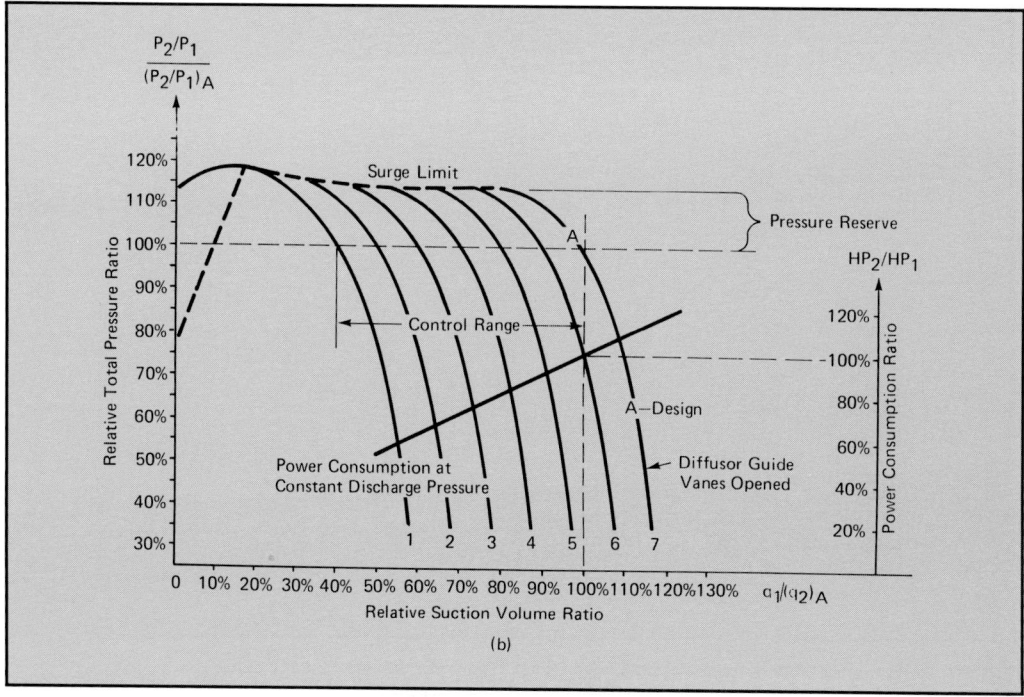

Fig. 5-8. (b) Typical Characteristic of a Constant Speed Compressor with Adjustable Diffuser Guide Vanes[9]

there is also a surge line for each vane position.[8] The surge curve, as discussed in Unit 3, will normally go through the peak or the zero slope of the performance curve. However, in operation, the surge curve becomes a surge point for constant speed operation. The surge curve shown on Fig. 5-7 is the locus of the surge points for the various vane angle performance curves.

Figure 5-7 shows the inlet guide vanes to provide a rangeability of almost 3:1 for the combination static-friction load curve shown. Rangeability would be much less for either an "all-friction" or an "all-static" load curve. In general, greater rangeability is provided by *diffuser guide vanes* than with inlet guide vanes. Figure 5-8 shows a comparison of compressor maps for inlet and diffusor guide vanes.

Figure 5-8a shows a compressor map for inlet guide vanes similar to Fig. 5-7. A rangeability of some 2.5:1 is also shown.

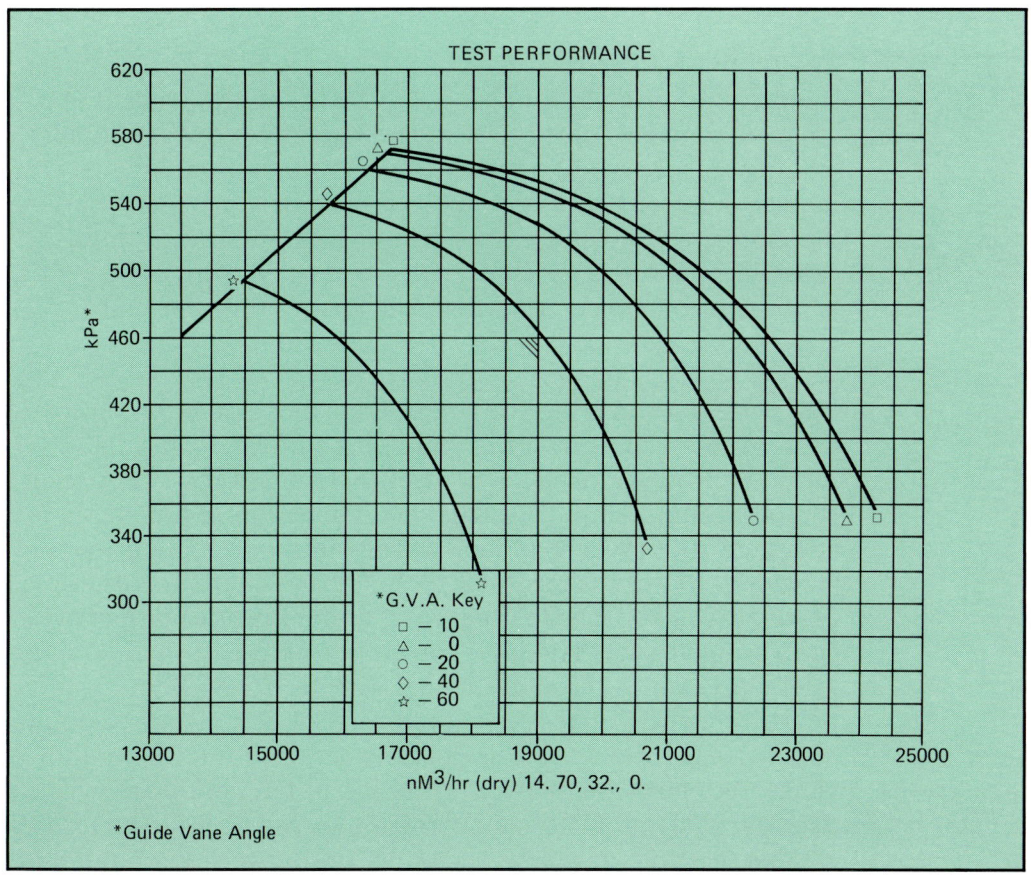

Fig. 5-9. Actual Characteristic of a Constant Speed Compressor with Adjustable Inlet Guide Vanes[25]

Atlas Copco Energas GMBH® recommends inlet guide vanes for loads where discharge pressure falls with flow or for an all-friction load line. A map for an actual compressor, Fig. 5-9, shows a more typical rangeability of 1.6 for an all-static load line.

Diffusor guide vanes are conceptually similar to inlet guide vanes, except that they are located at the discharge of the compressor stage. The compressor map showing their performance curves is shown on Fig. 5-8b. They are recommended for processes that have an all-static load line or a controlled discharge pressure.[9] The configuration of Fig. 5-8b shows the surge line movement to be substantially more for the diffusion guide vanes, with the locus of surge points being represented by a horizontal line that tracks the surge line for the farthest closed vane position.

5-2. Overload Protection Control

The inductive electrical motor, as discussed in Unit 4, runs at a synchronous, constant speed. Equation 4-3 indicates that a motor of a given horsepower (bhp) will slow down (s) as the loading, or torque (A), is increased. As the motor slows down, its resistance or "back emf" decreases, and the current through the motor windings increases. The increased current flow causes temperature to rise. If permitted to continue, the temperature increase will reach the set point of the high temperature interlock, which will shut the motor down.

The penalty for having the compressor unexpectedly remove itself from service is very large in terms of safety and lost production, so an override control that causes the compressor to continue operation at a degraded condition is preferable to complete shutdown. Figure 5-10 shows the override generally provided. Motor wattage will normally be well below the overload controller (XIC) set point, and controller output will go to a high value. The flow controller (FIC) will modulate the inlet guide vanes through the low selector relay. Should power demand increase past the overload controller set point, controller (XIC) output will decrease to close the inlet guide vanes, reducing the density of the gas in the compressor. The decreasing signal from the overload controller will decrease to a value less than the flow controller output at some point and take over the positioning of the guide vanes.

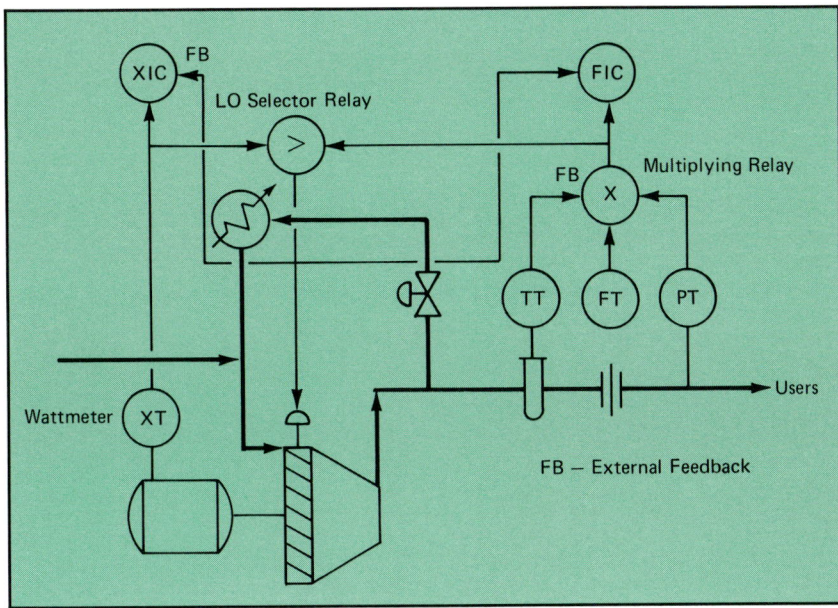

Fig. 5-10. Constant Speed Compressor with Adjustable Inlet Guide Vanes Providing Constant Mass Flow to Process, with Motor Overload Override

Figure 5-10 shows a compressor capacity control system that will accommodate a process downstream that requires a certain mass flow (or mol flow) of gas. The head-type flow measurement is compensated for pressure and temperature to provide a true mass flow measurement.[10] The compressor is basically a constant pressure, variable flow machine; it manipulates flow to regulate a controlled pressure quite well. It can also be used to regulate flow to a set point, as shown in Fig. 5-10. However, the sensitivity of flow to inlet guide vane position depends on the slope of the load curve, as shown in Fig. 5-7. An "all friction" load curve results in good sensitivity of flow to vane angle but indicates a large change in discharge pressure as the flow set point changes. An "all static" load curve (necessarily) requires a constant pressure but has an extremely high sensitivity of flow to vane angle. Reference to Fig. 5-8 shows diffusor guide vanes to better accommodate the latter situation.

A three-phase watt-measuring system is provided to transmit wattage to the overload controller. A typical system is shown in Fig. 5-11. The system generates a signal by the use of several components associated with the electrical motor and the motor starter. Two current transformers, two potential (voltage) transformers, and a watt transducer located in the motor starter generate a variable millivolt signal proportional

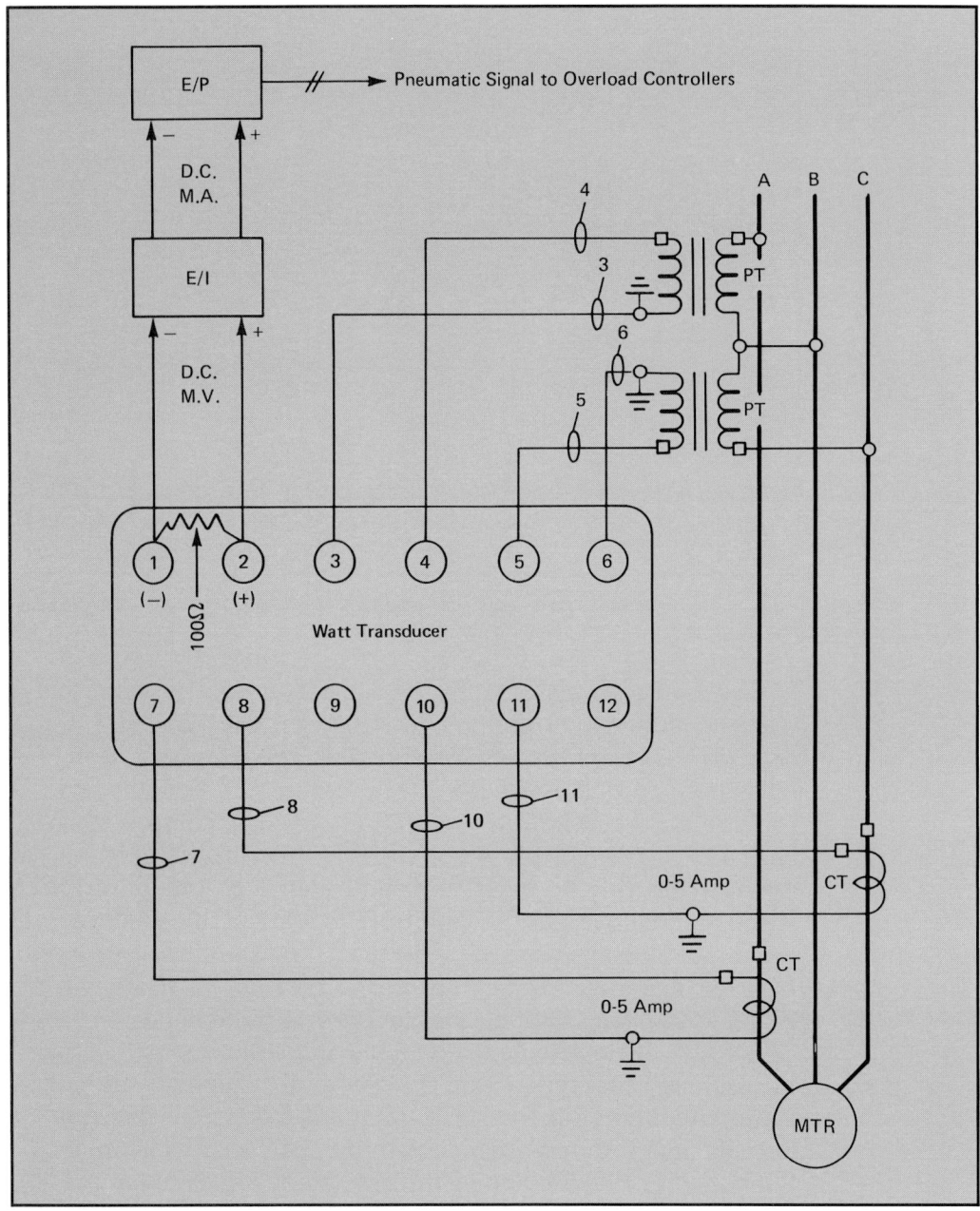

Fig. 5-11. Watt Measuring System Showing Individual Components of the System[11]

to the motor horsepower. Span of the millivoltage for this system is 0–100 mV.[11]

The system shown on Fig. 5-11 consists of two current transformers (CT), two potential transformers (PT), one watt transducer (WT), one millivolt-to-current transducer (E/I), and one current-to-pneumatic transducer (I/P). These components measure the kilowattage (kW) in lines A, B, C and produce a 3–15 psi signal proportional to kW.

The current transformers (CT) are used to measure current flows in each of lines A and C. The output span is always 0–5 amps ac. Current transformer ratings can always be determined from markings on the transformer, which denote the maximum primary/secondary current range. For example, 100/5 means that the output will be five amps when the input is 100 amps. Care must be taken never to open the secondary circuit of a current transformer when the primary is powered. Extremely high voltages will be generated by this action, imperiling both man and machine.

The potential transformers (PT) are used to step the line voltage down to a level compatible with standard instrument ranges. The primary sides of the transformers are connected to lines A-B and B-C in Fig. 5-11. The transformer ratios are such that they produce an output with a maximum voltage of 110 V ac.

The watt transducer receives as inputs the signals from the CTs and the PTs and essentially multiplies voltage by amperage to produce wattage. Inputs are 0 to 5 amps ac and 0 to 135 volts ac. Transducer span is 0 to 1000 watts, and the output span is 0 to 100 mV. The E/I transducer merely converts the 0 to 100 mV to 4 to 20 mA, which can be used directly in electronic control systems. The E/P transducer converts the milliamperage to a 3–15 psi pneumatic signal for use in pneumatic control systems. An A/D converter could be used to convert to a digital signal for microprocessor or computer use.

Example 5-1. Calibration of Wattage Transmitter[11] Consider the following electrical motor. Nameplate data:

Motor rated horsepower (mrhp)	—700 hp
Motor full load amperage (FLA)	—153 amps
Motor output horsepower at two load conditions (hp)	
@ 100% of compressor rated hp	—680 hp
@ 75% of compressor hp	—510 hp
Current transformer ratio (CTR)	—200/5
Potential transformer ratio (PTR)	—20/1

Remember the constant—746 watts/hp.
Assume motor efficiency of 94.4%.

Then,

75% Load

$$i = \frac{(hp)(FLA)}{(mrhp)}$$
$$= \frac{(510)(153)}{(700)}$$
$$= 111.5 \text{ amps}$$
$$mV = \frac{(79)(hp)}{(CTR)(PTR)}$$
$$= \frac{(79)(510)}{(40)(20)}$$
$$mV = 50.4 \text{ mV}$$
$$mA = [(mV)(10) - 125](0.032)$$
$$mA = [(50.4)(10) - 125](0.032)$$
$$mA = 12.1 \text{ mA}$$
$$\text{psi } 0.75 \text{ mA}$$
$$\text{psi} = (0.75)(12.1) = 9.1 \text{ psi}$$

100% Load

$$i = \frac{(hp)(FLA)}{(mrhp)}$$
$$= \frac{(680)(153)}{(700)}$$
$$= 148.6 \text{ amps}$$
$$mV = \frac{(79)(hp)}{(CTR)(PTR)}$$
$$= \frac{(79)(680)}{(40)(20)}$$
$$mV = 67.2 \text{ mV}$$
$$mA = [(mV)(10) - 125](0.032)$$
$$mA = [(67.2)(10) - 125](0.032)$$
$$mA = 17.5 \text{ mA}$$
$$\text{psi} = 0.75 \text{ mA}$$
$$\text{psi} = (0.75)(17.5) = 12.8 \text{ psi}$$

The foregoing equations can be solved in another order to solve for other parameters, such as the transformer ratios.

The output of the watt transmitter is transmitted to the overload controller, as shown on Fig. 5-10, for comparison to the overload set point.

Overload control is especially applicable to air compressors with interstage cooling, which are sensitive to variations in incoming air and cooling water.[12] Cooler air will increase density and, thus, will increase mass flow for the same volumetric flow. Additional mass flow requires more horsepower. Cooling water temperature affects the degree of intercooling and compression efficiency. Lower cooling water temperatures will raise the compressor performance curve and produce more pressure at a given volumetric flow. Or, moving out on the curve, a given controlled pressure will give a greater volumetric flow. This phenomenon is shown by the performance curves in Fig. 5-12.

The electrical motor driver must be sized to supply the power required at the lowest ambient temperature. Figure 5-12 shows the difference in power requirement between the temperature extremes is 20 to 25%. To require that the compressor produce sufficient compressed air in hot conditions while the driving equipment be sized to meet extreme cold conditions is to risk inefficient operation. If the compressor is running under summer conditions, yet the motor is sized to winter

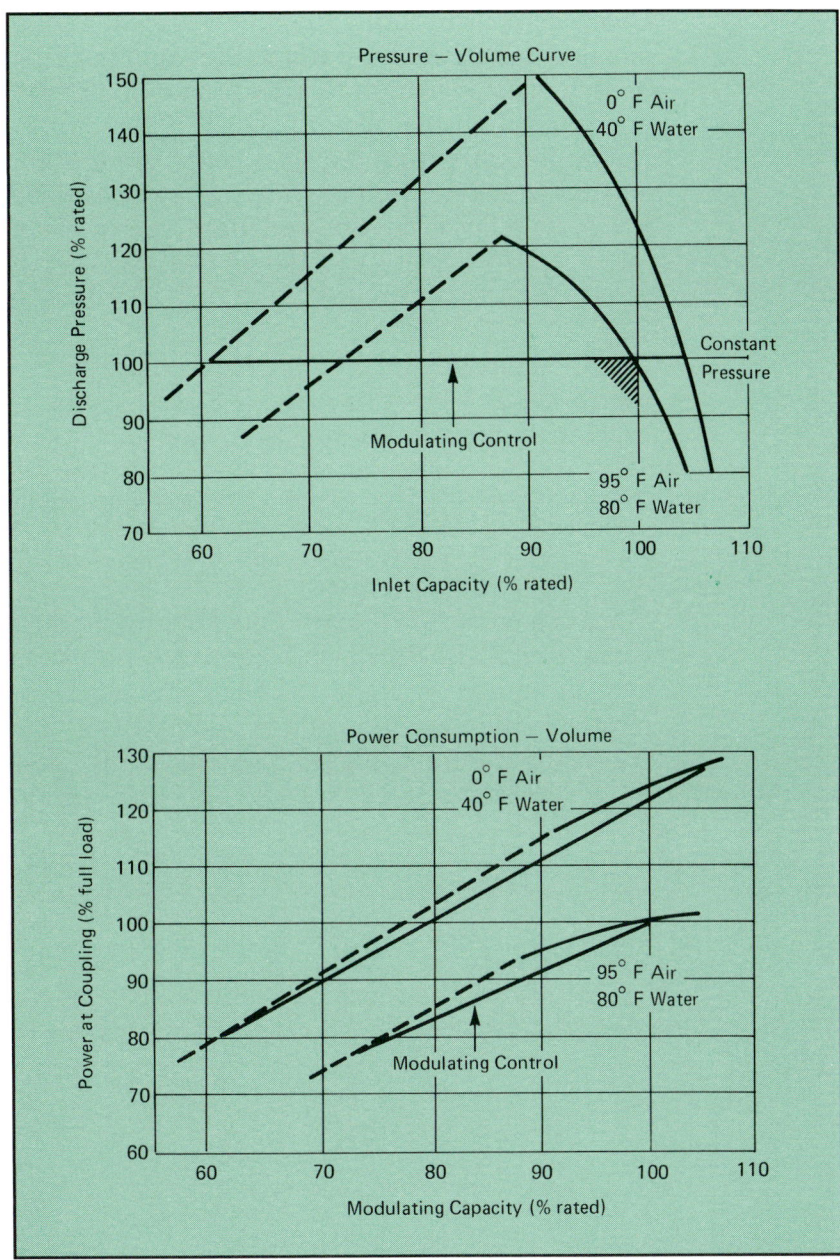

Fig. 5-12. Movement of Surge, Performance, and Power Curves with an Increase in Inlet Temperature[12]

conditions, the motor operates at three-quarters load, with consequent lower efficiency and power factor. Under winter conditions, the compressor supplies more air than the process requires and must be throttled back. Throttling is inefficient in itself, and even greater inefficiency might be introduced by venting air to prevent surge.

Overload control limits the motor load to a safe value, so a smaller electric motor designed for summer conditions can be chosen. The override prevents production of the additional compressed air that is not needed in the winter. Savings accrue not only from the smaller motor cost but also from smaller inrush current at start-up (electrical load peaking), lower switchgear ratings, and a less robust electrical system.

5-3. Variable Speed Compressors

As discussed in Unit 3, centrifugal compressors and blowers follow the Fan Laws or Affinity Laws regarding variation in capacity and head as a function of speed:

$$\frac{s_1}{s_2} = \frac{q_1}{q_2} = \frac{\sqrt{H_1}}{\sqrt{H_2}} \qquad (5\text{-}3)$$

Since the compressor map plots volumetric throughput (q) versus head (H), the speed (s) appears as a parameter on the map, as shown on Fig. 5-13. The variable speed feature is the most effective way to match the compressor characteristic to the required output. Figure 5-13 shows the compressor to be

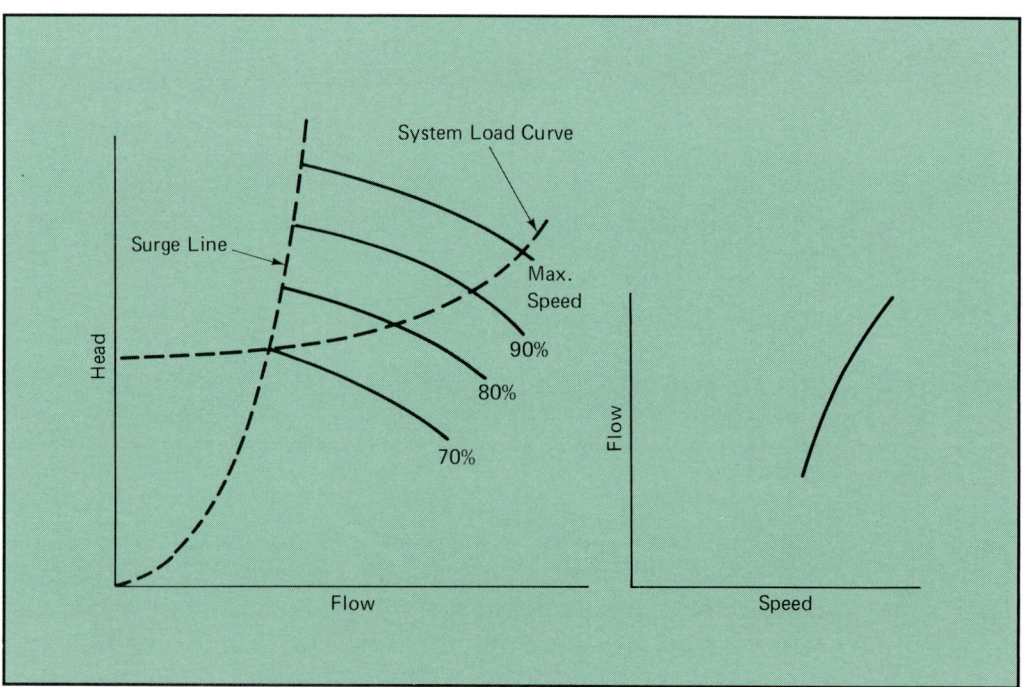

Fig. 5-13. Performance Curves for a Variable Speed Compressor[24]

able to easily deliver constant capacity at variable pressure, variable capacity at constant pressure, or a combination of variable capacity and variable pressure.

The Fan Laws are at best an approximation and are a "law" only for a fan handling air, a light gas. At the small head produced by a fan, the effect of changing conditions on the volumetric flow are quite small, and excellent accuracy can be obtained by the dimensional approach on which the Fan Law is based. The accuracy of the Fan Law diminishes with higher heads, heavy gases, and impeller design. For example, the head output of a backward-leaning blade impeller handling 10 percent more inlet acfm at 10 percent higher speed will be somewhat more than the 21% increase predicted by the Fan Laws.[3,13,14]

Figure 5-14 shows that the effect of speed on capacity of the compressor varies with the system resistance, or the load curve. An "all static" system requires constant pressure, and a small speed change will effect a large capacity change as shown by increment "A". On the other hand, the same speed change in a flow resistance circuit would produce a smaller capacity change as shown by increment "C".[4]

Proper application of variable speed regulation depends on the compressor design and the process characteristics. Speed control of discharge pressure is generally somewhat less than satisfactory with rapid disturbances and a small discharge volume (piping and vessels). Figure 5-14 shows the performance curve is also a factor. As the curve becomes flatter, note that a given speed change makes a much greater change in capacity. Should the curve become so flat that a gain of 10% capacity change per 1% speed change is attained, pressure control will not be feasible. Much better control could be attained by using flow as the controlled variable, if flow control is compatible with the process.[14]

Figure 5-15 shows the gas recovery compressor for a fluid catalytic cracking unit.[4] The compressor operates against an essentially constant discharge pressure system. The inlet pressure is controlled by varying the speed of the turbine driver. In this case, location of the recycle cooler must be placed on the suction side of the compressor. Placing the cooler in the compressor discharge would cause the "heavy ends" (gas components with high molecular weights) to

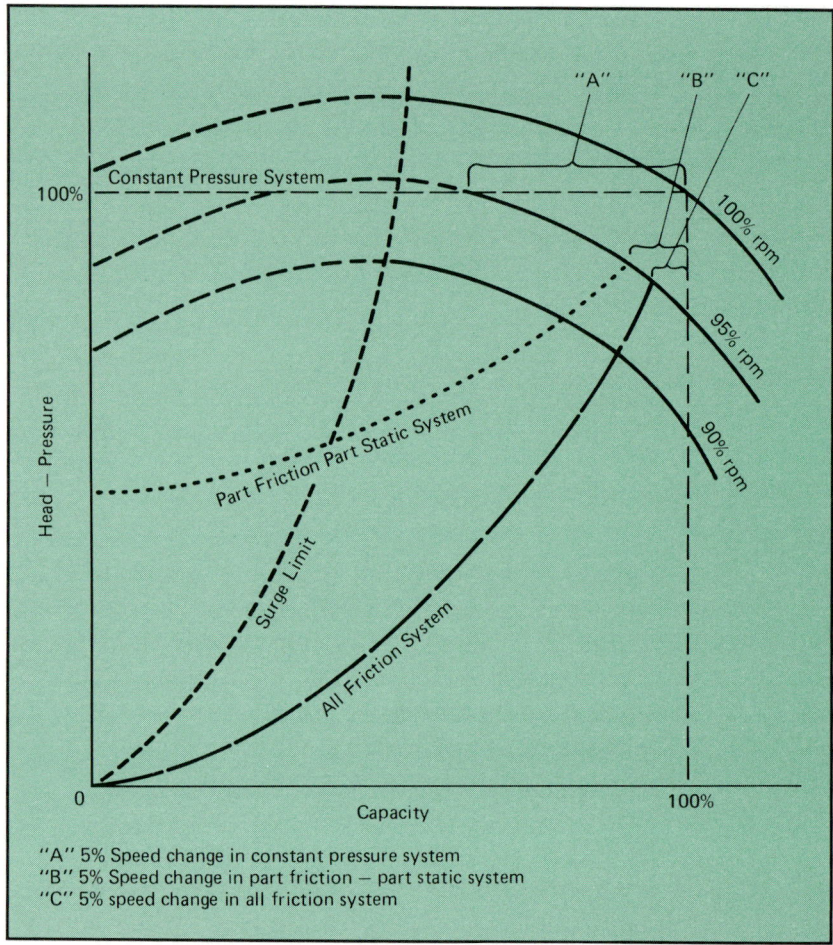

Fig. 5-14. Performance Curves for a Variable Speed Compressor, Showing Effect of Load Curve on Gain of Capacity to Speed[4]

condense out. Thus, a lighter molecular weight gas would be returned to the compressor inlet. The pressure ratio of the compressor would be reduced if the molecular weight of the inlet gas were lower than the design molecular weight.

Figure 5-16 shows a similar application in a turbine-driven alkylation unit refrigeration system.[15] The function of the system is to remove reaction heat in the contactor with refrigerant. Proper reaction rate is maintained by controlling pressure in the knockout drum. The pressure is regulated by varying the speed of the compressor. This process includes the typical cascade of process variable on speed (the governor discussed in Unit 4). Figure 5-16 also shows the complexity of control applications to a "real-world" process.

Unit 5: Capacity Control

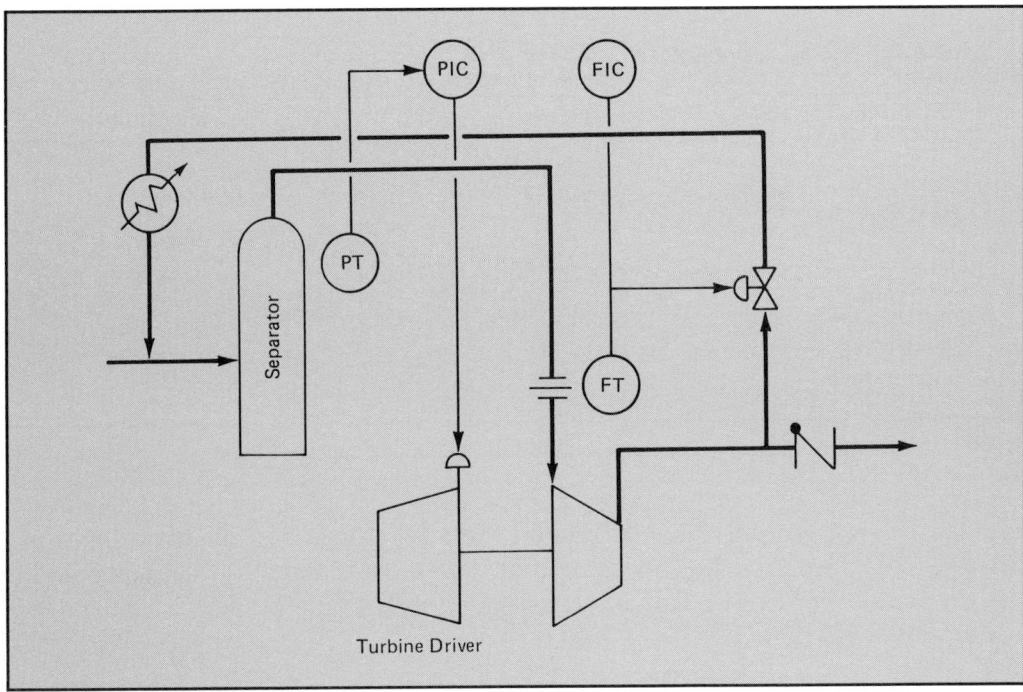

Fig. 5-15. Gas Recovery Compressor in a Fluid Catalytic Cracking Unit[4]

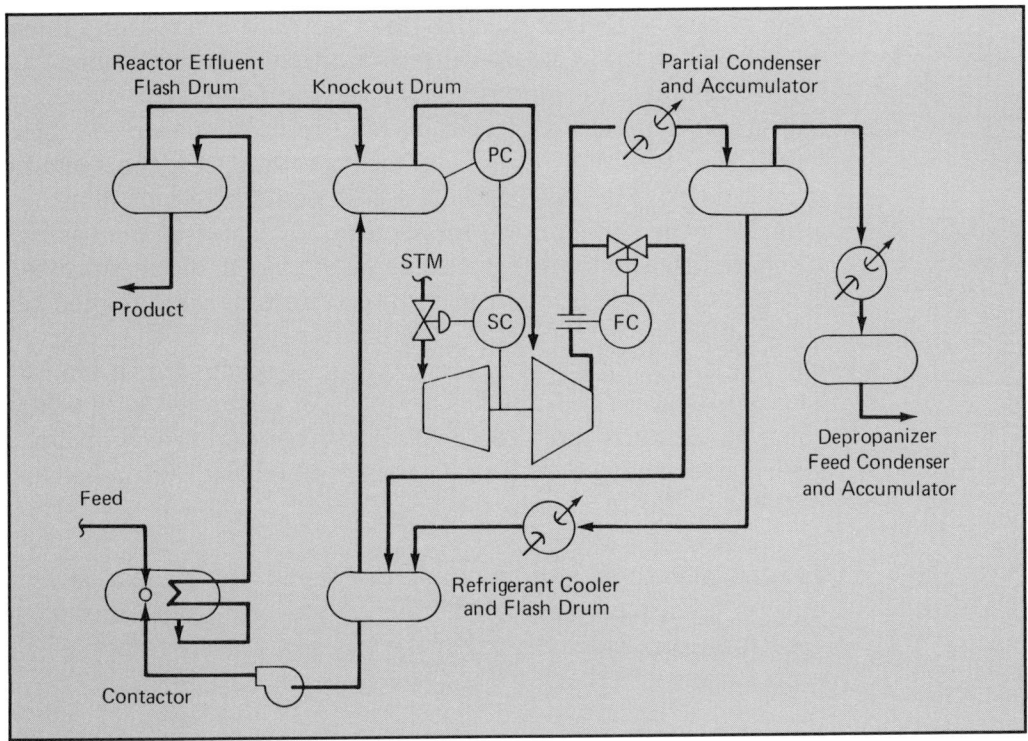

Fig. 5-16. Compressor in an Alkylation Unit Refrigeration System[15]

Manipulative Variable	Energy Efficiency	Ease of Implementation	Reliability	Affect on Stability
Recycle flow	Low	Very easy	High	Minimum
Speed	Very good	Moderately difficult	Moderate	Usually none
Suction throttling	Moderate	Very easy	High [except for P_s near atmospheric]	Moderate
Discharge throttling	Very low	Easy	Good	Potentially high
Guide vanes	Good	Difficult	Low	Moderate
Suction temperature	Moderate	Expensive	Good	Minimum

Table 5-1.[5] Advantages and Disadvantages of Manipulative Variables for Capacity Control

Nisenfeld[5] has tabulated the advantages and disadvantages of each of the foregoing manipulated variables for capacity control in Table 5-1.

5-4. Multistage Compressors

The pressure ratio required by most industrial processes is typically between 3:1 and 15:1, while the ratio of a single stage (single wheel) is usually less than 2:1. Therefore, almost all industrial applications require multistage compressors. Multistage machines are preferred over single-stage compressors in series since they require only one driver, take less space, and require less capital investment for the same net pressure rise. Multistage compressors contain two or more wheels in the same casing (or section). The size of the casing is limited by the temperature rise of the gas being compressed. The gas must be removed from the compressor and cooled by intercoolers as it reaches the temperature limit of the mechanism. (The effect of intercooling is discussed in Unit 3.) The temperature limit of the compressor is some 450°F, while some gases and vapors require a lower temperature limit for safety. For example, chlorine gas cannot exceed 300°F because of the possibility of chlorine fire.[3,16,17,18]

The manufacturer will assemble or "stack" the stages for optium compressor overall efficiency and minimum power consumption. Assembly is on the basis of the efficiency of each stage plotted against the normalized variable q/s (volumetric throughput divided by speed). Each stage operating flow point is determined by establishing its inlet condition, which is the outlet condition of the preceding stage altered by temperature and pressure changes through interstage piping and intercoolers. The compressor inlet design condition at which power is required will result in individual stage flows that are to the right or left of their best efficiency flows.

A stage efficiency curve with a broad, flat shape in the area of best efficiency provides the manufacturer with more flexibility in applying that stage to a wider range of operating conditions. The operating speed of the common shaft powering all the stages is then selected as a compromise to permit operating the wheels as near as possible to their highest efficiency. The final selection of speed will favor the highly loaded impeller, since poor efficiency on that stage will have the greatest effect on overall efficiency.[19]

The stacking of stages permits the manufacturer to provide compressor maps for each stage as well as an overall compressor map, as shown in Fig. 5-17. The overall map is a composite of the surge and stonewall points of the individual stage curves, so it will generally have a narrower operating range between surge and stonewall than do the individual stages.[19,20] The individual compressor maps for each stage must be referred to for an assessment of the rangeability of the compressor.

The volume reduction of the gas due to pressure increase is less per stage when operating at speeds lower than design. Under low-speed conditions, the stages near the discharge are forced to handle more than their rated volumetric flows (approaching stonewall), while those nearer the inlet handle less than normal (approaching surge). It is common for surge to be initiated by one of the latter stages when the compressor is operating at design speed, but at an earlier stage at lower speeds. A distinct change in the surge line is evident where this occurs. This effect, seen on Fig. 5-17c, introduces a severe difficulty into characterizing the surge line.

Multistage compressors have the process flexibility of being able to accept inlet gas at each point where intercooling takes place and to discharge compressed gas at these same locations. Operations can also be performed on or by the gas stream while it is removed from the compressor for intercooling. At

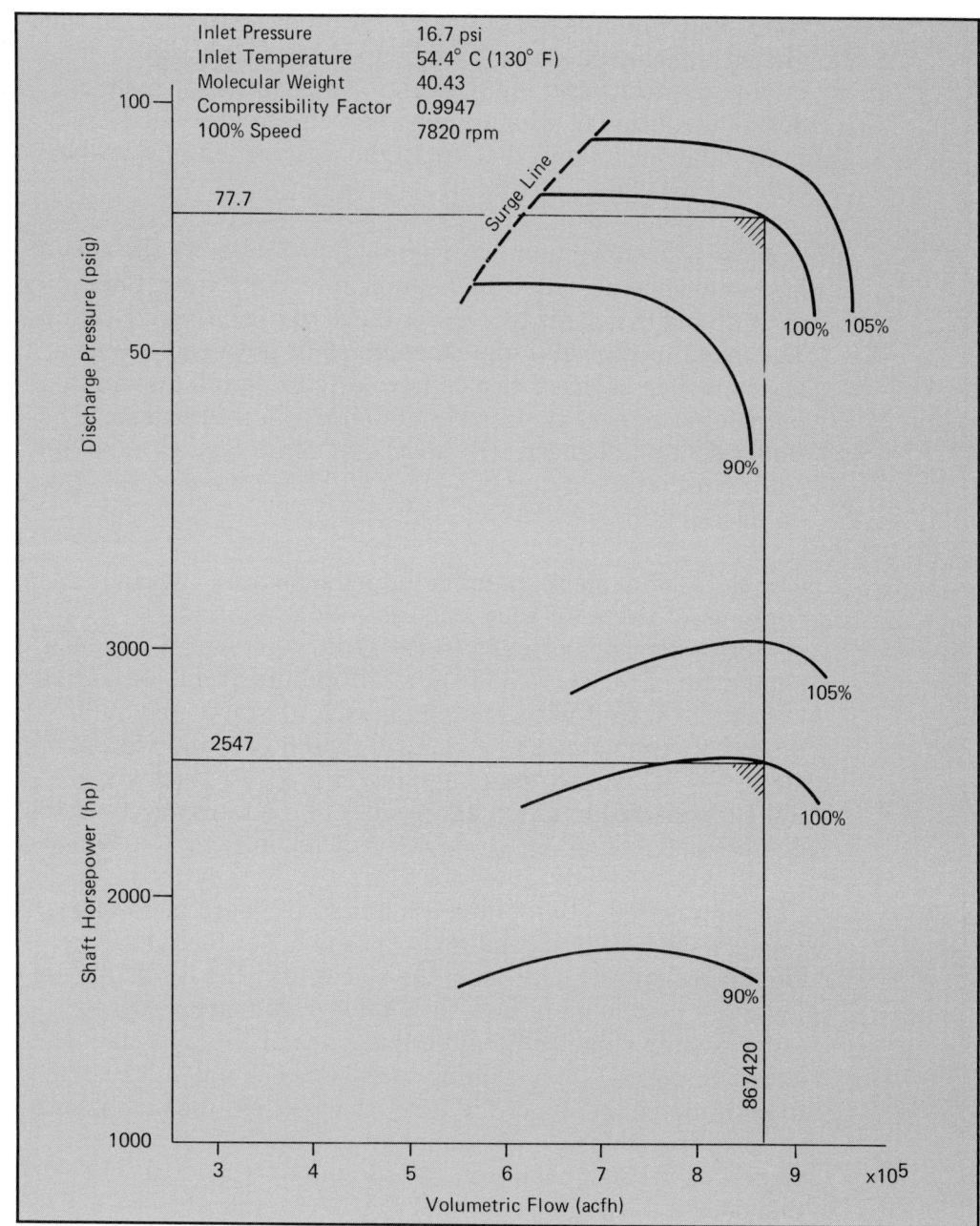

Fig. 5-17. (a) Compressor Map for the First Stage of a CO_2 Compressor

Unit 5: Capacity Control

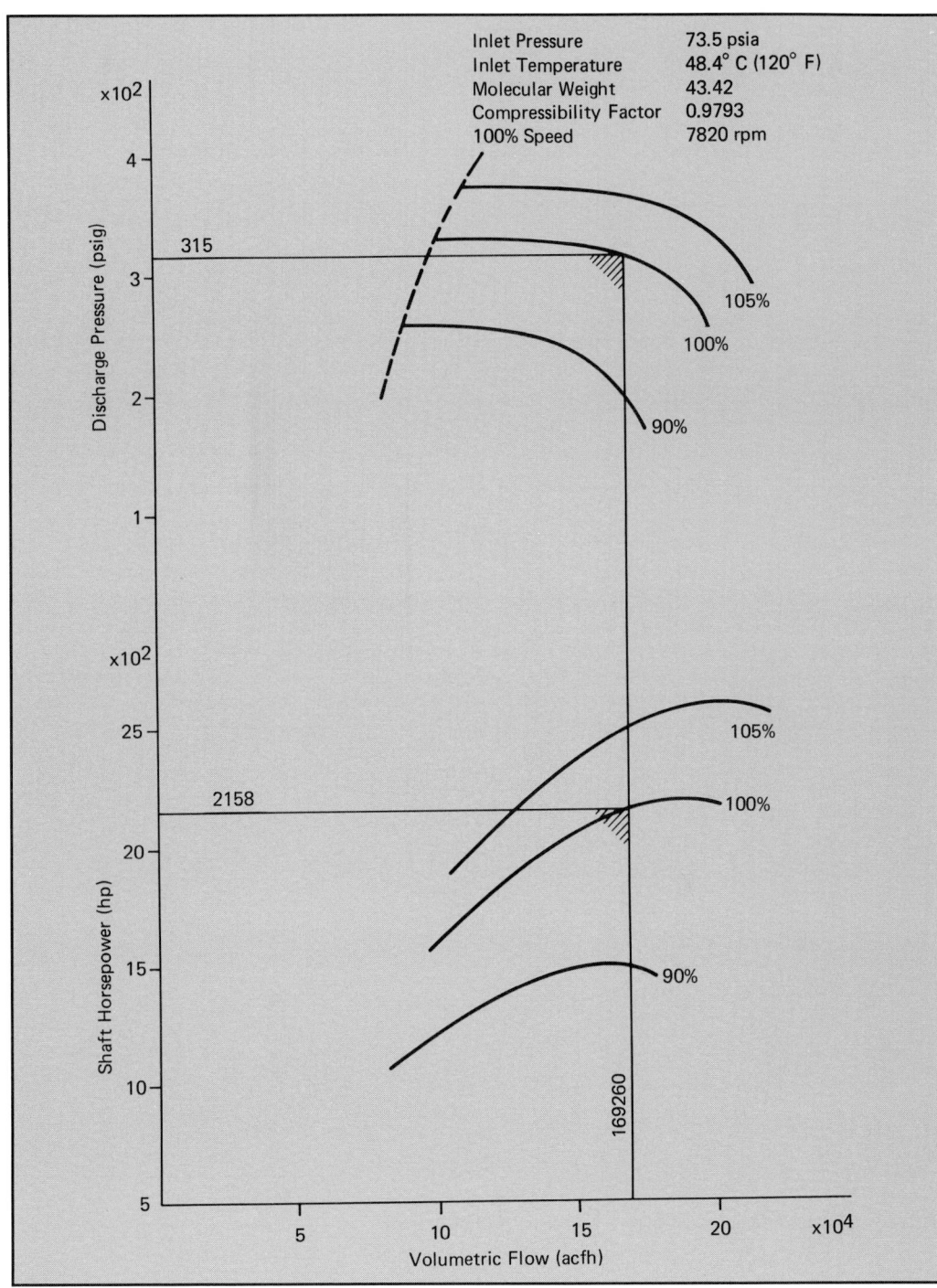

Fig. 5-17. (b) Compressor Map for the Second Stage of a CO_2 Compressor

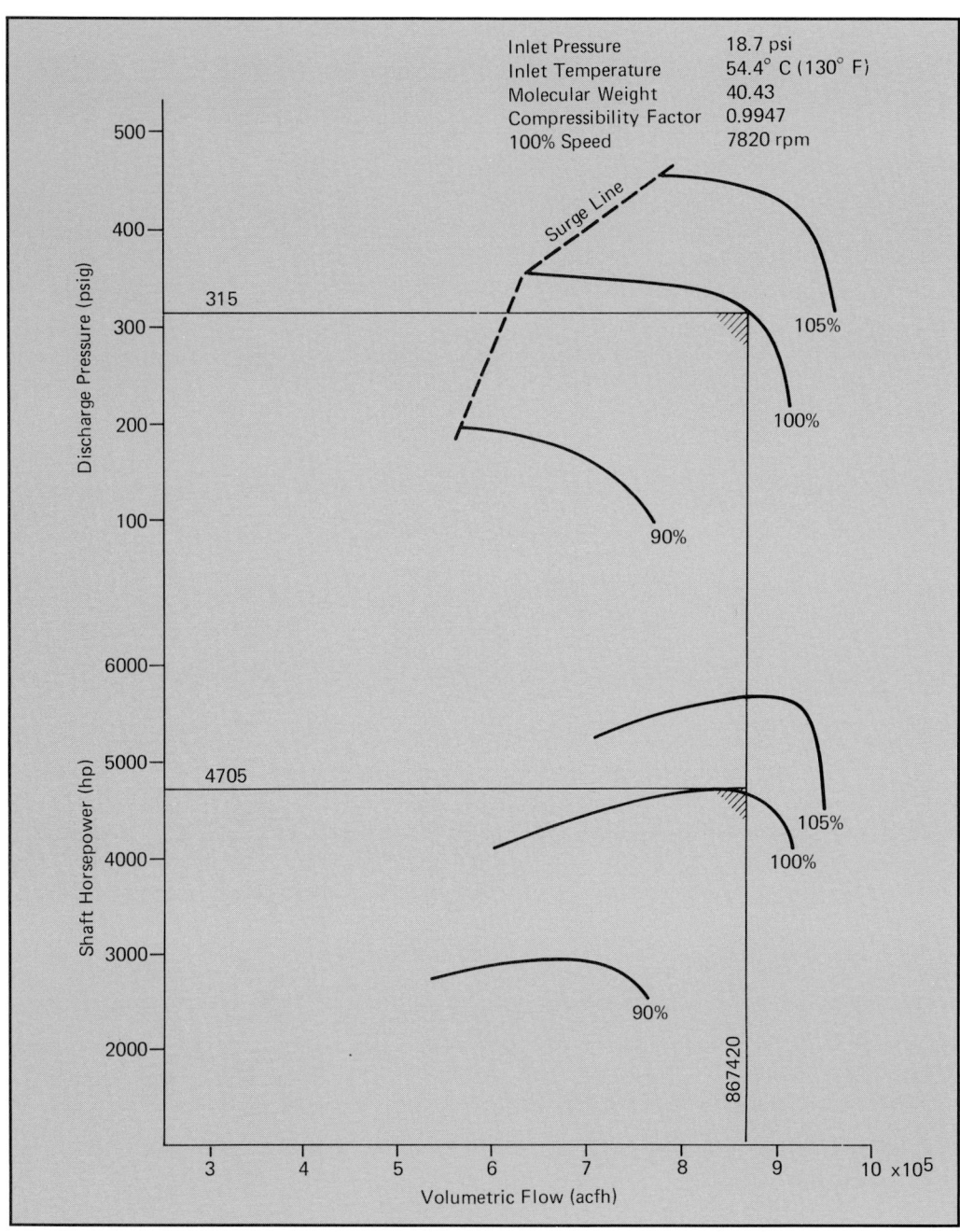

Fig. 5-17. (c) Overall Compressor Map for a 2-Stage CO_2 Compressor

the very least, condensate can be removed by the intercooling and knockout operations, reducing the flow to the next stage and possibly changing the molecular weight. This type of multistage operation can essentially be described as a series of separate compressors performing different duties but driven by a single shaft. Therefore, manipulating speed to accommodate changes in one stage would upset all of the other stages and could possibly drive them into surge. Variable speed is no longer a viable option as a manipulated variable under these conditions.[21]

Boyd[22] has described the complications of multistage operation. His experience with a 4-stage compressor underlined the requirement for adequate instrumentation and monitoring to observe the status of the compressor.

Finally, the compressor must be started up. Temperatures and molecular weights could be entirely different from those during normal operation. Simulation is a useful tool for designing start-up strategy,[23] while microprocessors can be used to insure the repeatability of a complex strategy. In some cases, instruments and process equipment dedicated to start-up alone are required to safely bring the compressor on line.

Exercises

1. (a) *What is the advantage of an inlet throttling valve over inlet guide vanes for capacity control? The disadvantage?*

   (b) *What is the advantage of inlet guide vanes over the inlet throttling valve? The disadvantage?*

2. *Why are diffuser guide vanes preferable to inlet guide vanes for flow control with an "all static" load curve?*

3. *Is variable speed a viable manipulated variable for the compressor providing constant header pressure in the compressor map shown here?*

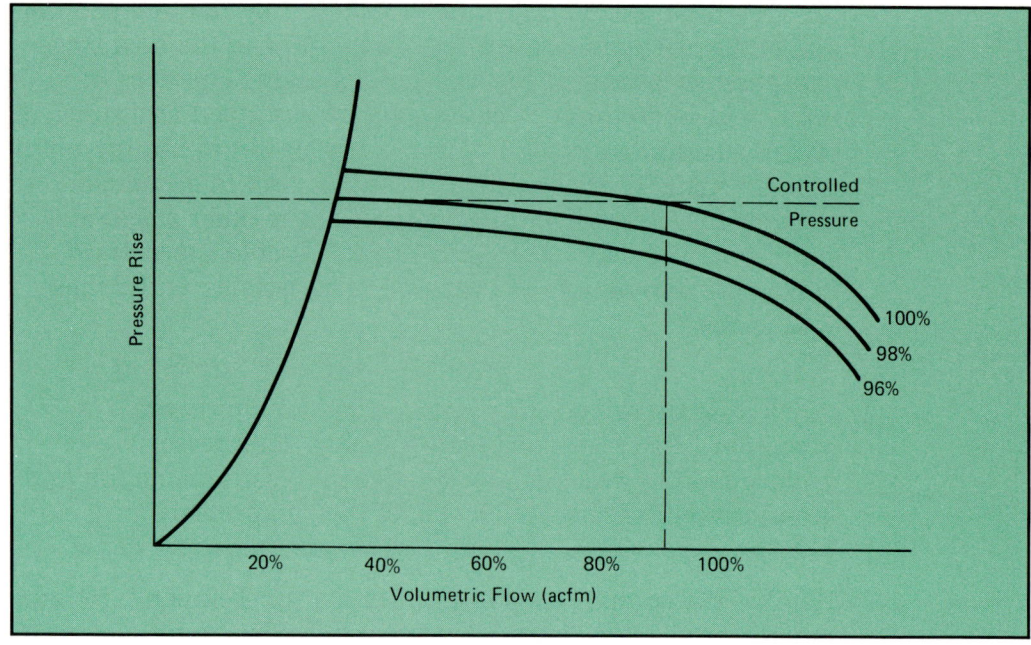

4. What is the ideal location for the primary flow element in Fig. 5-1? Is it as shown, or between valve and compressor, or in the discharge of the compressor? Why?

5. What are the disadvantages of flow elements and control valves in the discharge of the compressor?

6. Characterize the load curve shown in Fig. 5-7.

7. Partially closing a valve in the compressor discharge affects which part of the load line in Fig. 5-7?

8. Why does the surge line in Fig. 5-13 have the shape of a parabola, while the surge line in Fig. 5-7 "bends" the other way?

References

1. Buzzard, W. S., and G. Carlo-Stella, "Control of Dynamic Compressors," Fischer and Porter Application Bulletin 91-53G-06, 1973.
2. McMillan, G. K., *Centrifugal and Axial Compressor Control*, Instrument Society of America, Research Triangle Park, NC, 1983.
3. Neerken, R. F., "Compressor Selection for the Chemical Process Industries," *Chemical Engineering*, January 20, 1975, p. 79.
4. Daze, R. E., "How to Instrument Centrifugal Compressors," *Hydrocarbon Processing*, October, 1965, p. 125.

5. Nisenfeld, A. E., *Centrifugal Compressors, Principles of Operation and Control*, Instrument Society of America, Research Triangle Park, N.C., 1982.
6. Moore, R. L., *The Dynamic Analysis of Automatic Process Control*, Instrument Society of America, Research Triangle Park, NC, 1985.
7. B. F. Sturtevant Co., "Sturtevant Vane Control for Mechanical Draft Fans".
8. White, M. H., "Surge Control for Centrifugal Compressors," *Chemical Engineering*, December 25, 1972, p. 54.
9. Bulletin, Atlas Copco Energas GMBH, Am Rheinofer 20 Post Box 501150, D5000 Koein 50 (Suerth), Germany.
10. Moore, R. L., "Flow Measurement," *Measurement Fundamentals, Basic Instrumentation Lecture Notes and Study Guide*, Third Edition, Instrument Society of America, Research Triangle Park, NC, 1982.
11. Dresser Clark Division, "Watt Measuring System Functional Description and Troubleshooting Procedure," Dresser Industries, Inc., Olean, N.Y.
12. Nolar, R. L., and N. H. Robson, "Compressors Load Limit Controls," *Compressed Air*, Ingersoll Rand Co., Phillipsburg, NJ, June, 1982, p. 20.
13. Tezekjian, E. A., "How to Control Centrifugal Compressors," *Hydrocarbon Processing*, Gulf Publishing Co., Houston, TX, July, 1963.
14. Gaston, J. R., "Centrifugal Compressor Control," *Proceedings of the Fifteenth Annual ISA Chemical and Petroleum Instrumentation Symposium*, Instrument Society of America, Research Triangle Park, NC, 1974.
15. Schultz, H. M., R. K. Miyasaki, T. B. Leim, and R. A. Stanley, "Analog Simulation of Compressor Systems, A Straightforward Approach," *Instrumentation in the Chemical and Petroleum Industries, Volume 10*, Instrument Society of America, Research Triangle Park, NC, 1974.
16. Haden, R. D., "Chlorine Caustic Compressors," Lake Charles, LA
17. Gaston, J. R., "Centrifugal Compressor Operation and Control," *Centrifugal Compressor Operation and Control*, Instrument Society of America, Research Triangle Park, NC, 1976.
18. Weiner, S., "Multi-Stage Centrifugal Compressor Controls," *Proceedings of 1983 Annual Conference and Exhibit*, Instrument Society of America, Research Triangle Park, NC.
19. Stadler, E. L., "Understand Centrifugal Compressor Stage Curves," *Hydrocarbon Processing*, August, 1986, p. 51.
20. Baggio, J. L., and Y. Pouradier, "Method Forecasts Multistage Compressor Performance," *Orland Gas Journal*, September 1, 1980, p. 118.
21. Waggoner, R. C., "Process Control for Compressors," *Centrifugal Compressor Operation and Control*, Instrument Society of America, Research Triangle Park, NC, 1976.
22. Boyd, D. M., "How Accurate are the Surge Curves?" *Proceedings of the Fifteenth Annual ISA Chemical and Petroleum Instrumentation Symposium*, (San Francisco, CA) 1974, Instrument Society of America, Research Triangle Park, NC.
23. Marks, A. F., "Dynamic Analysis Can Eliminate Start-Up Problems," *Proceedings of Annual Conference and Exhibit*, 1973, Instrument Society of America, Research Triangle Park, NC.
24. Buzzard, W. S., "Controlling Centrifugal Compressors," *Instrumentation Technology*, November, 1973, p. 39.
25. Specification, Joy Mfg. Co., 3101 Broadway, Buffalo, NY, 14225.

Unit 6:
Surge Control

UNIT 6
Surge Control

Control of a centrifugal compressor must meet two fundamental requirements. First, the throughput must be matched to the demand, as described in Unit 5. Second, the compressor must be kept out of surge. Of these two objectives, surge prevention must take precedence over capacity control because of the disabling damage that can result from a few seconds of surging. Repair of the damage can cause the compressor (and the process) to be out of service for months. Well designed anti-surge control is mandatory. However, most anti-surge systems are energy intensive and must not operate before needed because of the cost.

Learning Objectives — When you have completed this unit, you should:

A. Understand the objectives of surge control.

B. Be aware of the various techniques for accomplishing anti-surge control.

C. Appreciate the judgment factor involved in the anti-surge system that is designed on fuzzy data, must perform flawlessly, yet is extremely expensive to operate.

6-1. General

Centrifugal compressors are energy intensive, very expensive machines, costing up to several million dollars. Monitoring and control can yield significant cost savings. The economics of pipeline compressors serve as an illustration of the potential. Hamell[1] describes a 12,500-hp gas pipeline compressor that consumes fuel costing from $1.5 to $2.9 million each year and requires operation and maintenance expenditures of over $350,000. Costs like these provide strong economic incentives for well designed compressor control systems. The controls must allow the compressor to operate to meet process requirements without undue hazard, while simultaneously avoiding the energy waste represented by venting or recycling.

Surge control is effected by the following:

- Increasing the throughput flow
- Decreasing the head required
- Decreasing the speed

The most dependable, and thus the most widely used, method of surge control is increasing the throughput by opening a bypass valve that either recycles gas around the compressor or blows gas off to the atmosphere. Since bypassing or venting gas wastes power, it is desirable to determine surge flow as accurately as possible to avoid bypassing gas unnecessarily while maintaining safe operation. However, determining surge flow is not a simple matter. No detector measures surge directly. Inlet volumetric flow in acfm is not a particularly convenient variable to measure since it is affected by a number of other variables. Unit 3 showed that the intersection of the performance curve and the surge curve can be denoted by many combinations of the measurable variables' inlet pressure (p_i), discharge pressure (p_d), inlet temperature (T_i), speed (s), and the unmeasurable variables' ratio of specific heats (k), compressibility (Z), and molecular weight (M). Thus, surge conditions can be defined completely in terms of variables other than flow, and, in fact, surge control systems are calibrated to an operating envelope defined on a graph of chosen variables. The problem of defining surge conditions has led to the development of a wide variety of systems for determining surge conditions from different measurements.

The simplicity (or complexity) of the surge control system for a particular application will depend on a number of factors, including:[2,3]

- *Kind of compressor*—A constant speed compressor has one less variable affecting surge conditions than a variable speed machine, and determination of surge conditions is simpler. Similarly, more complex controls are required on a multistage compressor than on a single-stage machine.
- *Load changes*—Applications where load conditions, such as inlet temperature or molecular weight, exhibit little variation are amenable to simpler surge systems where these variables are treated as constants.
- *Ease and availability of measurements*—The simplest and most economical surge control system will use variables already being used for other purposes. Similarly, the

measurement of some variables is not feasible, e.g., the long straight run of large pipe for an inlet flow measurement is sometimes prohibitively expensive, and a surge control system utilizing other measurements is desirable.
- *Accuracy of surge control required*—A rudimentary control system is adequate for small compressors with safety being insured by controlling well above surge flow and the knowledge that the horsepower is small enough that extensive surging will have to take place to do discernible damage. The price to be paid is bypassing or venting more flow than is necessary. With a larger compressor, the power loss would be more serious, and a more accurate control that would permit operating closer to the minimum flow would be required.

Because of the large number of possible variations in compressor surge control systems to meet process objectives and measurement limitations, some of the more widely applied systems are described. Several surge control systems are designed specifically for use with limited conditions such as constant compressor speed or constant discharge pressure. Other examples are applicable to systems for control of different process variables such as discharge pressure or gas flow. The systems described by no means exhaust the possible configurations for surge control loops, but the principles illustrated suggest additional systems to accommodate other process objectives.

6-2. Constant Speed Compressor with Inlet Throttling

A typical compressor installation is an air compressor driven by a constant speed electrical inductance-type motor. The compressor is required to maintain pressure in a discharge piping header that transports air throughout the operating plant complex. Frequently, the header will be supplied by a number of compressors bought at different times from different manufacturers as demand grew. The optimum operation of multiple compressors is considered in later units. For the single compressor shown in Fig. 6-1, discharge pressure is used to regulate capacity. Constant speed dictates that the compressor will always follow the performance curve shown on Fig. 6-1(b). Thus, a point can be chosen on the performance curve at a predetermined distance from the surge curve (the surge margin) at which the vent valve will open and prevent flow from decreasing past that point. This surge

Unit 6: Surge Control

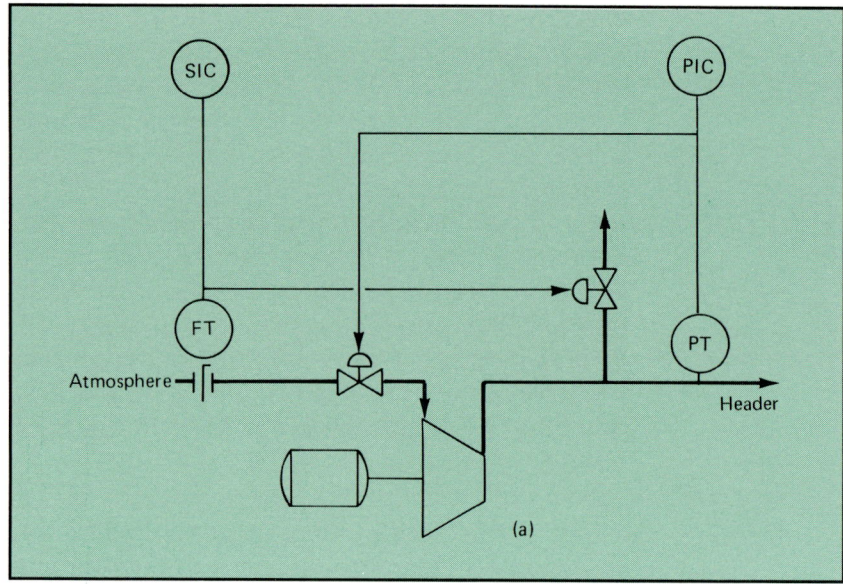

Fig. 6-1. (a) Typical Capacity and Surge Control Systems on a Constant Speed Air Compressor

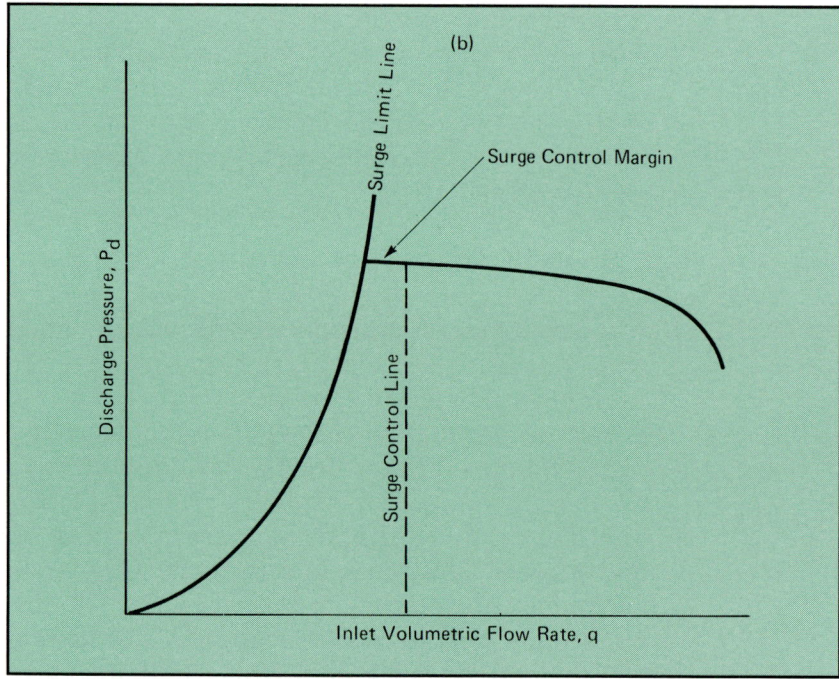

Fig. 6-1. (b) Compressor Map for Air Compressor Showing Constant Flow Surge Control Line

Unit 6: Surge Control

control point will be the set point of the anti-surge controller (ASC).

Actually, the performance and surge curves are not single lines but will move with inlet (ambient) pressure and temperature, as shown on the map for an actual compressor, Fig. 6-2. Given that the compressor must provide header pressure (220 psig), the mass flow of air that can be delivered at the controlled pressure varies with suction pressure and temperature. Pressure varies with the opening of the inlet throttling valve, as shown in Fig. 6-2. Since temperature varies from summer to winter, operation will change from

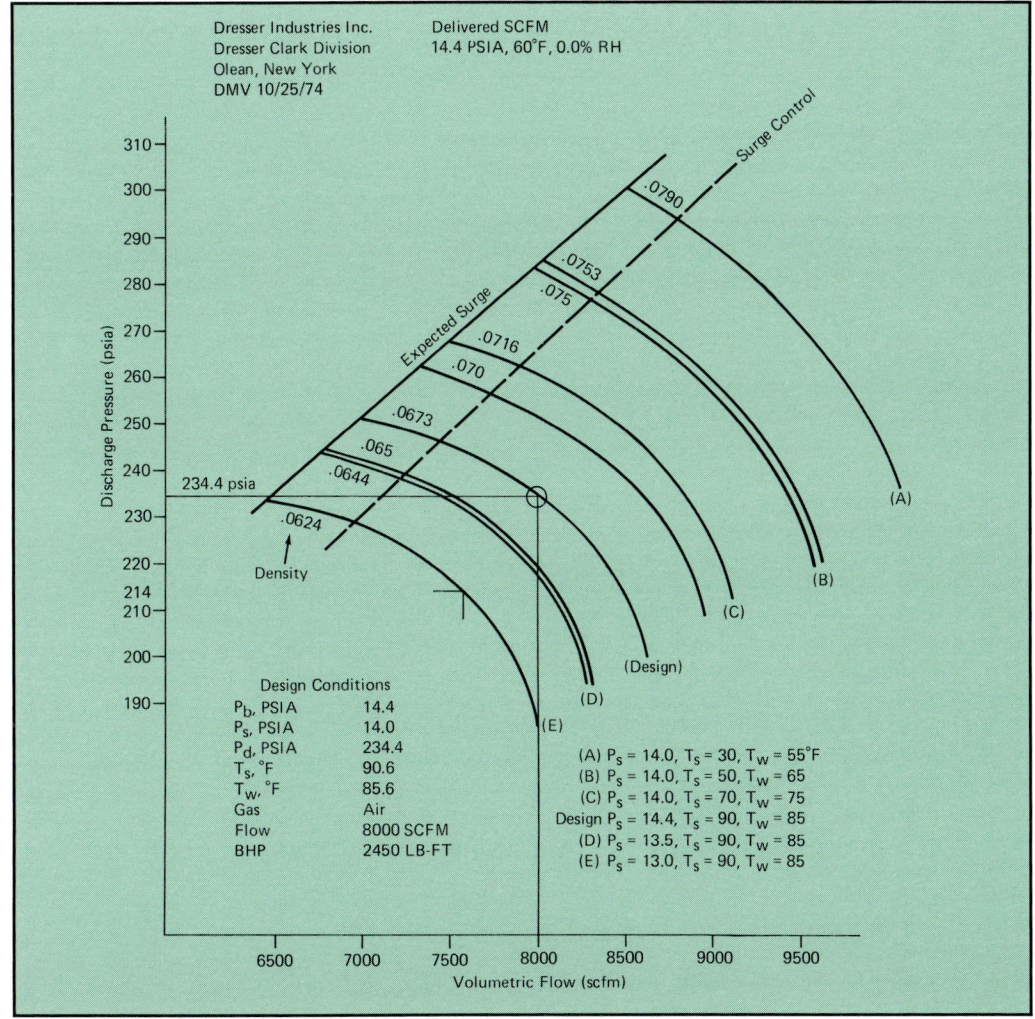

Fig. 6-2. Compressor Map for Dresser Clark ISOPAC™ Constant Speed Air Compressor

performance curve to performance curve with flow demand and season.

Each performance curve will have its own surge curve or surge point for a single-speed machine. The locus of those surge points is shown as the "expected surge" line on Figure 6-2. The expected surge line indicates that the surge point will change during operation and that this change can be quantified by information on the compressor map. Using this information, the surge control system can be compensated for changes in surge point by using the existing measurement of discharge pressure. The compensated system, known as pressure-compensated surge control, is illustrated by the following example.

Example 6-1. Consider the diagram Fig. 6-3 and the compressor map Fig. 6-2. For pneumatic instruments, let

p_d = Discharge pressure, psi
θ_p = Pressure transmitter output, 3-15 psi
a = Gain and bias relay output, 3-15 psi
θ_q = Flow transmitter output, 3-15 psi
q = Flow, scfm
mv = Measured variable to controller, 3-15 psi

Transmitter Ranges:

Pressure transmitter: 160 to 260 psig
Flow transmitter: 0 to 10000 scfm

For the gain & bias relay (R)

$$a = \alpha \theta_p + B$$

From the compressor map, on the surge control line

q = 8000 scfm, θ_q = 12.6 psi, p_d = 250 psi, θ_p = 13.8 psi
q = 7120 scfm, θ_q = 11.5 psi, p_d = 220 psi, θ_p = 102 psi

Then

$$12.6 = 13.8 \alpha + B$$
$$11.5 = 10.2 \alpha + B$$

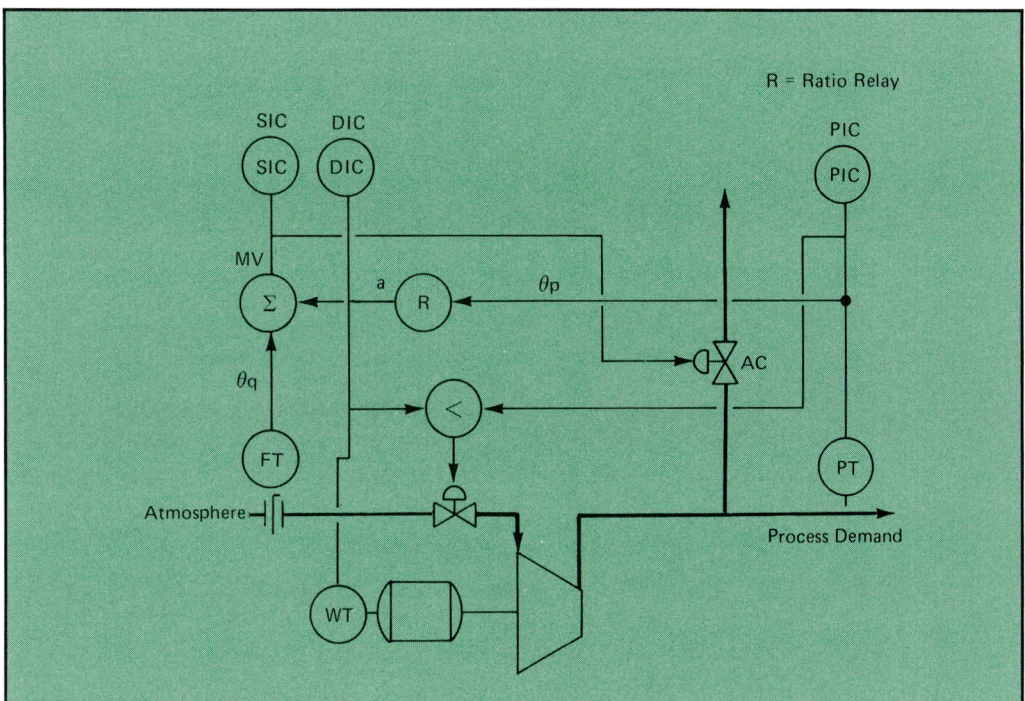

Fig. 6-3. Pressure-Compensated Surge Control System for Clark Compressor

and

$$\alpha = 0.305$$
$$B = 8.38$$

But, let bias (B) be zero for ratio relay, and put all required bias in the summer.

The summer must calculate:

mv = θ_q − a ± k = SP
 k = Summer bias
 SP = Controller set point in 3-15 psi signal span
    = 7120 scfm = (0.712)(12) + 3 = 11.5 psi
 p_d = 220 psi, θ_p = 10.2 psi, a = 3.11 psi
 q = 7120 scfm, θ_q = 11.5 psi

For design condition, mv = SP

$$11.5 = 11.5 - 3.11 \pm k$$
$$k = 3.11$$

And, for the summer,

$$mv = \theta_q - a + 3.11$$

Check for point at p_d = 250 psi, mv = SP = 11.5 psi

$$11.5 = \theta_q - (0.305)(13.8) + 3.11$$

$$\theta_q = 12.6 \text{ psi}$$

$$q = 8000 \text{ scfm}$$

Example 6-1 uses pneumatic instrumentation as a vehicle. Equations would have to be changed to accommodate the pneumatic or electronic instruments of a chosen manufacturer. Similarly, the formulation could be changed for programming into a microprocessor-based controller.

Figure 6-4 shows a discharge pressure control system with surge protection based on *controlling minimum electrical power to the constant speed motor*. Horsepower can be defined:

$$hp = \frac{(K)(p_d - p_i)(m)}{\rho} \tag{6-1}$$

Discharge pressure (p_d) is controlled and limited by the performance curve, and inlet pressure (p_i) varies below atmospheric with the closing of the inlet throttling valve. The major effect on horsepower (hp) is the mass flow (m), which is:

$$m = \frac{(K_1)(p_i)(M)(q)}{(T_i)(Z_i)} \tag{6-2}$$

Thus, the horsepower (hp) is essentially proportional to volumetric flow rate (q) for controlled discharge pressure (p_d) and constant inlet temperature (T_i). Surge control by minimum power is largely limited to air compressors powered by constant speed electrical motors and pumping into pressure-controlled headers.

Warnock[3,4] has offered the system shown in Fig. 6-5 to compensate for varying inlet temperature (T_i) and discharge pressure (p_d). The ratio relay sets the slope of the wattage-

Unit 6: Surge Control

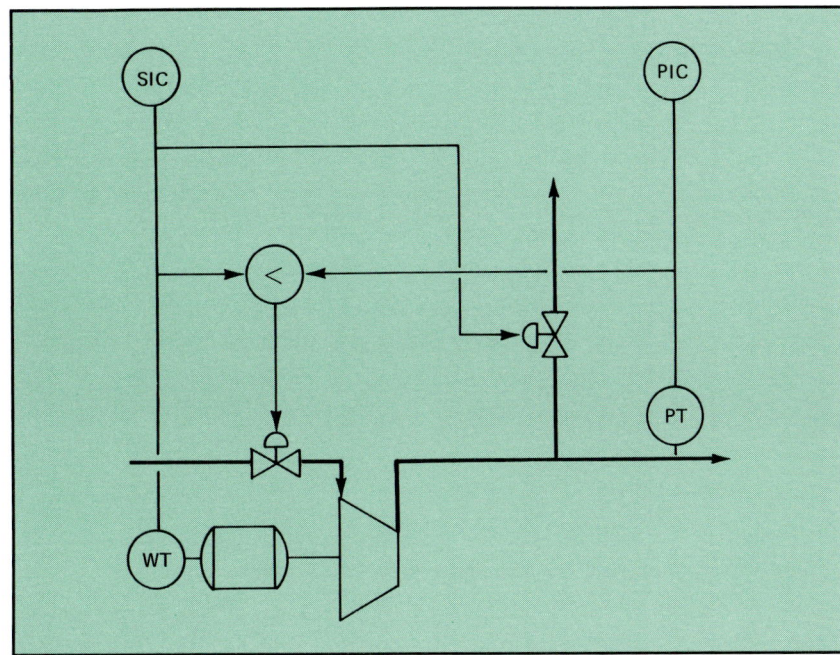

Fig. 6-4. Surge Control System Based on Motor Wattage with Wattage Also Used To Prevent Motor Overload

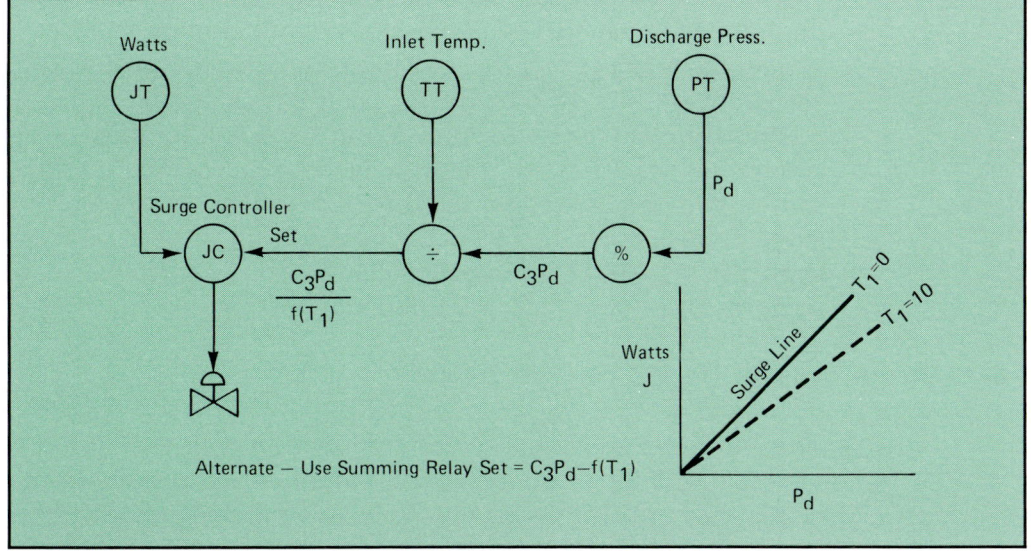

Fig. 6-5. Minimum Power Surge Control with Compensated-for Inlet Pressure and Temperature Changes[3]

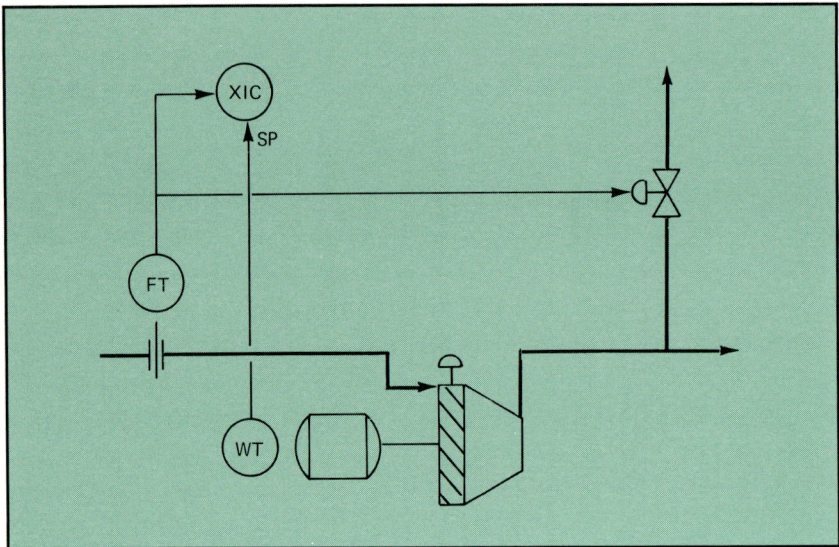

Fig. 6-6. Minimum Power Surge Control System

discharge pressure surge line. A dividing relay shifts the slope of the surge line with temperature. This compensation is an approximation that is useful for limited temperature ranges.

Buzzard[5] has used surge control by minimum power for significant simplification in an actual installation. The surge line horsepower used by a constant speed blower operating with atmospheric inlet and throttled by inlet guide vanes was observed to be linearly related to the differential produced on the inlet orifice meter, regardless of inlet temperature variations of from $-5°F$ to $+95°F$ and guide vane position variations of from 10% to 100% "open". A square law relationship between horsepower and head is found by combining eqs. (6-1), (6-2) and (3-34) for volumetric flow rate. This is apparently compensated by the "squared" locus of surge points shown in Fig. 5-7. The resulting surge system is shown in Fig. 6-6. Serendipity sometimes appears in compressor control, but don't depend on it.

6-3. Surge Control Based on Fan Laws[3,4,6]

The Fan Laws are discussed in conjunction with eqs. (3-37), (3-38), and (3-39) and are shown in another form in eq. (5-2). While they are strictly true only for low head machines like fans, they serve as a reliable engineering approximation for larger compressors. A simple anti-surge system for a variable speed compressor results from the application of the Fan Law, q/s equals a constant, to the surge curve. Thus, the surge control is based on the linear relationship:

$$q = Ks \qquad (6-3)$$

The system is shown in Fig. 6-7. The surge control is the minimum flow system shown in Fig. 6-1(a). When process demand reduces inlet flow below the set point, the surge controller opens the bypass valve to maintain minimum flow through the compressor. Controller set point is calculated from the speed measurement and reduces the surge set point as speed is reduced; if flow decreases because of a speed reduction, the surge control valve begins to open at a lower flow.

The adjustable gain and adjustable bias in the system are provided to (1) allow an adjustable surge margin between surge limit line and surge control line; and (2) adjust the calculated surge control line to accommodate actual conditions in the field as found by experimentation.

The effect of the Mach number (the ratio of gas velocity to the velocity of sound) is very important as the gas velocity at the

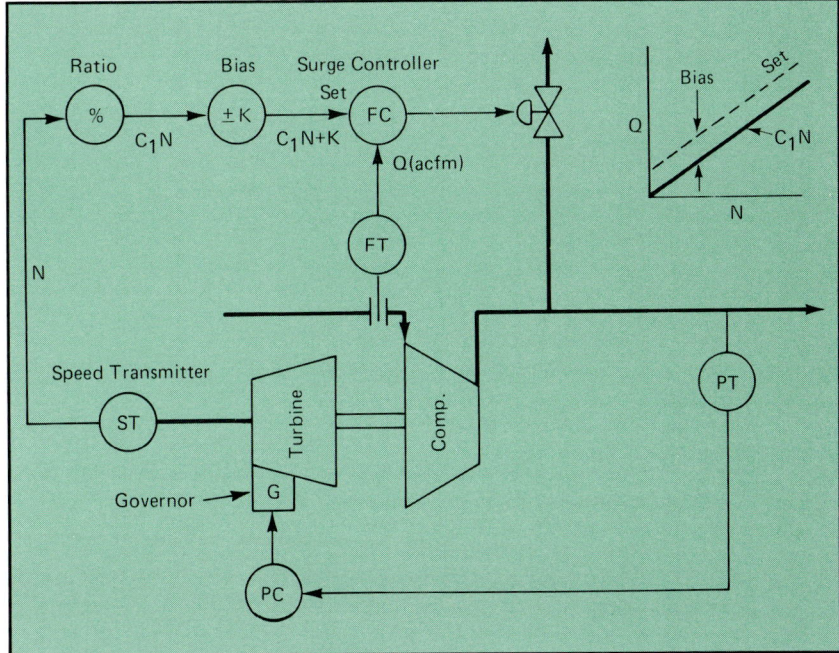

Fig. 6-7. Constant Flow/Speed Surge Control System[3]

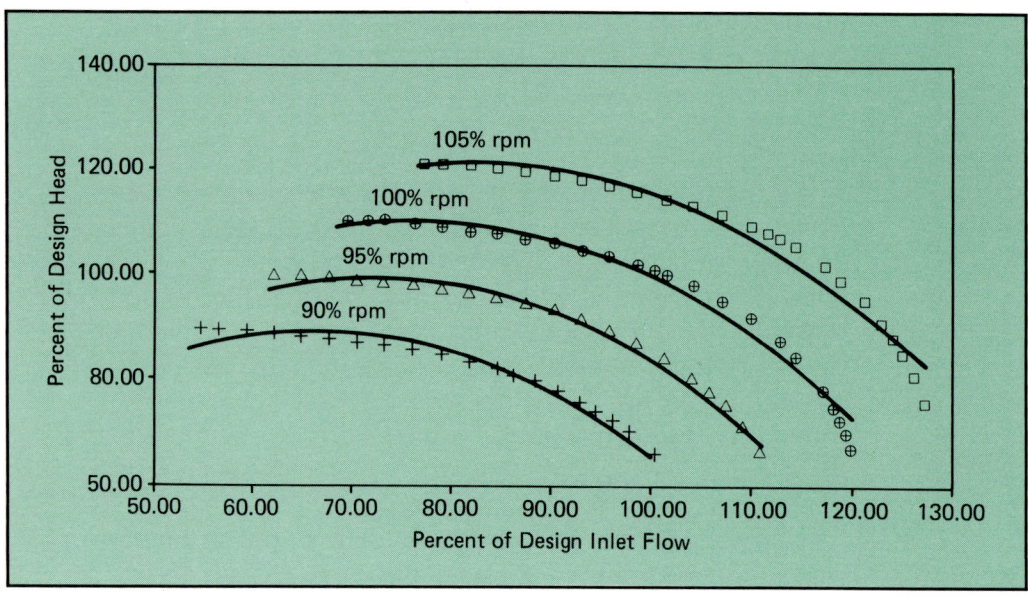

Fig. 6-8. (a) Compressor Performance Curves Developed from Correlations of Manufacturer's Data

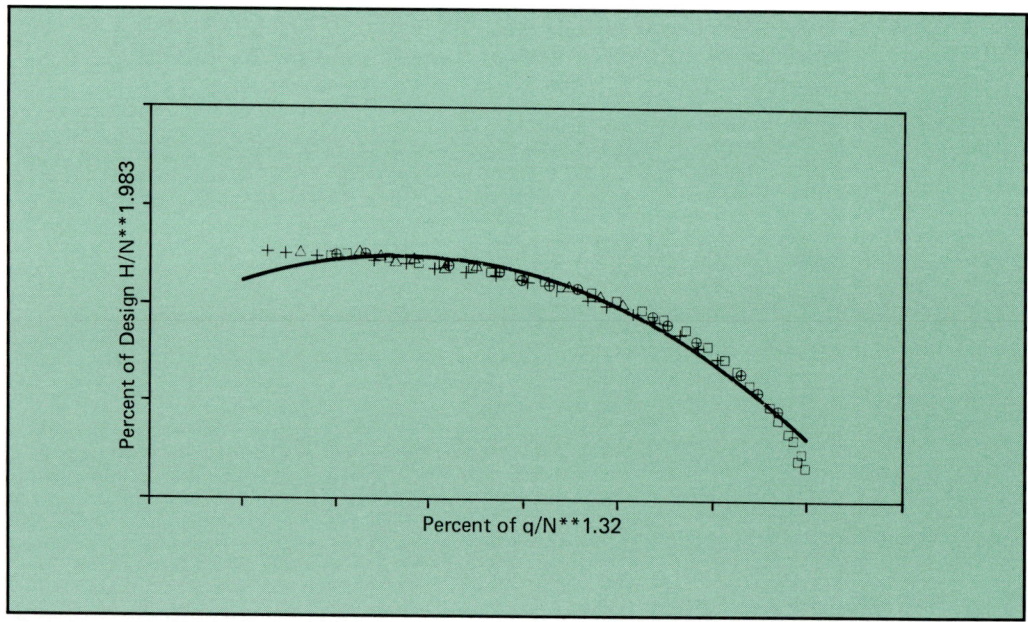

Fig. 6-8. (b) Reduced Performance Curve

compressor entrance approaches sonic velocity (Mach No. = 1) in the gas medium. This limits the volumetric flow rate through the compressor and is identified as the "stonewall" effect formerly described (Unit 3). The Mach effect is assumed negligible in the operating region of the compressor map. Under these conditions for a given compressor, the results of dimensional analysis indicate that if the quantity H/s^2 is plotted against q/s, the curves for different speeds of a variable speed compressor will collapse into a single curve on the compressor map. This technique is widely used in simulating[7,8] and rerating compressors and is merely a restatement of the Fan Laws.

This simplifying procedure has been applied to a number of performance curves for compressors supplied by various manufacturers. The data points were scattered in each case. This result is to be expected, since relationships developed for fans are being applied to compressors while ignoring compressib ity effects represented by the Mach number. Curve-fitting techniques result in the following relationships for one of the compressors analyzed:[8]

$$x = q/s^{1.32}$$
$$y = H/s^{1.983}$$

The "goodness of fit" of these relationships is shown in Fig. 6-8.

This analysis indicates that a surge control based on the Fan Laws will require (1) a final adjustment while in service, and (2) a relatively wide surge control margin because of uncertainties.

6-4. Surge Control Based on Other Variable Pairs

Surge control is so difficult to implement, both because of the "fuzzy" data on which it is based and the complexity of any control configuration that accommodates all possible variables; yet failure is so devastating that calibrated anti-surge systems have been based on many different pairs of variables. The various pairings have had varying degrees of success. Systems based on the following variables have been presented as viable combinations:

- Brake horsepower versus mass flow[9]
- Pressure ratio versus Mach number squared[9]
- Pressure ratio versus flow element differential pressure[10]
- Polytropic head versus speed[10]

Figure 6-9 shows the extreme sensitivity of the location of the surge limit line to an error in head measurement. The figure shows that the same relative error tolerance in a head-measuring system will produce a much larger error in fixing the surge point when close to the surge limit line than would be corresponding flow error. Regardless of the scaling factor for head and flow, the lines of constant speed approach a point of tangency to lines of constant head at surge, making the location of the surge point by speed and head extremely

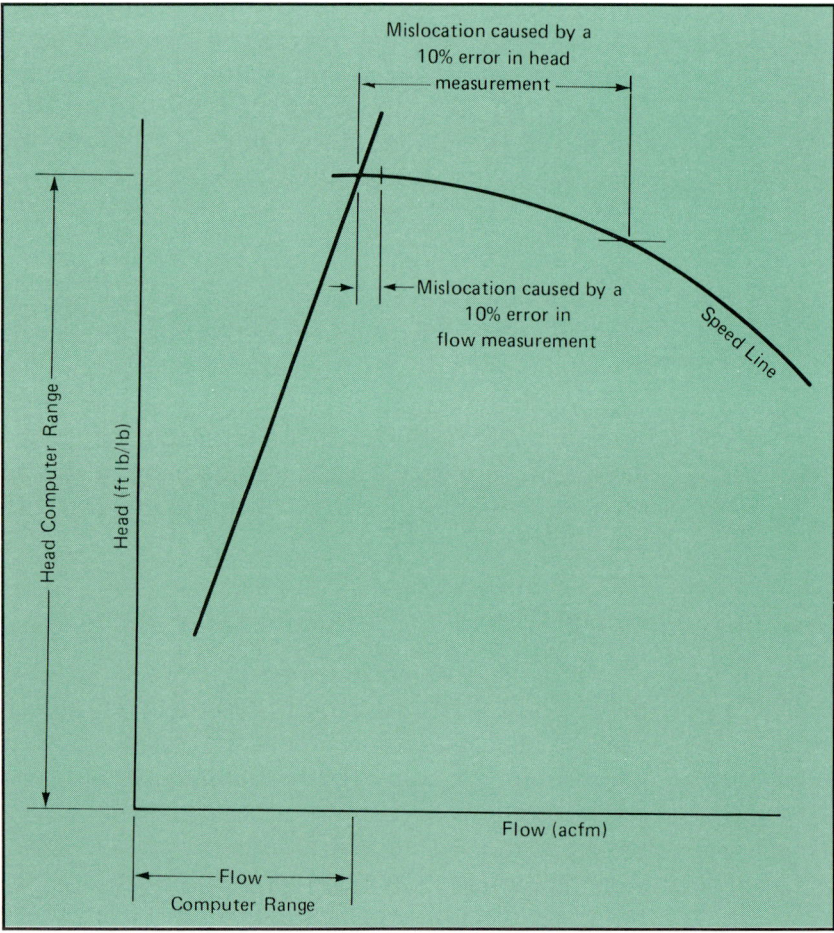

Fig. 6-9. Relative Errors Caused by the Same Inaccuracy of Measurement of Head and Flow[10]

difficult. This type of system is thus limited to machines having very steep performance curves.[10]

In addition to systems based on measured or computed physical variables, empirical equations of the surge limit line have also been found to be useful.[11] Consider the empirical equation

$$R_c = \frac{Ks^2}{T_f} + 1 \qquad (6\text{-}4)$$

Each side of this equation is computed and compared in a microprocessor-based controller. When the left-hand side of the equation (the compression ratio, R_c) increases in magnitude to be greater than the right-hand side, the controller opens the anti-surge control valve. The equation was written in the following form in the actual application:

$$\frac{\Delta p}{p_s} = \frac{Ks^{2.2}}{T_f} \qquad (6\text{-}5)$$

An advantage of the use of eq. (6-5) is that it eliminates the need for a flow measurement. The equation provided a very close match to the surge curve on the Solar Saturn compressor and permitted operation close to the surge limit line with safety.[11]

6-5. Simplified Surge Control System

The simplified surge control system for variable speed compressors developed by M. H. White in 1972[12] has found general application in industry. The concept has been widely discussed since that time.[3,4,10,13,14,15,16] The design philosophy of the system is that of matching a surge prevention control to the requirements of both the machine and its associated system with emphasis on the economics of control selection.

The simplified system assumes that the Fan Laws are valid along the surge limit line:

$$q = K_1 s \qquad (6\text{-}6)$$

$$H = K_2 s^2 \qquad (6\text{-}7)$$

Solving simultaneously to eliminate the speed (s),

$$H = K_3 q^2 \qquad (6\text{-}8)$$

Then consider the volumetric flow equation,

$$q = K_4 \sqrt{\frac{h Z T_i}{p_i M}} \quad (6\text{-}9)$$

Then, combining eqs. (6-8) and (6-9),

$$H = K_5 \, h \left[\frac{Z T_i}{p_i M}\right] \quad (6\text{-}10)$$

But, head (H) is difficult to measure. Consider a more convenient measurement—differential pressure (Δp),

$$\Delta p = p_i (R_c - 1) \quad (6\text{-}11)$$

and

$$H = \frac{1545 \, Z T_i}{M \sigma} (R_c^\sigma - 1) \quad (6\text{-}12)$$

where $\sigma = (n - 1)/n$. For simplicity, assume that:

$$R_c^\sigma - 1 = K_6 (R_c - 1) \quad (6\text{-}13)$$

Figure 6-10 shows the nonlinearity that must be linearized in order to implement this assumption for adiabatic compression ($n = k$). The approximation is quite accurate for gases with large ratios of specific heats and also for low pressure ratio (R_c) machines handling a fluid with a specific heat ratio of 1.3 or above.[10] Figure 6-11 shows the linearization to be within $\pm 5\%$ for a machine with a ratio of 1.5 and a low compression ratio.

Then, obviously,

$$R_c - 1 = \frac{\Delta p}{p_i} \quad (6\text{-}14)$$

Given that eq. (6-13) is reasonably valid, then combining eqs. (6-10), (6-11), (6-12), (6-13), and (6-14) results in

$$K_6 \left[\frac{1545 \, Z T_i}{M \sigma}\right] \left[\frac{\Delta p}{p_i}\right] = K_5 h \left[\frac{Z T_i}{p_i M}\right] \quad (6\text{-}15)$$

Cancelling the common variables on both sides of the equation

Unit 6: Surge Control

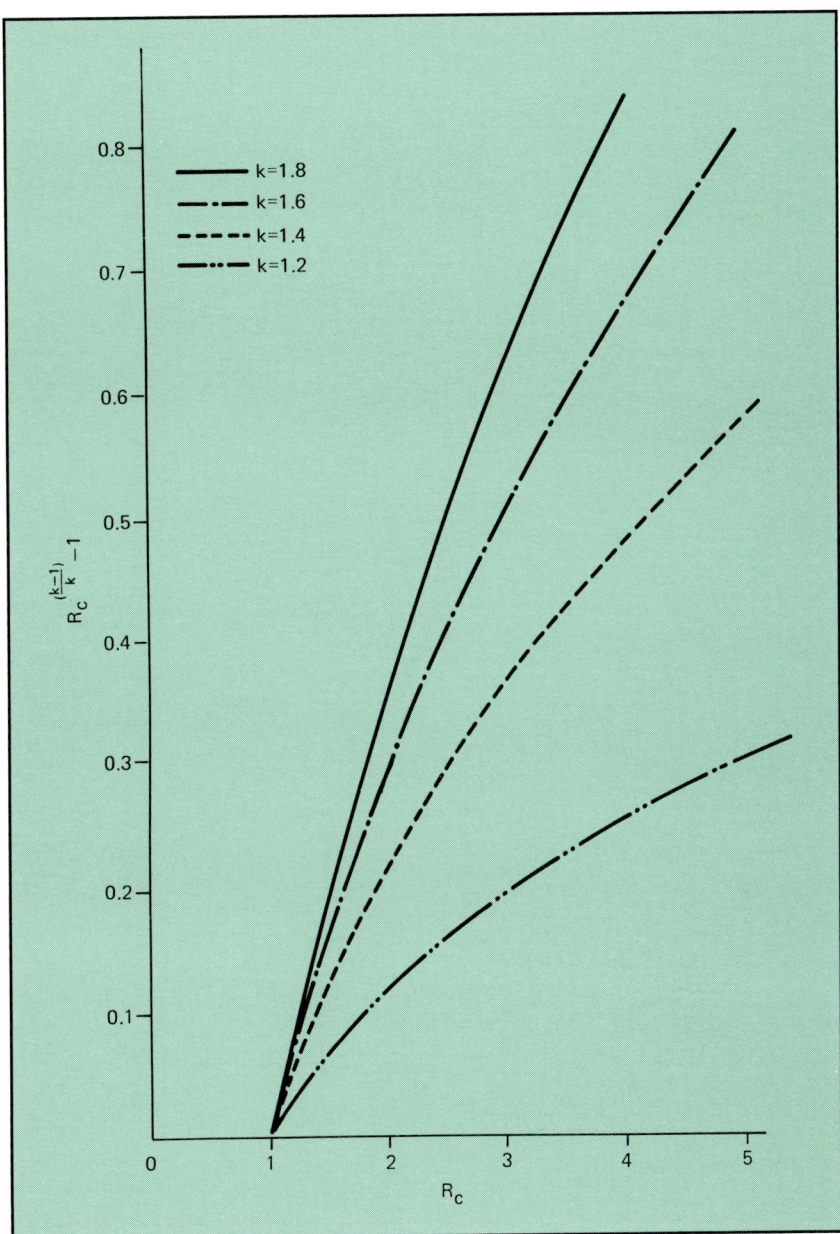

Fig. 6-10. Relationship of Function $R_c \frac{k-1}{k} - 1$ for Various Values of R_c and k^{10}

and collecting all the constants into a single constant, this reduces beautifully to

$$K_7 \, \Delta p = h \tag{6-16}$$

This relationship is independent of compressibility (Z), inlet temperature (T_i), inlet pressure (p_i), and molecular weight (M).

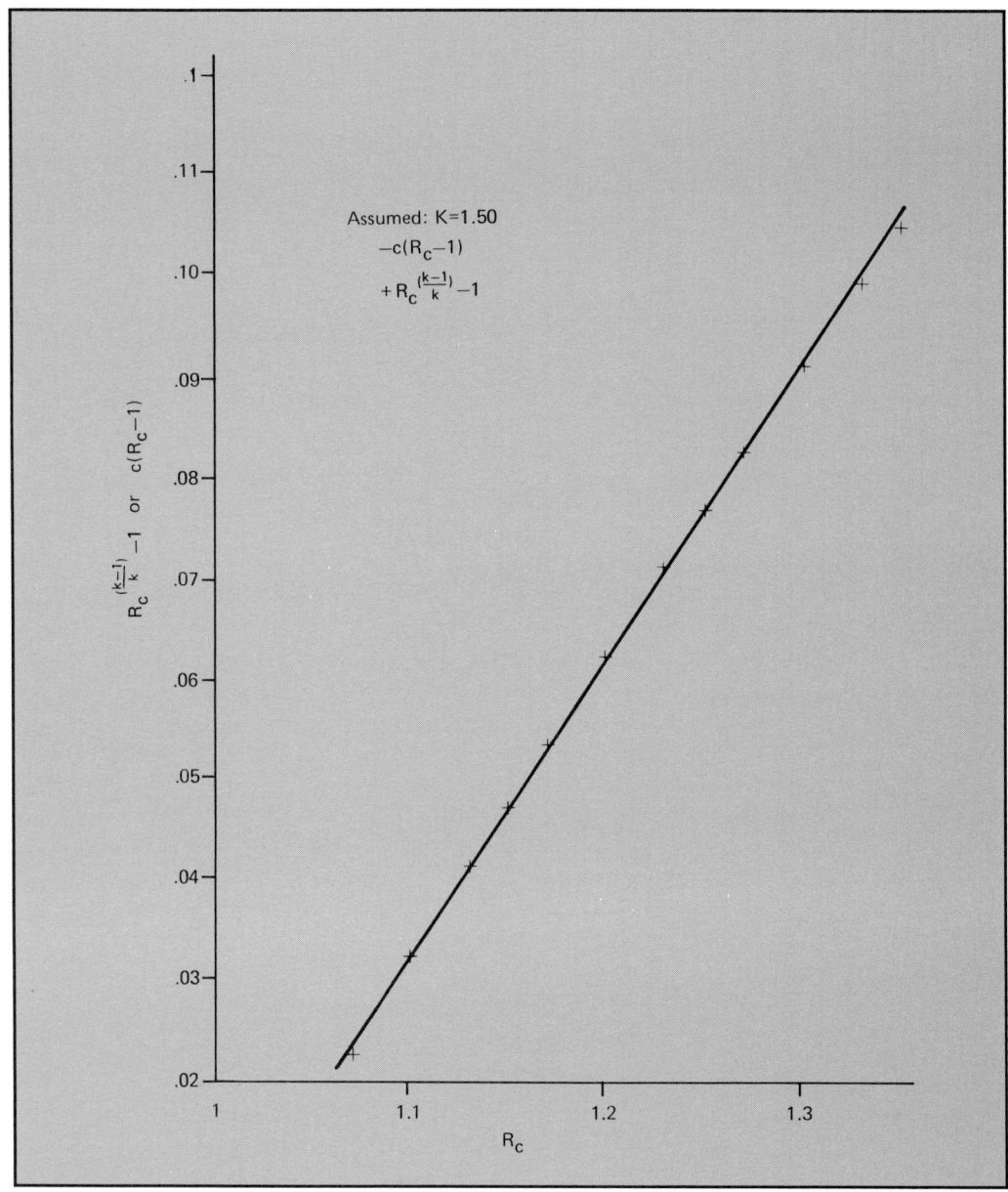

Fig. 6-11. Function Comparison of the Difference between $R_c \frac{k-1}{k} - 1$ and $C(R_c - 1)$ at Various Values of R_c[10]

It thus represents a simple, linear expression with an analytical foundation that is easily implemented. The control configuration used to implement the foregoing relationship is shown in Fig. 6-12. The surge controller (SIC) has as its measured variable (mv) the flow measurement signal from the flow transmitter (FT). The compensated compressor pressure

Unit 6: Surge Control

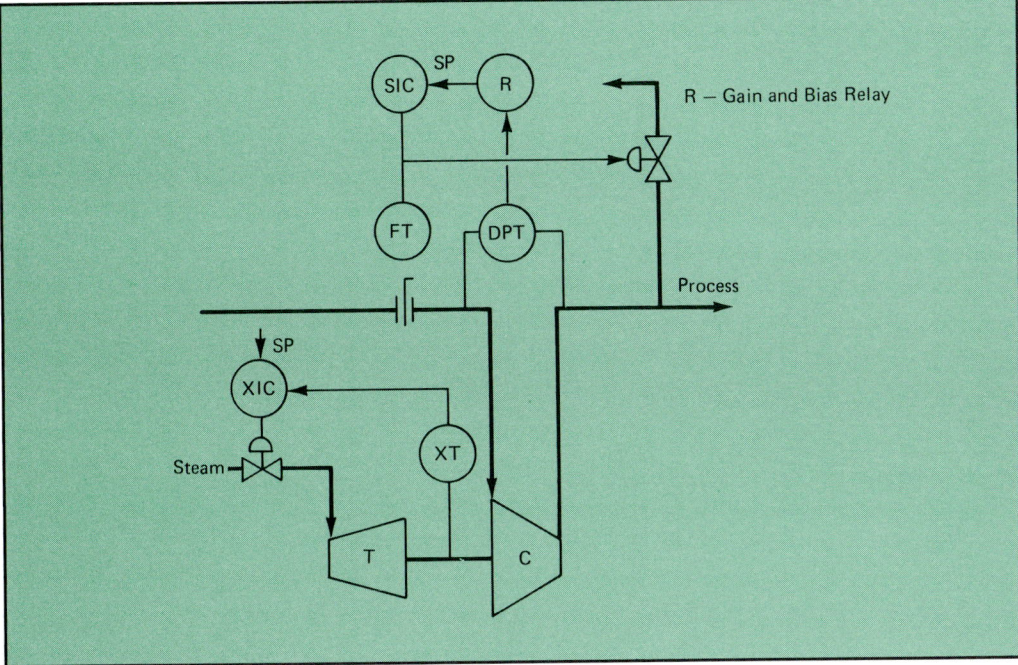

Fig. 6-12. Surge Control System for a Variable Speed Compressor Based on Compressor Δp and Flowmeter h

rise signal from the differential pressure transmitter (DPT) serves as the controller set point. Quite obviously, the signals (mv and SP) to the controller can be reversed. However, this arrangement appears as a flow control loop with a calculated set point when being serviced in the field, thus simplifying the troubleshooting function.

Figures 6-13a, b, and c show the positioning of the surge control line when plotted on the orifice head (h) versus compressor pressure rise (Δp) with the gain and bias relay on Fig. 6-12 calibrated to the following:

$$SP = (a)(DPT) + b \qquad (6\text{-}17)$$

Then from Fig. 6-13,

- Figure 6-13a—Relay calibrated to a = 0 and b = SP. This rudimentary system, discussed earlier, has the disadvantages of the danger of surging at higher pressure increases and excess circulation at lower speeds.
- Figure 6-13b—Relay calibrated to a = constant and b = 0. This system has the advantage of a simpler ratio relay but

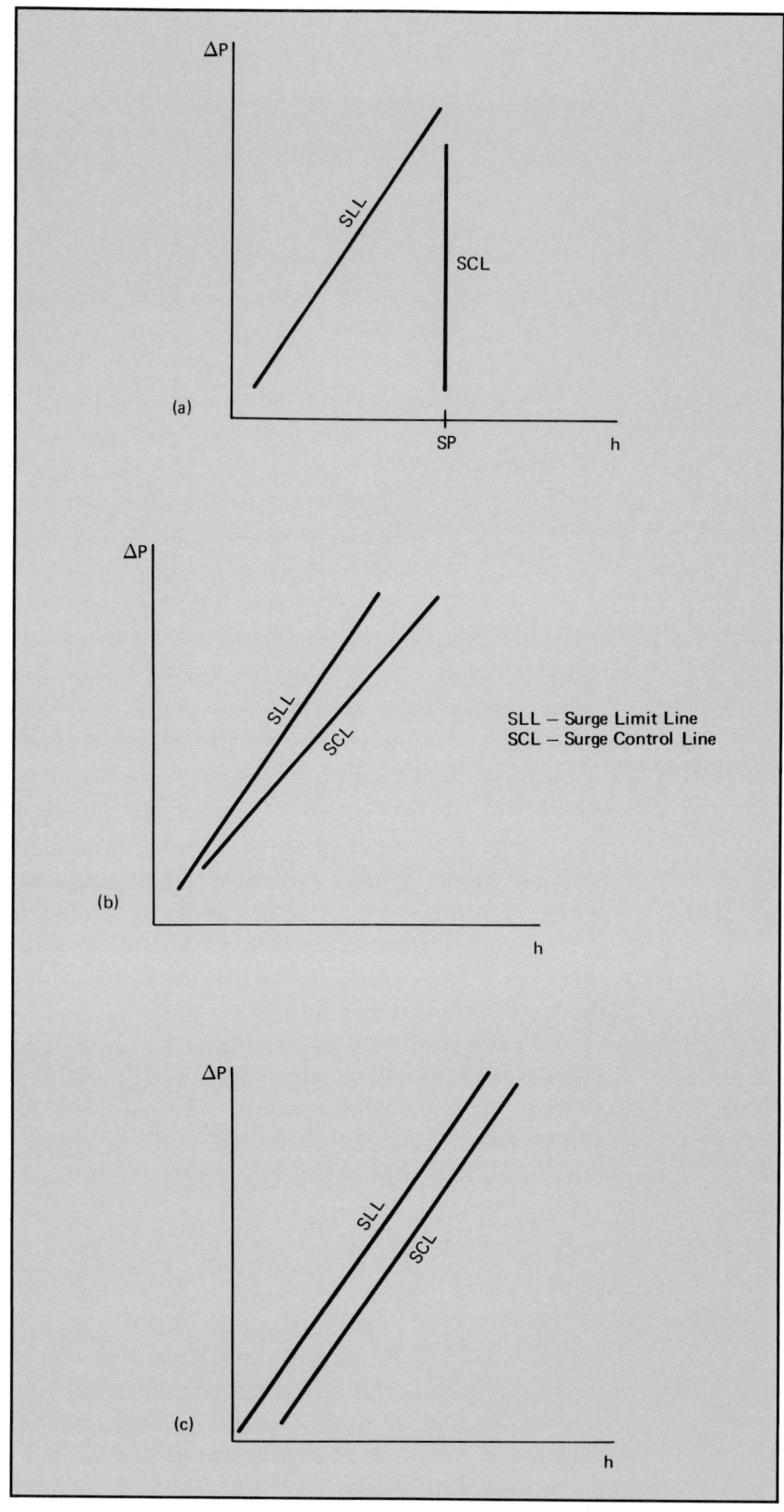

Fig. 6-13. Locus of Surge Control Line with Adjustment of Gain and Bias Relay

has the disadvantage of providing less surge protection at lower speeds.
- Figure 6-13c—Relay calibrated to gain and bias, as per eq. (6-16). This system provides minimum adequate surge protection across the speed range.

Example 6-2. Calibrate the gain and bias relay shown in Fig. 6-12 to provide anti-surge protection as shown in Fig. 6-13c for the compressor described by the map shown as Fig. 6-14c.

Basic Data:

Flow Measurement = 0-25 inches water for 0-100,000 LB/HR
Δp across compressor = 0-400 psi
Surge point @ 100% speed = 7820 rpm
$\quad \Delta p = 315$ psia $- 18.7$ psia $= 296.3$ psi
$\quad q = 635000$ acfh (67945 LB/HR)

Calculate surge control point with 8% surge control margin:

$$q = (1.08)(635000) = 685800 \; (73381 \text{ lb/hr})$$

Calculate flow element pressure difference (h) at surge line:

$$\frac{67945}{100000} = \sqrt{\frac{h}{25}} \quad h = 11.54 \text{ inches}$$

And at the surge control line:

$$\frac{73381}{100000} = \sqrt{\frac{h}{25}} \quad h = 13.46 \text{ inches}$$

Convert to pneumatic scale (3-15 psi span)

$$\Delta p = \left(\frac{296.3}{400}\right)(12) + 3 = 11.89 \text{ psi}$$

$$h_{SL} = \left(\frac{11.54}{25}\right)(12) + 3 = 8.54 \text{ psi}$$

$$h_{SC} = \left(\frac{13.46}{25}\right)(12) + 3 = 9.46 \text{ psi}$$

Calibrate A gain and bias relay with the formula:

$$h = a(\Delta p) + b$$

Note that the formula will vary with relay manufacturer. Calculate a:

$$h = a(\Delta p)$$

$$a = \frac{8.54 - 3}{11.89 - 3} = 0.623$$

This is the slope of the surge control line (SCL) shown in Fig. 6-13c.

Then,

$$b = 9.46 - 8.54 = 0.92 \text{ psi}$$

and

$$h = 0.623 \, \Delta p + 0.92$$

The compressor described by Fig. 6-14 is shown in Fig. 6-15. It is a two-stage (two-section) machine that removes water in the interstage cooler and could accommodate gas removal or addition between the stages. Thus, it could be described as two separate compressors performing different duties but driven by a single shaft. Each stage has its own set of operating curves and surge lines, as shown by Figs. 6-14a and 6-14b. The machine is designed for compatibility of the stages at 100% speed. Since the stages are different, manipulating speed to accommodate changes in one stage will upset the other stage and may even put it into surge. Surge control on multistage compressors must be even more stringent than that on a single stage. Frequently, a separate surge control system is required on each stage to protect that stage from surge, as shown in Fig. 6-16.[13] More stages will only serve to amplify the complexity of the system to protect against surge.

Figure 6-14c shows the surge limit line departing from the shape predicted by the Fan Laws. This results from the volume reduction per stage being less when operating at speeds lower than design. Generally, stages near the discharge are forced to handle more than their rated volume, while those nearer the inlet handle less than normal. The accumulated effect is that the surge line shape tends to depart from the Fan Law predictions as more stages are added. Surge is commonly initiated at one of the latter stages when the compressor is operating at design speed but by an earlier stage at lower

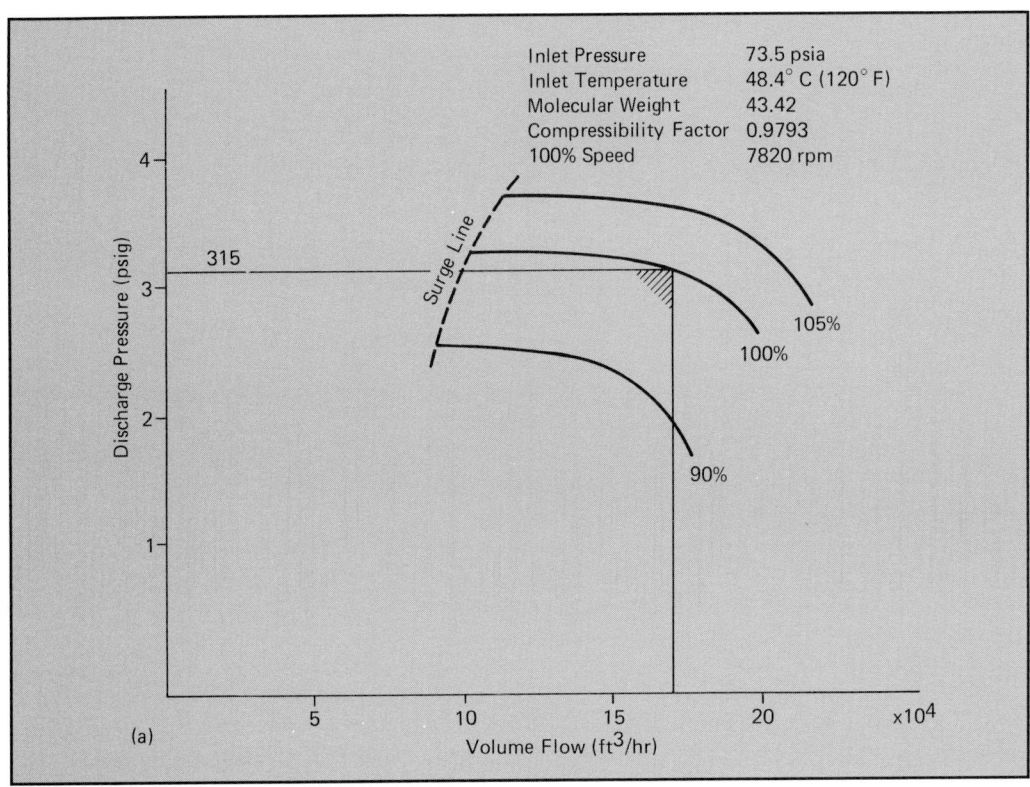

Fig. 6-14. (a) Performance Curves for a Variable Speed Compressor, First Stage

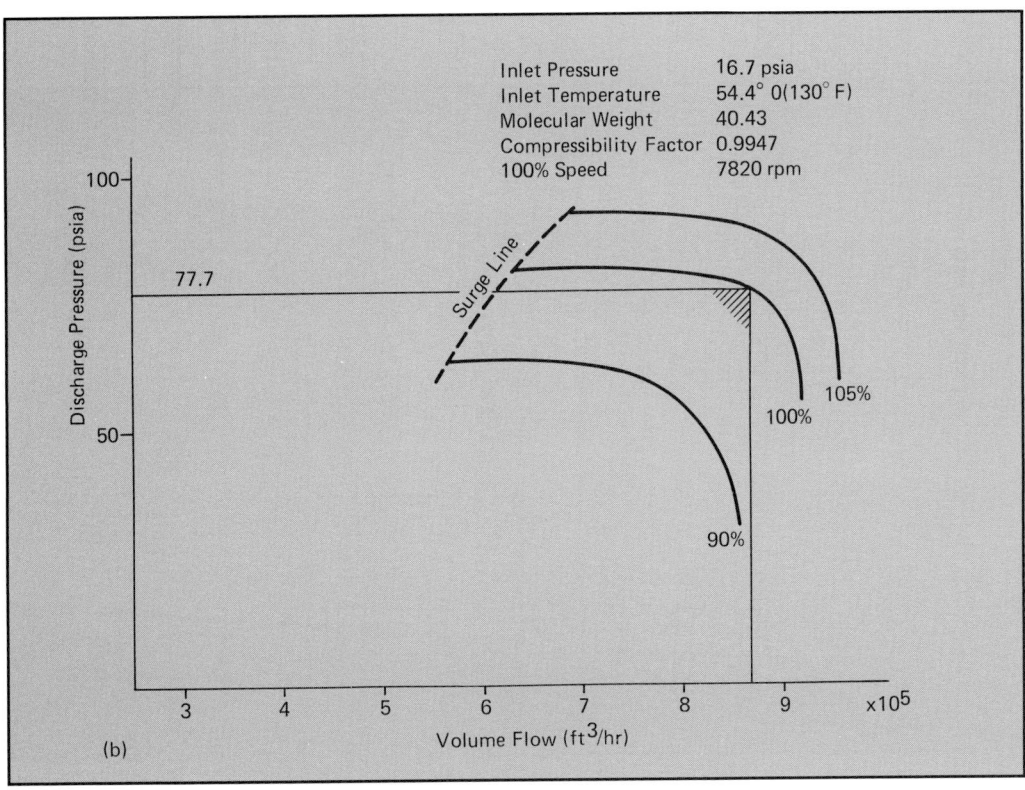

Fig. 6-14. (b) Performance Curves for a Variable Speed Compressor, Second Stage

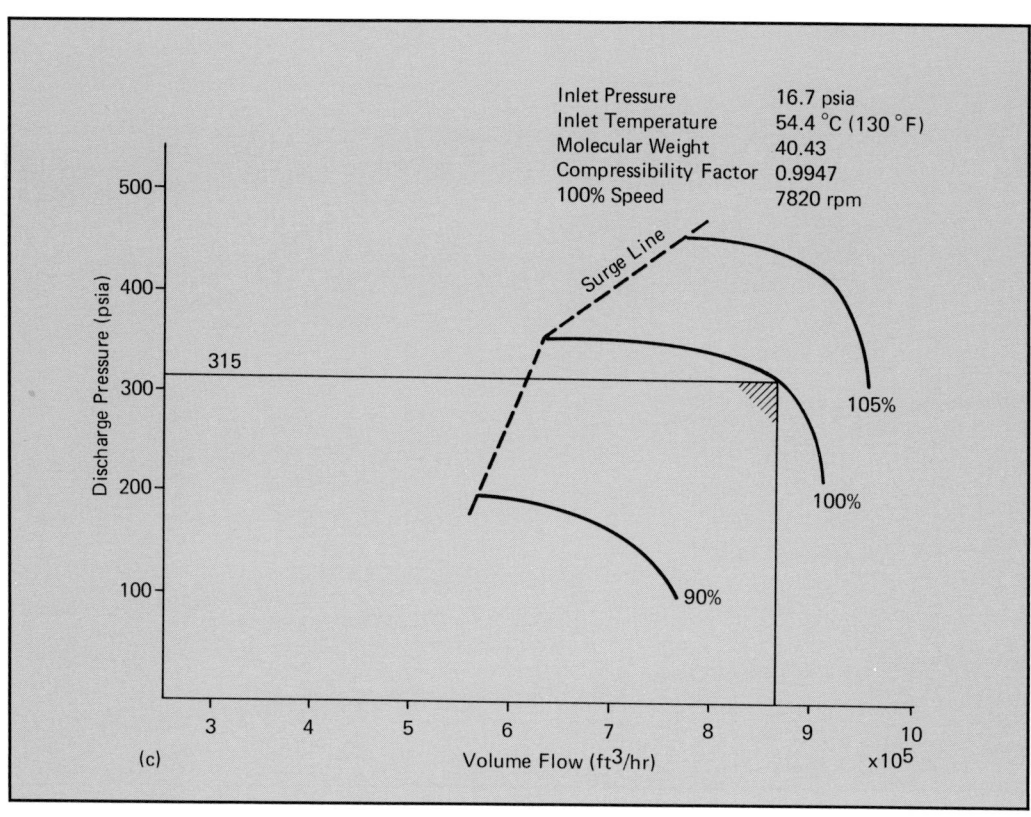

Fig. 6-14. (c) Performance Curves for a Variable Speed Compressor, Overall (Stages 1 and 2)

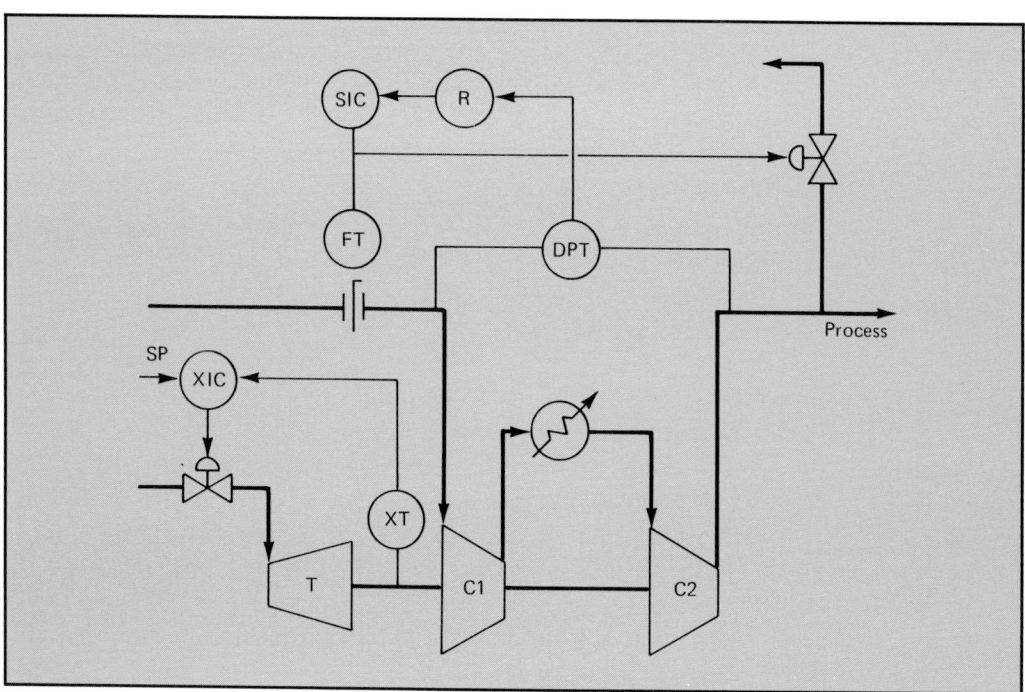

Fig. 6-15. Multistage Variable Speed Compressor with Anti-surge Valve Protecting Entire Compressor

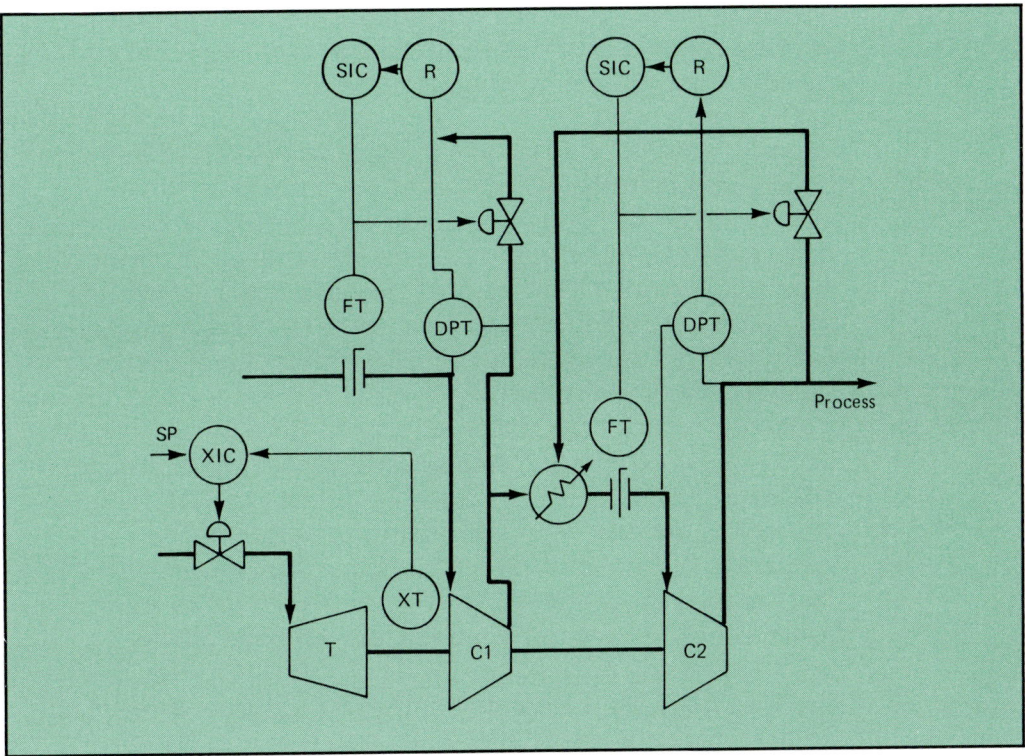

Fig. 6-16. Multistage Variable Speed Compressor with Anti-surge Valve for Each Stage

speeds. A distinct change in the surge line is evident at the point where this occurs. The concept of surge control for each stage shown in Fig. 6-16 accommodates this phenomenon quite well, while a more conservative (larger) surge margin is required with the single system shown in Fig. 6-15 to compensate for the mismatch in form between the surge limit line and the surge control line.

6-6. Manual Surge Control for Start-Up

The mismatch in pumping capacities among stages at lower speeds will sometimes incur surging of the first stage predictably in a given speed range during start-up. During start-up, the first stage will not yet have reached design pressure ratio; density entering the second stage is less than design; and even though the second stage is pumping the expected volumetric flow, it is not pumping away the expected mass flow. For this reason, volumetric flow through the first stage is less than expected, and surge ensues. Should the surge be severe enough to cause damage, a manual anti-surge valve can be installed in a recycle around the first stage, as shown in Fig. 6-17. The remote-operated valve in the recycle is opened upon starting, then gradually closed and left closed during normal operation.

The calculation of the anti-surge valve size is a trial and error procedure. Referring to Fig. 6-17, both anti-surge valves are assumed to be wide open, and compressor inlet is assumed to be at constant pressure. Further, the compressor is assumed to be on total recycle. The pressure rise and flow through the first stage can be determined from the first stage compressor map at a given speed; then enough information is available to calculate the flow through the manual anti-surge valve for the chosen valve size (C_v); the remainder of the flow goes to the second stage; reference to the second stage compressor map with flow, speed, and inlet pressure defines outlet pressure; and outlet pressure completes the information needed to calculate the flow through the automatic anti-surge valve. The flow through the automatic valve must converge to equal the flow through the second stage for this iterative process.

The load lines for the manual anti-surge valve as it relates to the first stage are essentially linear, as shown in Fig. 6-18.

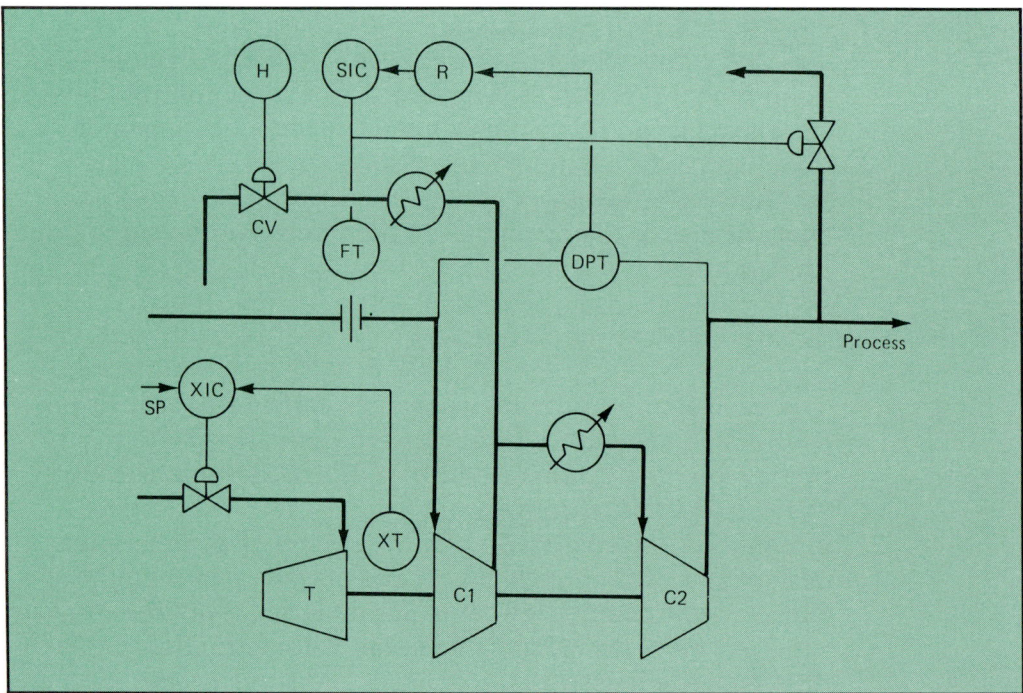

Fig. 6-17. Multistage Variable Speed Compressor with Manual Surge Control on the First Stage

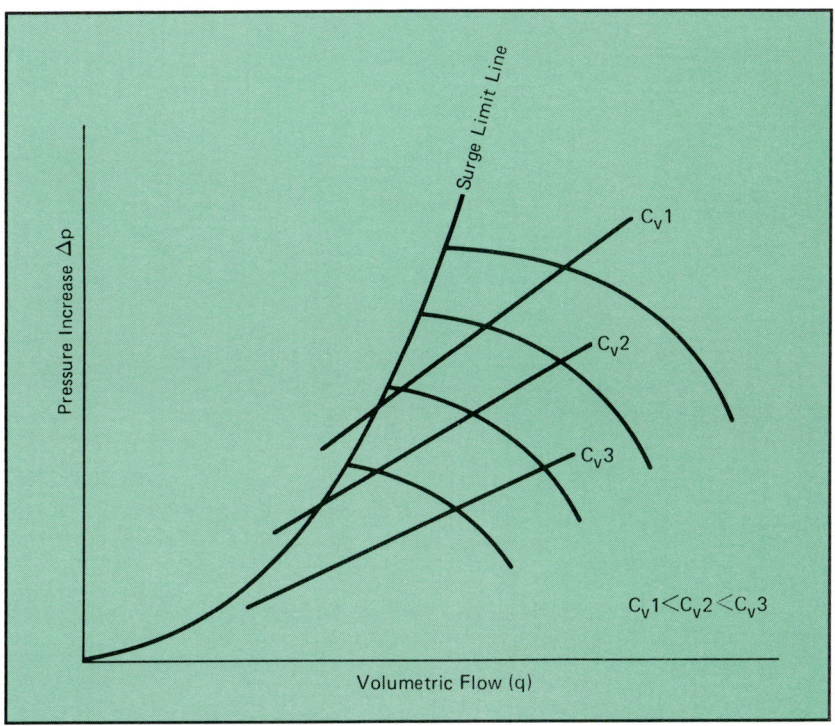

Fig. 6-18. Compressor Map Showing Load Lines Represented by Manual Anti-surge Valve on First Stage

Having constructed a series of load lines for valves of different sizes, a valve can be chosen that does not cross over the surge limit line.

6-7. Sophisticated Surge Control System

Modern microprocessor-based controllers provide extensive process modeling capability.[17] This capability permits use of nonlinear analytical expressions and eliminates the need for simplification. Therefore, the simplification of eq. (6-13) is not required, but eqs. (6-10) and (6-12) can be equated:

$$\frac{1545 \, ZT_i}{M\sigma} (R_c^\sigma - 1) = K_5 h \frac{ZT_i}{p_i M} \qquad (6\text{-}18)$$

where

$$R_c = \frac{p_d}{p_s} \qquad (6\text{-}19)$$

The compressibility (Z), inlet temperature (T_i) and molecular weight (M) terms again cancel, leaving the flowmeter pressure

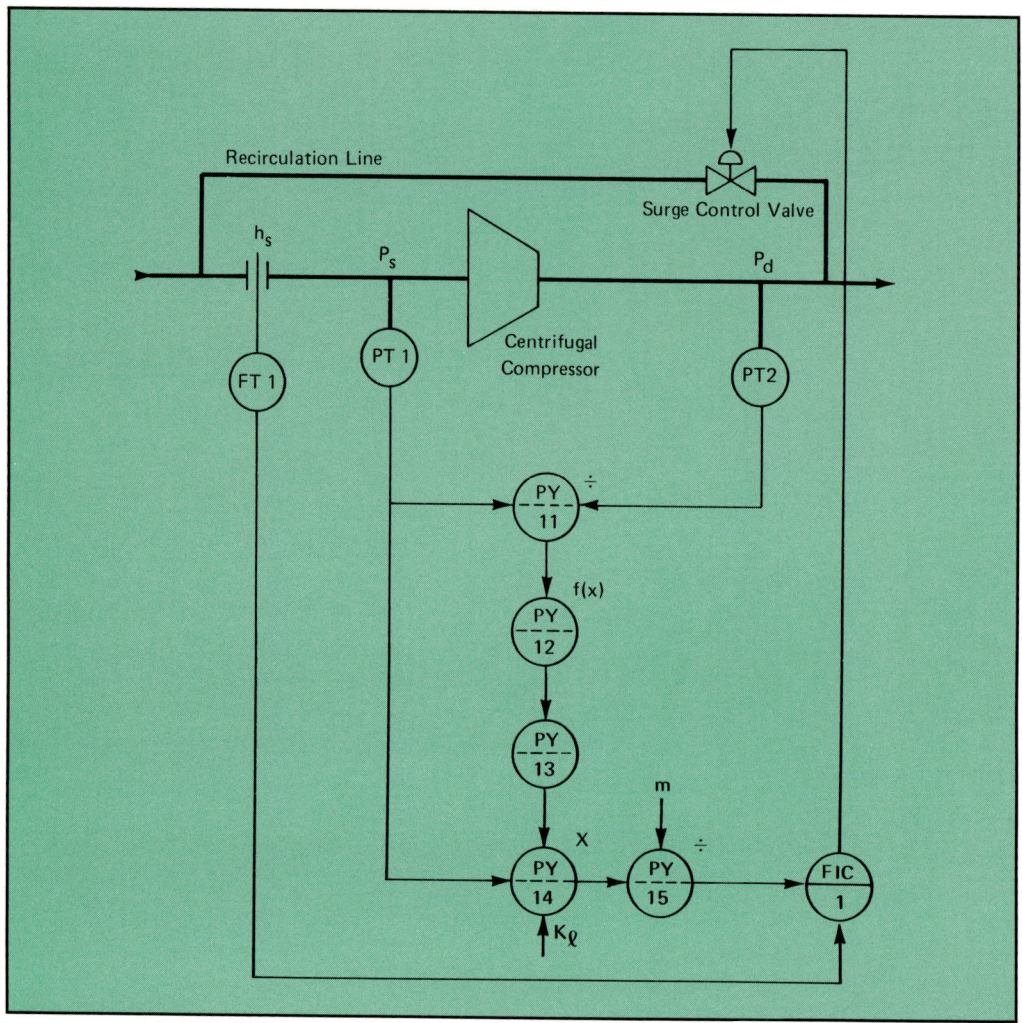

Fig. 6-19. Variable Speed Compressor with Surge Control Based on Flowmeter h and Compression Head (H)[18]

differential (h) in terms of suction pressure (p_s), discharge pressure (p_d), and ratio of specific heats (σ).

The flowmeter differential (h) calculated in eq. (6-18) is the set point value of the controller that actuates the control valve to hold the h measured by the flowmeter equal to the computed value of h. The resulting control system is shown as Fig. 6-19.[18]

6-8. Surge Control for Complex Systems

Energy costs and a competitive climate have renewed interest in efficiency and rangeability in centrifugal compressor installations. Rangeability is severely limited, and efficiency decreases significantly away from the design operating point. Variable speed drives and guide vanes are becoming more common to enhance these characteristics.

Figure 6-20 shows a compressor installation with both variable speed drive and diffuser guide vanes. The guide vanes on each stage are operated with constant relative opening to each other. The turbine governor acts as a speed control system, which receives its set point from the capacity control system. Speed and vane position are then adjusted in coordination to provide the best efficiency at the required throughput.

Compressor map Fig. 6-21 shows the flexibility of this arrangement. Performance curves only are shown. The dual number on each performance curve delineates the positions of the two sets of guide vanes. Changing the vane angle moves the performance curve horizontally, as shown in Fig. 5-8b. Changing the speed moves the performance curve vertically, as shown by the locus of speed points on the speed scale. Thus, many operating conditions can be attained by changing either

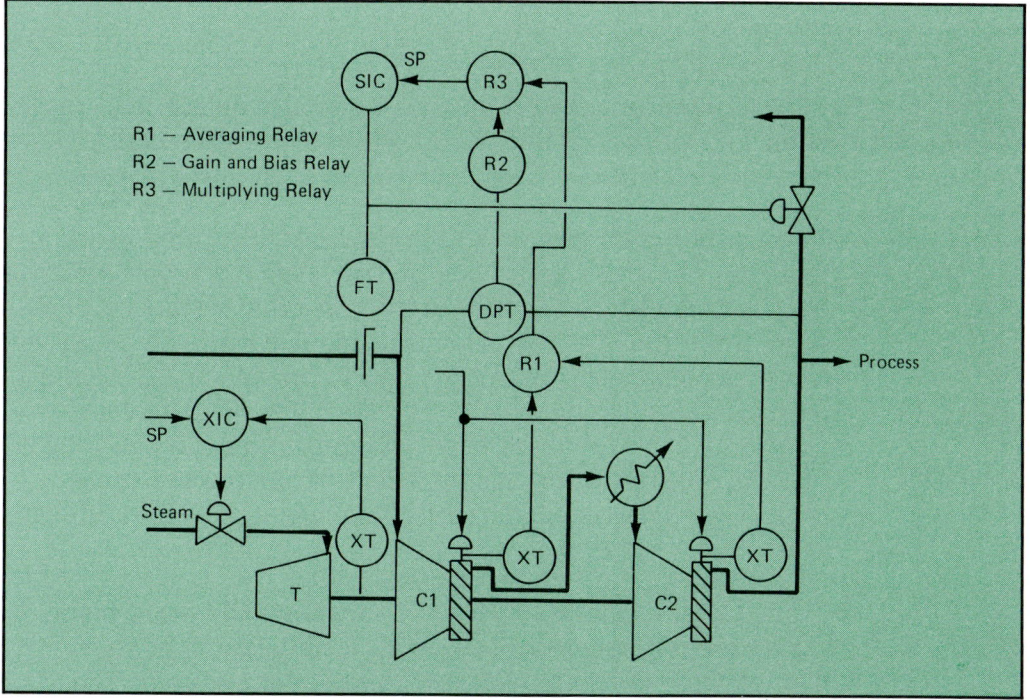

Fig. 6-20. Surge Control System for Variable Speed Compressor with Adjustable Diffuser Vanes

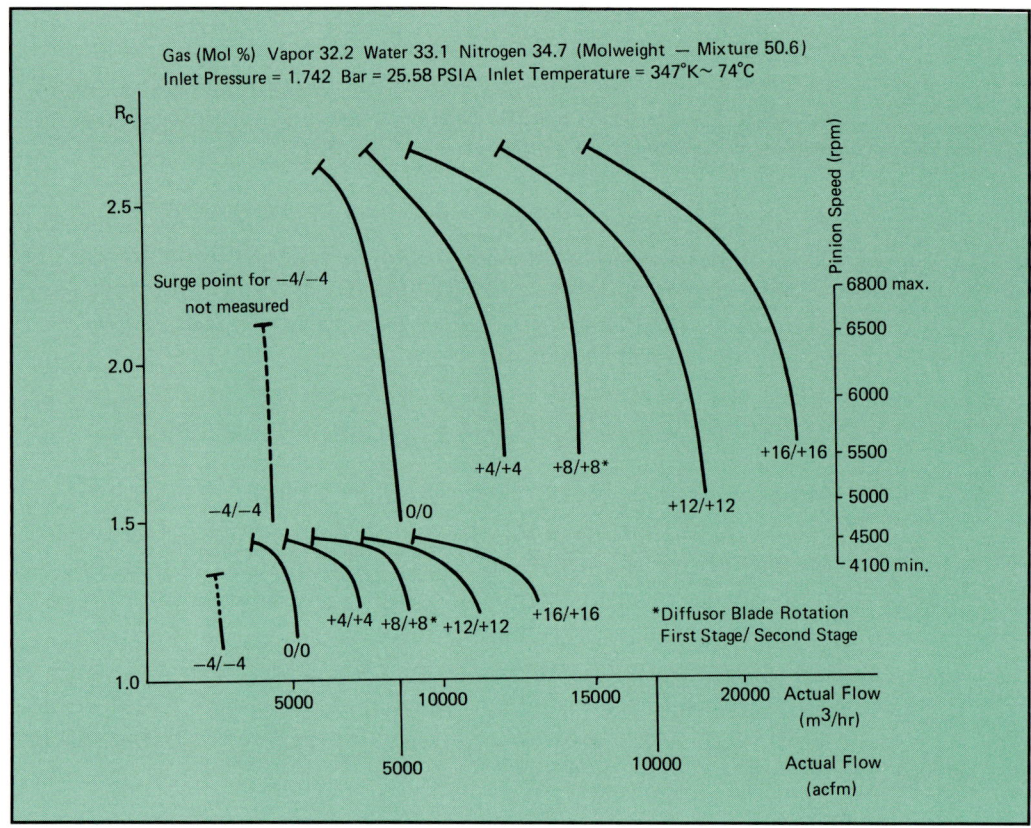

Fig. 6-21. Compressor Map for Variable Speed and Diffuser Vanes[30]

vane position or speed. The experimentally determined surge point for each performance curve is shown by a small perpendicular line at the upper end of each curve.

Two control concepts have been developed to provide surge protection for this type of compressor. One is shown in Fig. 6-20. It uses a third parameter to calculate the position of the surge limit line. The surge control line is calculated using the compressor pressure rise (Δp) and the flowmeter differential pressure (h) to calibrate gain and bias relay R2. The vane position is measured and used as a multiplier through relay R3. The multiplication has the effect of moving the surge control curve horizontally with the movement of the diffusor vanes.

A comparison of Figs. 6-21 and 5-8b shows the performance curves are not equally spaced with vane position as they theoretically are, so a nonlinear function is required. The required function is supplied in the Compressor Controls Series II™ Controller, which is designed for this purpose.[19]

Another notable feature is the increasing slope of the performance curves as the vanes close, as shown in Fig. 6-21. The slope is so steep at the minimum vane opening that the surge and stonewall points are very close together, making calibration that avoids surge yet does not demand running in the stonewall area very difficult.

An alternative surge control configuration depends on the observation that the surge points at a given speed are all at the same compression ratio (R_c) or head (H). Further, as formerly discussed, the performance curves are steep enough to dependably measure pressure ratio. Nonlinearity must be introduced to accommodate the surge point at lower compression ratio at minimum vane opening. McMillan[20] suggests that the nonlinearity is conveniently provided by two square root extractor relays in series with the flowmeter, resulting in a curve that is proportional to the square of the flowmeter differential pressure (h) signal. This arrangement can also be programmed on microprocessor-based controllers.

Having based the anti-surge system on compression ratio (R_c), the surge control line must be moved vertically to accommodate speed changes. The Fan Law, eq. (6-7), is used for this purpose.

The differences between Fig. 5-8b, showing theoretical performance curves, and Fig. 6-21, showing experimental performance curves, emphasizes the need to adjust the parameters of the surge control system to meet observed requirements.

By this time, it is apparent that the needs of the process engineer in increasing rangeability and minimizing energy usage are a challenge (read that, "nightmare") to the control engineer.

6-9. Surge Control System Components

The surge control systems described in this unit have been based on typical measurements, calibrated to identify a surge limit line depicted graphically either by the manufacturer or by field testing. Thus, the instruments do not require a bandwidth sufficient to measure surge itself but need only be adequate to open an anti-surge valve to compensate for process disturbances. Unfortunately, the only method that quantitatively identifies the required speed of response and the required surge control margin is computer simulation. Fortunately, experience has shown that petrochemical plants move slowly enough that even pneumatic instruments (which respond in fractions of a second) are adequate for most compressor control installations. Unit 8 will discuss control systems that operate on the basis of measured surge. These systems demand instruments of much higher frequency response.

The inlet volumetric *flow measurement* is governed by eq. (6-9). The measured variable (h) is a function of not only volumetric flow (q) but also of inlet pressure (p_i), temperature (T_i), and molecular weight (M). Figure 6-22 shows the method for compensating for inlet temperature and pressure[4,21] when they are expected to vary significantly. Should molecular weight also vary, it is simpler to consider an alternative form of the equation

$$q = K_8 \sqrt{\frac{h}{\rho}} \qquad (6\text{-}20)$$

The compensation is then accomplished by a density transmitter in preference to the temperature and pressure transmitters.

Flow measurement in the discharge of the compressor is often preferable to measurement in the suction. This installation might be preferred because (1) the pressure gradient in the

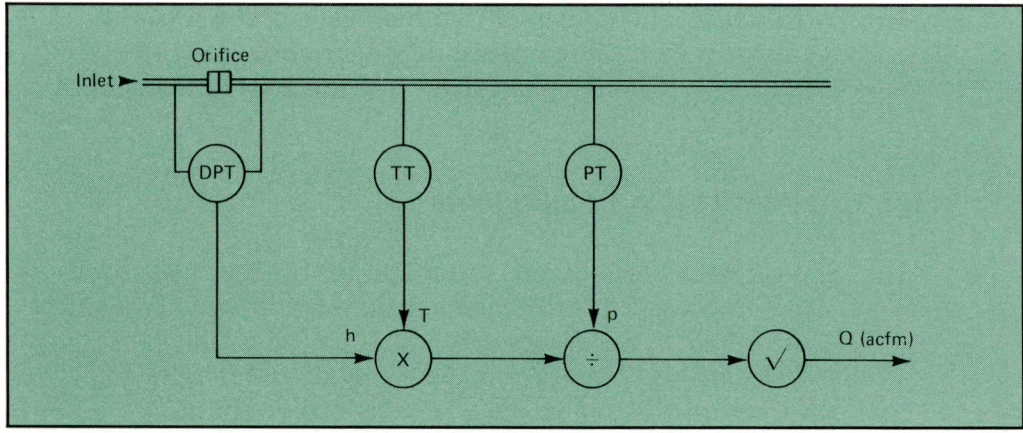

Fig. 6-22. Inlet Volumetric Flow Rate Measurement with Pressure and Temperature Compensation[4]

suction is too small to achieve a reliable flow signal, or (2) flowmeter permanent pressure loss is not as objectionable in the discharge, or (3) the inlet pipe diameter is so large that the required straight run would be impractically long, or (4) a discharge flow measurement is already required for process reasons. Then the discharge flow measurement should be corrected to inlet conditions, as shown in Fig. 6-23.[4,22] This configuration is based on the measurement of mass flow in the discharge and the knowledge that mass flow in equals mass flow out. The discharge flow is compensated to mass flow, and the mass flow is compensated to inlet volumetric conditions.

In addition to the usual valve selection guidelines, the *anti-surge control valve* must:

- Be in accordance with anti-noise regulations.
- Be amply large enough.
- Fully stroke in one second or less.

Most manufacturers can calculate the expected noise levels of their various valve configurations for design conditions.[23,24] Sometimes muffling devices (insulation, downstream mufflers) will reduce the noise level to within government mandated standards to prevent hearing loss. Installations with more

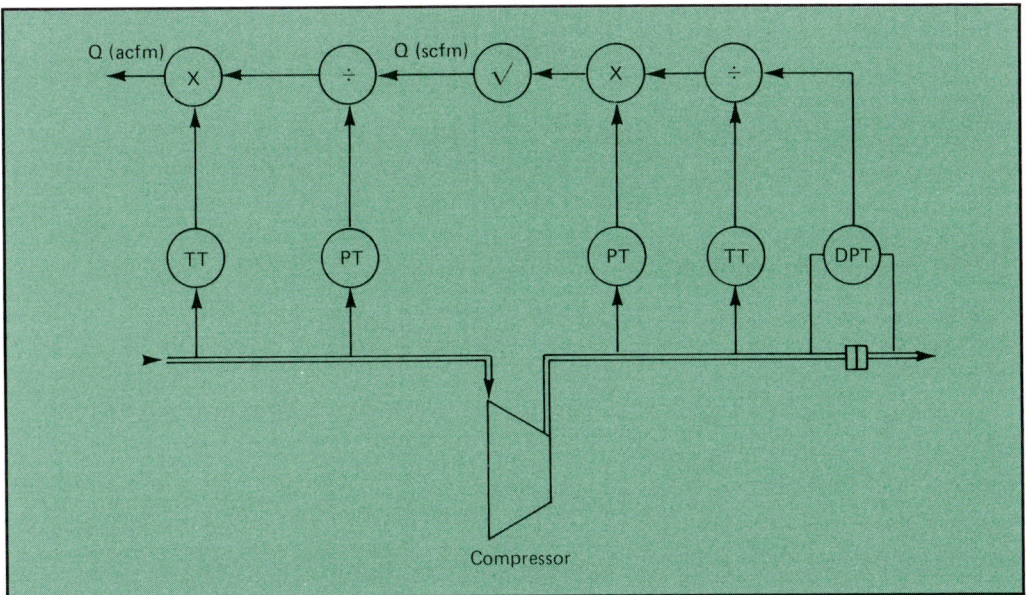

Fig. 6-23. Discharge Mass Flow Rate Measurement Compensated to Inlet Volumetric Flow Rate[4]

severe physical conditions require special low-noise valves with such exotic sounding names as the drag valve,[25] and dragon tooth valves.[26]

The anti-surge valve must be large enough. One conservative guideline for selecting the valve size is that the valve must pass full design flow at 70% of the pressure rise (Δp) across the compressor. Assurance that the anti-surge valve has adequate capacity for other operating conditions can be attained from the calculation in Example 6-3.

Example 6-3. Consider the compressor map Fig. 6-24 and the control valve sizing equation for gas service:

$$q_s = 963 \, C_v \sqrt{\frac{p_d^2 - p_s^2}{GT_d}}$$

where:

q_s = Standard volumetric flow rate, scfm
p_d = Discharge pressure, psia
p_s = Suction pressure = 362.7 psia
G = Specific gravity = 0.31
T_d = Actual discharge temperature = 515°R

and

$$q = \left[\frac{p_{std} T_s}{p_s T_{std}}\right] q_s$$

p_{std} = Standard pressure = 14.68 psia
T_{std} = Standard temperature = 520°R
T_s = Inlet temperature = 565°R

Then, substitute

$$q = 0.0559 \, C_v \sqrt{\Delta p (p_s + p_d)}$$

Evaluate this equation with pressures from compressor map Fig. 6-24 and for the two following control valves:

$1\frac{1}{2}$-inch control valve, C_v = 83.6

3-inch control valve, C_v = 155

The calculated flows for the series of chosen pressure conditions are shown on the compressor map. The $1\frac{1}{2}$-inch

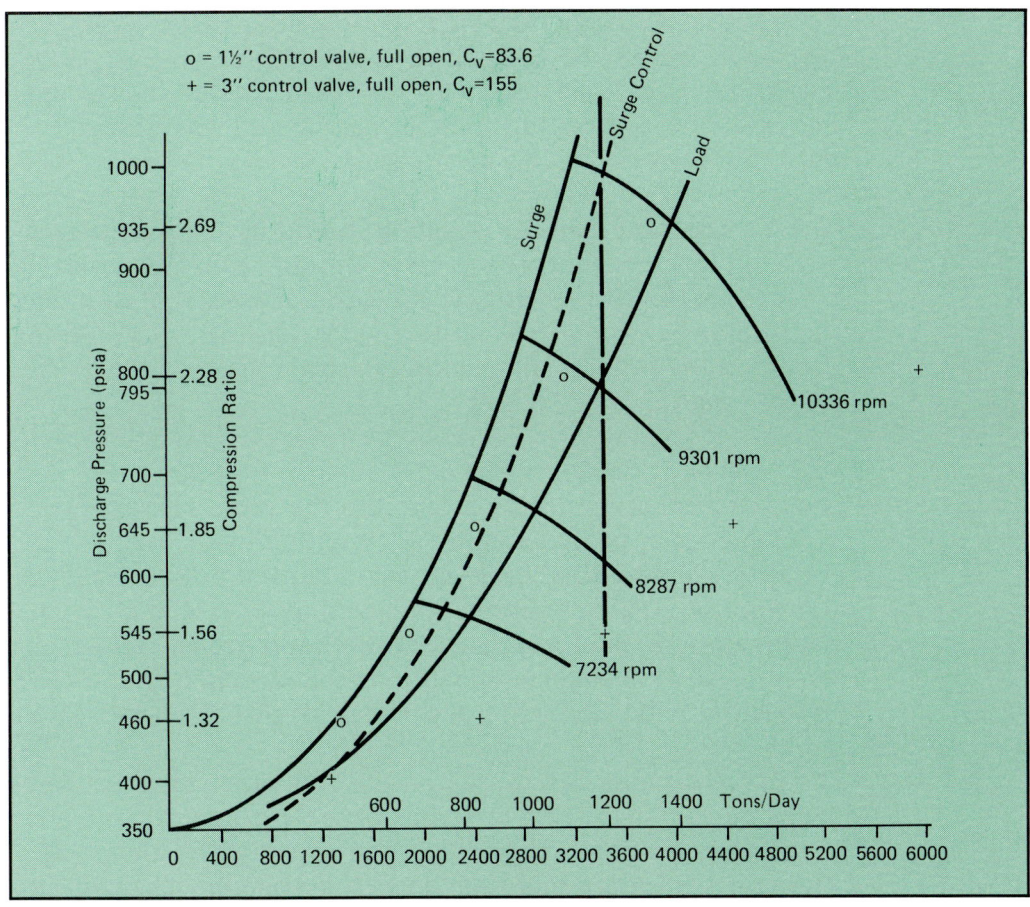

Fig. 6-24. Load Lines Representing Anti-surge Valves of Different Sizes

valve is too small because the valve load line crosses the surge control line at low flows.

The foregoing calculation assumes that there is no forward flow through the compressor and that the full-open anti-surge valve on each stage must be large enough to pass adequate flow to avoid surge at the pressure rise across the compressor stage. Pressure rise across the compressor stage is pressure drop across the anti-surge valve. The inlet is assumed to be at constant pressure for the first stage. Later stages are more complicated since the inlet pressure is varying because of the action of preceding stages.

The anti-surge valve is typically a large valve, especially if it is of a low-noise design, actuated by a pneumatic actuator. It will not normally stroke in the required one second. Quantitative verification of the required stroking time can only be attained from computer simulation. It is possible that such speed will not be required for petrochemical applications with large piping. However, the simulation to determine the exact requirements is expensive, while the guideline of one second has been found to be adequate for just about every known application and is widely accepted.

The 1-second time for full stroking should not be confused with the time constant of the valve and actuator. The time constant is likely to be around a second. However, it describes the response of some 10% of the total stroke. Velocity limiting occurs in the full stroke, making the stroking time much longer than the time constant.

The 1-second stroking speed can be realized by the incorporation of volume boosters into the valve actuator.[20] A volume booster is a pneumatic relay with a large flow capacity. The stroking speed criterion applies only to opening the anti-surge valve to protect the compressor. Generally, no harm results from closing more slowly, and, in fact, a fast opening-slow closing combination helps minimize the interaction with other control systems, as discussed in Unit 7.

A valve positioner is required on the anti-surge valve because it is a large valve with a high and variable pressure drop, and precise positioning is required (hunting cannot be tolerated). The positioner does represent a cascaded servomechanism in the control loop and, as such, could require some detuning of the surge controller.[20,27] However, advantages outweigh disadvantages in almost every application.

The installed valve flow characteristic of the anti-surge valve will be the inherent characteristic, since little equipment causing pressure drop is in series. For this reason, a linear valve flow characteristic is used.[27] The linear characteristic has the additional advantages of representing a linear element in the control system and of providing greater flow at a small opening, thus relieving surge more quickly. It has the disadvantage of a very small stem movement at low flows, where Murphy's Law will unquestionably require operation.

The *surge controller* can be pneumatic, electronic, or microprocessor-based. The microprocessor-based controller has tremendous flexibility and can be reprogrammed in the field to accommodate conditions observed during operation. However, if it is part of a distributed control system, its response is dependent on the system sampling rate. A slow sampling rate (say, of the inlet flow measurement) can negate all of the other efforts to provide fast surge system response. A sampling rate of 10 samples/second is required to approach analog system performance.[28]

Two conditions must be considered in the design of the surge control configuration: (1) the surge control system will be off-line (the valve will be closed) most of the time, and (2) the controller cannot be tuned on-line. If the compressor application is well designed, the anti-surge valve will be closed under normal operating conditions to conserve energy. With the valve closed, the measured variable does not equal the set point, and the resulting error drives the controller into saturation. Then, as throughput flow falls toward surge, time is required for the controller to come out of saturation before it can begin opening the anti-surge valve.

Conceptually, the batch switch[29] provided for pneumatic controllers is an ideal solution. It forces saturation from the end of the span where it occurs to the other end of the span. This action has the effect of shifting the controller proportional band (the gain that causes the valve to open) closer to the changing measured variable (the flow). In this manner, quicker response is attained.

A more typical installation is shown in Fig. 6-25. Modulation of the control valve can be taken from the surge controller (SIC) by an emergency interlock through the solenoid valve or by the override of another variable through the selector relay (R). Normally, the controller would saturate for either of these conditions, as described above. However, feeding back the signal proportional to valve position into the reset mechanism keeps the controller always balanced at the valve attitude. Then, in the case where the controller tends to saturate when the controller is connected to the anti-surge valve but the valve is closed, limits are placed on the controller output that permit the output to go only a small amount past the ends of the span. Thus, the signal being fed back cannot saturate very far. Further, the controller will always be tuned like a flow controller with a very fast reset, which will bleed down the saturation very quickly. While the system shown in Fig. 6-25 does not have the conceptual advantage of the batch switch, it has proved adequate in many applications.[27]

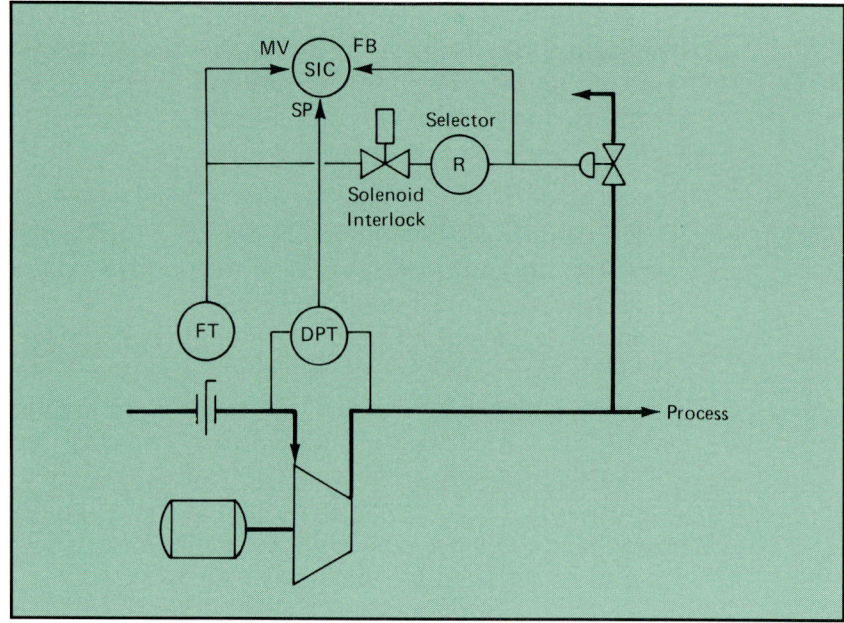

Fig. 6-25. Surge Controller with External Reset Feedback

The surge controller cannot be tuned with the compressor operating because of the danger of damaging the compressor while experimenting with the control. Therefore, the field tuning techniques[29] by which final tuning is traditionally attained are not available. It is necessary to put "good" tuning parameters on the surge controller prior to start-up. The practitioner now finds himself with a surge control system on which the safety of a multimillion dollar compressor depends but which has been designed on fuzzy data, has been developed on analytical approximations (e.g., pressure increase (Δp) for head (H)), and now has "guestimated" controller tuning! Fortunately, the surge control system closely approximates a flow control system, and adequate tuning is known:[29]

$$PB = 180\%$$

$$\text{Reset} = 0.5 \text{ minutes}$$

Then, in operation, the proportional band can be gingerly narrowed if the system doesn't seem to respond quickly enough.

Alternatively, McMillan[20] provides formulae by which the controller tuning parameters can be calculated:

$$T_u = 4 \cdot (TD_v + TC_v + TD_m + TC_m + TD_c + TC_c) \quad (6\text{-}21)$$

$$TC_c = 0.6 \cdot T_s \text{ (0.03 second for analog controller)} \quad (6\text{-}22)$$

$$TD_c = 0.5 \cdot T_s \quad (6\text{-}23)$$

$$T_i = 0.8 \cdot T_u \quad (6\text{-}24)$$

$$TC_p = 60 \cdot \{(V \cdot P)/(P_a \cdot F)\} \quad (6\text{-}25)$$

$$PB = \{(4 \cdot T_u)/(2\pi \cdot TC_p)\} \cdot K_o \cdot 100\% \quad (6\text{-}26)$$

where:

PB = proportional band of the controller (percent)
K_o = open-loop gain (percent/percent)
F = flow through the control valve (acfm)
P = pressure upstream of control valve (psia)
P_a = atmospheric pressure (psia)
TC_c = time constant of controller measurement filter (sec)
TC_m = time constant of transmitter (sec)
TC_p = time constant of pressure response (sec)
TC_v = time constant of control valve (sec)
TD_c = dead time of digital controller (sec)
TD_m = dead time of transmitter (sec)
TD_v = prestroke dead time of control valve (sec)
T_i = integral time setting (sec/repeat)
T_s = sample time of digital controller (sec)
T_u = ultimate period of the control loop (sec)
V = volume of the piping and equipment (cuft)

Exercises

1. Why is surge control mandatory on large compressors?

2. Why not make surge control very conservative, providing double assurance that the machine will not go into surge?

3. Venting is so much more economical than recycling for surge control. Why not use it all the time?

4. What are the limitations on the simplified surge control system described in Section 6-5?

5. Why incorporate variable speed and guide vanes into a compressor when they only (1) increase purchase cost, (2)

increase maintenance cost, and (3) increase surge control system complexity?

6. Where inlet temperature and pressure vary widely, should the inlet flowmeter be compensated for pressure and temperature (Fig. 6-25) when using the h vs. Δp (Section 6-5) surge control configuration?

References

1. Hamell, E., "Compressor Control: Surging Ahead," *InTech*, Instrument Society of America, Research Triangle Park, NC, December, 1982, p. 9.
2. Anon., "Selecting a Surge Control System," *Bailey Control Systems*, Bailey Controls, Wickliffe, Ohio, 1983.
3. Warnock, J. D., "Typical Compressor Control Configurations," *Centrifugal Compressor Operation and Control*, Instrument Society of America, Research Triangle Park, NC, 1976.
4. Warnock, J. D., "Methods of Control of Centrifugal and Reciprocating Compressors," Moore Products Co., Spring House, PA.
5. Buzzard, W. S., "Controlling Centrifugal Compressors," *Instrumentation Technology*, November, 1973, p. 39.
6. Bozeman, H. C., "Analog Computer Prevents Surge in Centrifugal Compressor," *Oil and Gas Journal*, December 1, 1964, p. 87.
7. Franks, R. G. E., *Modeling and Simulation in Chemical Engineering*, John Wiley & Sons, New York, NY, 1972.
8. Davis, F. T., and A. B. Corripio, "Dynamic Simulation of Variable Speed Centrifugal Compressors," *Instrumentation in the Chemical and Petroleum Industries*, Volume 10, Instrument Society of America, Research Triangle Park, NC, 1974.
9. Gupta, B. P. and M. F. Jeffrey, "Compressor Controls Made Easy with Microprocessors," *Proceedings of 1979 Maintenance Division Symposium*, Instrument Society of America, Research Triangle Park, NC, 1979.
10. Chandler, E. G., "Surge Prevention in Large Centrifugal Gas Compressors," *En-Tronic Controls Bulletin*, Cooper-Bessemer Co., Mount Vernon, Ohio.
11. Hatton, C., "Controlling a Gas Compressor Station Electronically," *Instruments and Control Systems*, May, 1967, p. 145.
12. White, M. H., "Surge Control for Centrifugal Compressors," *Chemical Engineering*, December 25, 1972, p. 54.
13. Waggoner, R. C., "Process Control for Compressors," *Centrifugal Compressor Operation and Control*, Instrument Society of America, Research Triangle Park, NC, 1976.
14. Anon., "Simplified Surge Control System for Centrifugal Compressor," Bailey Control Systems Application Bulletin 999-11, Bailey Controls, Wickliffe, Ohio, 1982.
15. Staroselsky, N., and L. Ladin, "Improved Surge Control for Centrifugal Compressors," *Chemical Engineering*, McGraw-Hill, New York, N.Y., May 21, 1979.
16. Magliozzi, T. L., "Control System Prevents Surging in Centrifugal-Flow Compressors," *Chemical Engineering*, McGraw-Hill, New York, N.Y., May 8, 1967.
17. Golla, J. T., "Centrifugal Compressor Variable Surge Line Computation and Control," ISA Paper No. 87-1100, Advances in Instrumentation, Vol. 42, Instrument Society of America, Research Triangle Park, NC, 1987.

18. Anon., "Centrifugal Compressor Surge Control System," Bailey Control Systems Application Bulletin 999-10, Bailey Controls, Wickliffe, Ohio, 1982.
19. Anon., "Series II Antisurge Controller," Bulletin PB12, Compressor Controls Corp., Des Moines, Iowa, December, 1984.
20. McMillan, G. K., *Centrifugal and Axial Compressor Control*, Instrument Society of America, Research Triangle Park, NC, 1983.
21. Moore, R. L., "Flow Measurement," *Measurement Fundamentals, Basic Instrumentation Lecture Notes and Study Guide*, Third Edition, Instrument Society of America, Research Triangle Park, NC, 1982.
22. Nisenfeld, A. E., "Anti-Surge Control—With Discharge Flow Measurement," *Instruments and Control Systems*, September, 1977, p. 166.
23. Arant, J. B., "How to Cope with Control Valve Noise," *Instrumentation Technology*, March 1973, p. 37.
24. Baumann, H. D. "On the Prediction of Acrodynamically Created Sound Pressure Level of Control Valves," ASME Paper WA/FE-28, American Society of Mechanical Engineers, United Engineering Center, 345 E. 47th Street, New York, N.Y., 10017.
25. Anon., "Why Velocity Control?" Bulletin 100, Control Components, Inc., 2567 SE Main Street, Irvine, California.
26. Anon., "The Silent Treatment," Bulletin No. 80:005, Fischer Controls, Marshalltown, Iowa, September, 1976.
27. Nisenfeld, A. E., *Centrifugal Compressors, Principles of Operation and Control*, Instrument Society of America, Research Triangle Park, N.C., 1982.
28. Chiu, K. C., and N. E. Pobanz, "Applicability of Distributed Control to a Fast Response Loop," *Proceedings of International Conference and Exhibit*, October, 1983, Instrument Society of America, Research Triangle Park, NC.
29. Moore, R. L., *The Dynamic Analysis of Automatic Process Control*, Instrument Society of America, Research Triangle Park, NC, 1985.
30. Bulletin, Atlas Copco Energas GMBH, Am Rheinofer 20 Post Box 501150, D5000 Koein 50 (Suerth), Germany.

Unit 7:
Control System Interaction

UNIT 7
Control System Interaction

All of the control systems on a single compressor or on a multiple compressor system affect the mass balance on the downstream piping or the pressure header. Thus, a change in one control valve requires changes by the other control systems to bring their measured variables back to the set points. These changes can require further readjustment. The potential for interaction among the control systems is great, and, in fact, interaction is frequently observed among the variables on operating compressors. As an oversimplification, two kinds of interaction are observed. First, a static interaction takes place when the movement of one control valve causes variations in several variables; or, second, a dynamic interaction occurs when two feedback control systems have the same natural response frequency, and the response of one upsets the other at its natural frequency. The potential for interaction is also very high, since all of the systems are typically controlling either flow or pressure, and these systems have essentially no capacitance resulting in similar response frequencies.

Learing Objectives — When you have completed this unit you should:

A. Understand the mechanism of interaction.

B. Be aware of techniques for estimating the possibility of interaction and its severity.

C. Have an appreciation of the decoupling techniques available.

D. Be aware of heuristic techniques for decoupling that have proved helpful on operating compressors.

7-1. General[1,2,3]

Cross coupling between controlled and manipulated variables affects control systems of two dimensions or higher. The cross coupling is evident in Fig. 7-1, where pressure and flow affect the same mass balance, and in Fig. 7-2, where the two flows affect the pressure in the same mass balance.

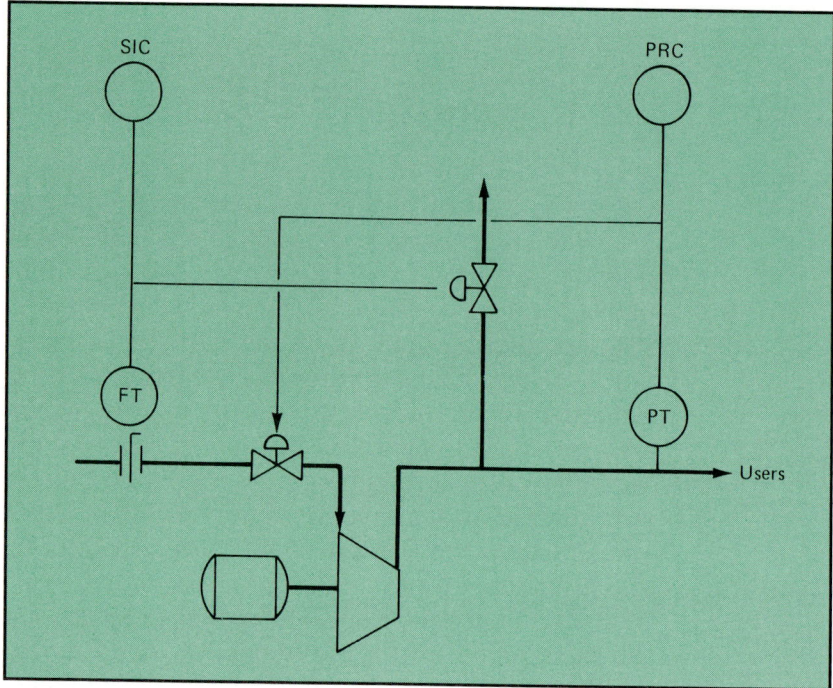

Fig. 7-1. Constant Speed Compressor with Capacity and Surge Controls

In single-variable applications, a process variable can be successfully controlled by the manipulation of one flow stream. Should the compressor in Fig. 7-1 be operating at the design point on the performance curve, the anti-surge valve would be closed, and the pressure control system would be an example of single-variable control. Manipulation of the stream through the compressor does not significantly interfere with the regulation of other process variables, and the pressure control loop is not greatly influenced by other controlled process variables.

By contrast, a multivariable control problem arises when no such exclusive relationship exists between controlled and manipulated variables. This situation is shown by Fig. 7-1 when the compressor is operating on the surge control line and the anti-surge valve is open and being actuated by the surge controller. Here, each manipulated variable influences both controlled variables. The operability of the system made up of the two single-loop processes will depend on the degree of cross coupling and on whether each of the controlled variables is paired with the most appropriate manipulated variable.

Unit 7: Control System Interaction

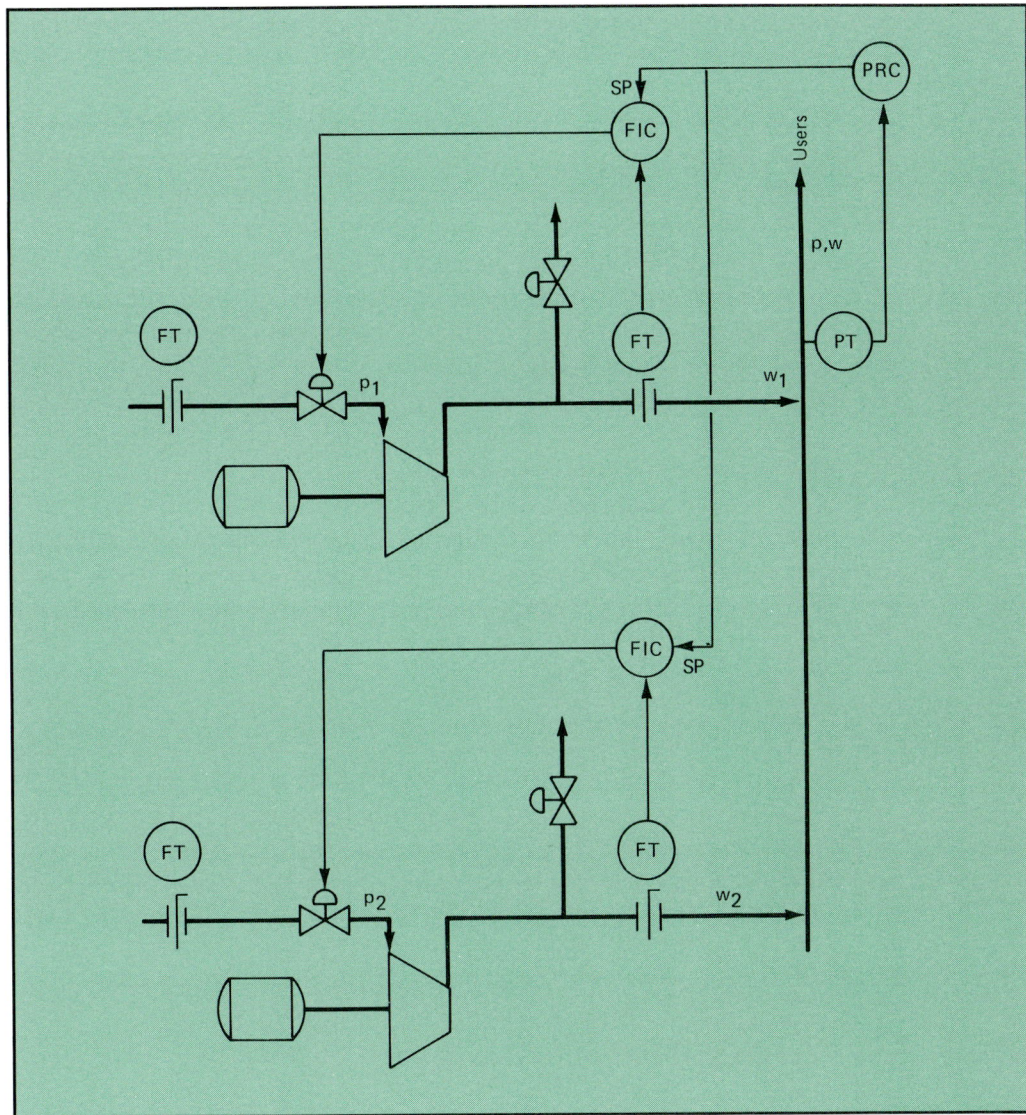

Fig. 7-2. Two Constant Speed Compressors Supplying a Pressure Header

Cross coupling (interaction) can be so severe that attempting to control with two single-loop controllers will result in an inoperable system. A workable control strategy for systems with severe interaction must take into account the multivariable nature of the problem. This is usually accomplished by devising a configuration that decouples the system. Block diagram and matrix approaches are used to develop a decoupling control strategy. A technique that measures the severity of interaction helps decide whether multiple single-loop or multivariable control should be used and also determines how to effectively pair the controlled and manipulated variables.

The general multivariable control problem is, of course, not restricted to a two-input, two-output system but covers all systems of two dimensions and higher. In Fig. 7-2, the order of the system and the rank of the matrix increase rapidly as capacity and anti-surge systems are considered on each compressor and as additional compressors serving the same pressure header are included. As might be expected, higher-order systems are frequently encountered throughout the fluid-processing industries. The complexity of the analysis to assess severity of interaction and to design decoupling strategies increases exponentially with the order of the system, and higher-order systems can be efficiently handled only by computerized matrix techniques. Since cost is proportional to complexity and since heuristic solutions have been found to be adequate if not optimal, such analyses are seldom made unless a readily identifiable problem exists.

7-2. Block Diagrams

The multivariable compressor control system shown in Fig. 7-1 can be shown as a conceptual block diagram as in Fig. 7-3. Here, it is evident that changing one control signal (or manipulated variable) causes changes in both controlled variables. This is a typical multivariable control problem. An increase in the surge controller set point causes anti-surge flow to increase and header pressure to decrease. Likewise, an increase in pressure controller set point causes pressure to increase and anti-surge flow to increase.

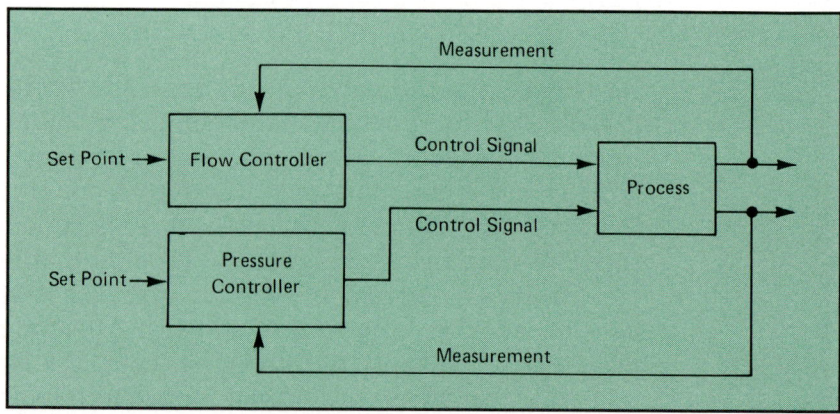

Fig. 7-3. Typical Interacting Pressure and Flow Control Systems[4]

The conceptual block diagram of Fig. 7-3 can be put in the form of Fig. 7-4 to quantitatively identify the interaction. Figure 7-4 emphasizes that no exclusive relationship exists between controlled and manipulated variables. *Each* manipulated variable influences *both* controlled variables. The success of treating the system as a two single-loop processes will depend on the degree of cross coupling. Interaction between control loops operating on the same process makes control difficult. Thus, interacting controls have another undesirable characteristic. The control loops cannot be tuned independently, leading to the repeated tuning of the process controllers in all loops. Satisfactory adjustment of controllers may never be achieved.[3,4]

The block diagram of Fig. 7-4 can be generalized to the diagram shown as Fig. 7-5a. The following notation is used to describe the system:

C—Conventional Process Controller

P—Process Characteristics

S—Set Point

O—System output, or controlled variable

Subscripts refer to the individual loops, as shown.

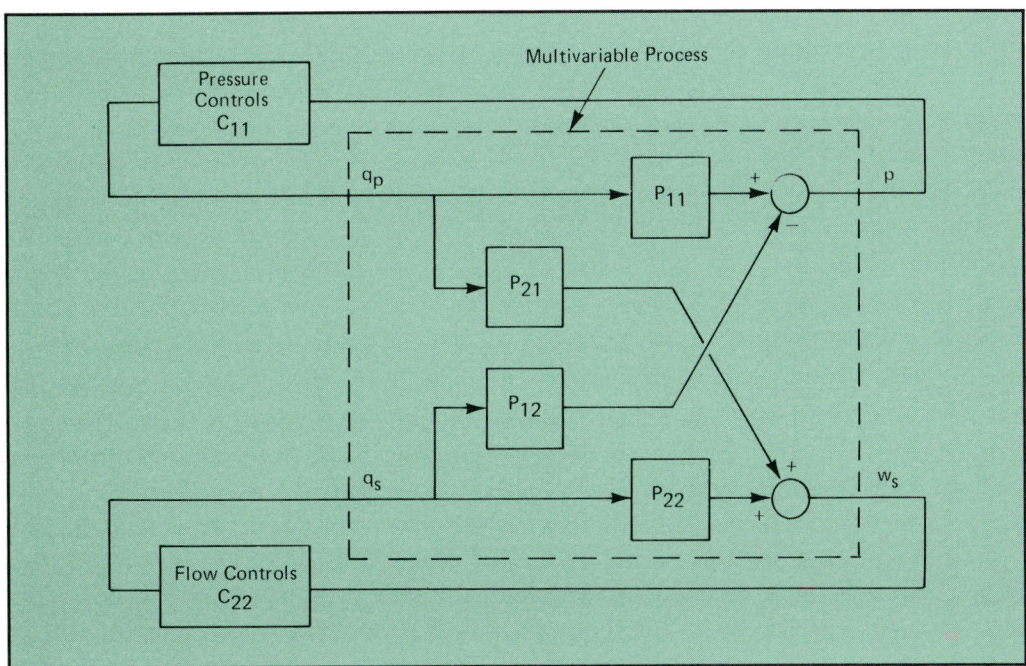

Fig. 7-4. Block Diagram of Two-Dimensional Control on a Compressor

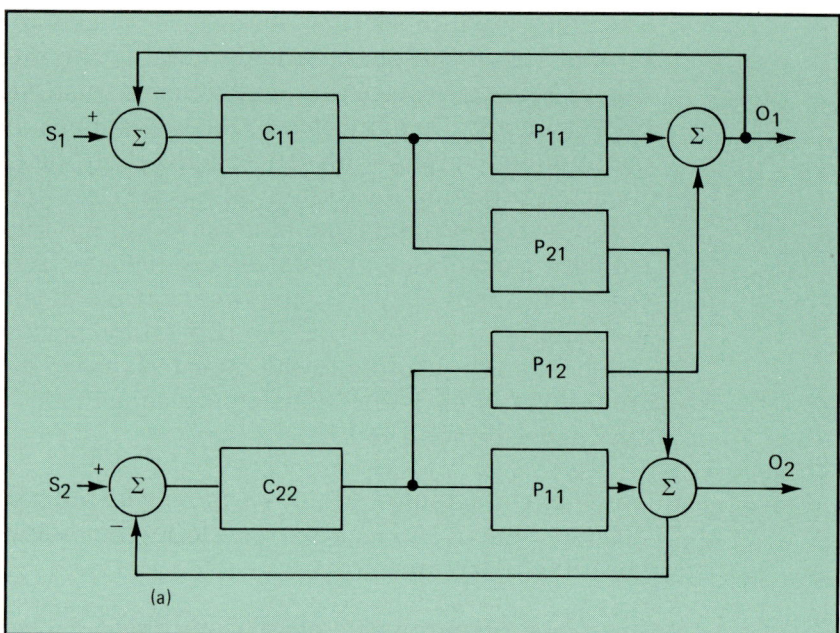

Fig. 7-5. (a) Conventional Diagram of an Interacting Process[4]

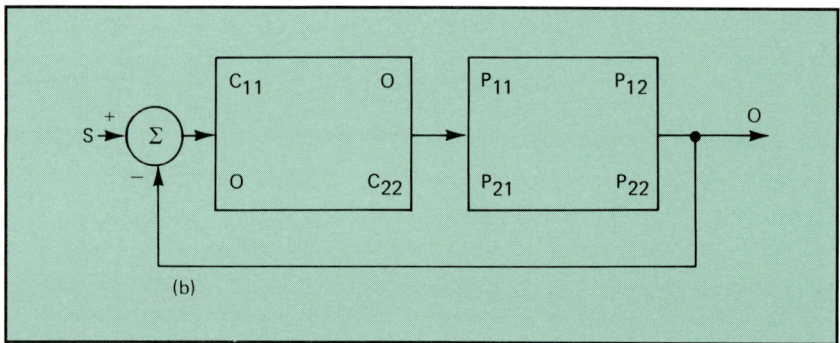

Fig. 7-5. (b) Matrix Notation for Control System of an Interacting Process[4]

Matrix notation is used to describe the control system configuration in Fig. 7-5b. The controllers themselves do not interact, and thus the off-diagonal terms in the controller matrix block are zero. The process does interact, and the interaction terms, P_{12} and P_{21}, appear in the matrix. This observation leads to a very useful generality about matrices; namely, the smaller the off-diagonal terms in comparison to the terms on the diagonal, the less interaction is present. This is the basis for the desirability of matrices with a strong (large) diagonal.

7-3. Matrix Block Diagrams

Figure 7-6 shows a matrix block diagram for three interacting control systems. The parallel lines of information for each loop are shown to emphasize the term-by-term multiplication denoted by the matrix. For example,

$$c_1 = o_1 + N_{11}u_1 + N_{12}u_2 + N_{13}u_3 + N_{14}u_4 \qquad (7\text{-}1)$$

Or, more elegantly,

$$c_i = o_i + \sum_{j=1}^{m} N_{ij}u_j \qquad i = 1, 2, \ldots m \qquad (7\text{-}2)$$

The relationships of the feedback system shown by Fig. 7-6 can be similarly developed. Then, the partial derivative of the rth control variable with respect to the tth error signal is the

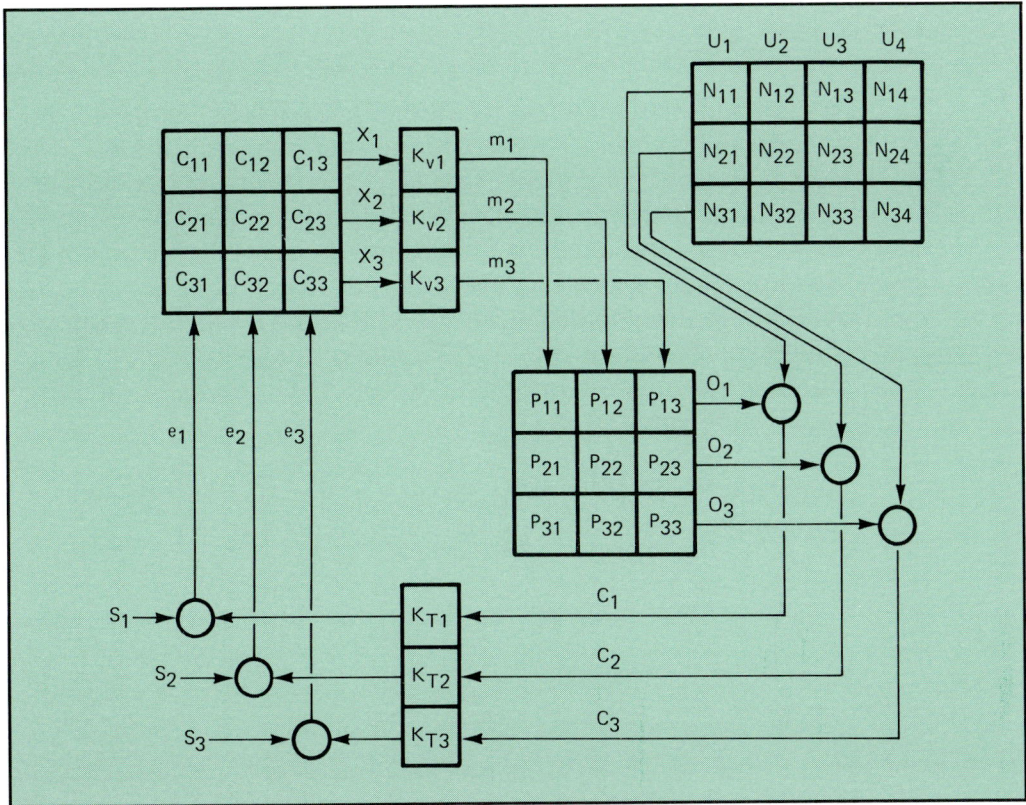

Fig. 7-6. General Block Diagram for a Three-Dimensional Control System

coefficient of e_t. Thus, if $i = r$ and $k = t$,

$$\frac{\partial c_r}{\partial e_t} = \sum_{j=1}^{m} P_{rj} K_{vj} C_{jt} \qquad (7\text{-}3)$$

The sufficient condition for non-interaction is that eq. (7-3) be identically equal to zero for $r \neq t$. This condition states that the change in c_r for any change in e_t (all other error signals considered fixed) is to be zero whenever r differs from t.[5]

Matrix techniques can be used to specify decoupling networks that reduce or eliminate interaction. Figure 7-7a shows the two dimensional interacting system discussed in conjunction with Fig. 7-5a, except that the decoupling elements I_{ij} have been added. Thus, standard non-interacting controllers are used, and the decoupling networks are external from the controller. Figure 7-7b shows the same control system in matrix notation. Here, network I acts on network P such that it looks to C as if C is controlling a process with no interaction. Therefore, C controls the combined effects of P and I. These two networks can be considered to be a single system component.

A common mathematical technique is to design I equal to the inverse matrix of P. This technique sometimes results in compensating components that are impractical or impossible to build. A better approach is to simplify the design of the decoupling components so that they are easily constructed. In this approach, I_{11} and I_{22} are set equal to a gain of one, and I_{12} and I_{21} are designed to make the off-diagonal elements zero. The result is shown in Fig. 7-7c, where interaction is eliminated, and conventional controllers C_{11} and C_{22} can be tuned independently.[4]

Referring to Fig. 7-7a, if $I_{11} = I_{22} = 0$, the required design conditions for the decoupling components are:

$$I_{12} = -\frac{P_{12}}{P_{22}} \qquad (7\text{-}4)$$

$$I_{21} = -\frac{P_{21}}{P_{11}} \qquad (7\text{-}5)$$

Matrix techniques make possible the application of multivariable control theory to systems larger than two

Unit 7: Control System Interaction

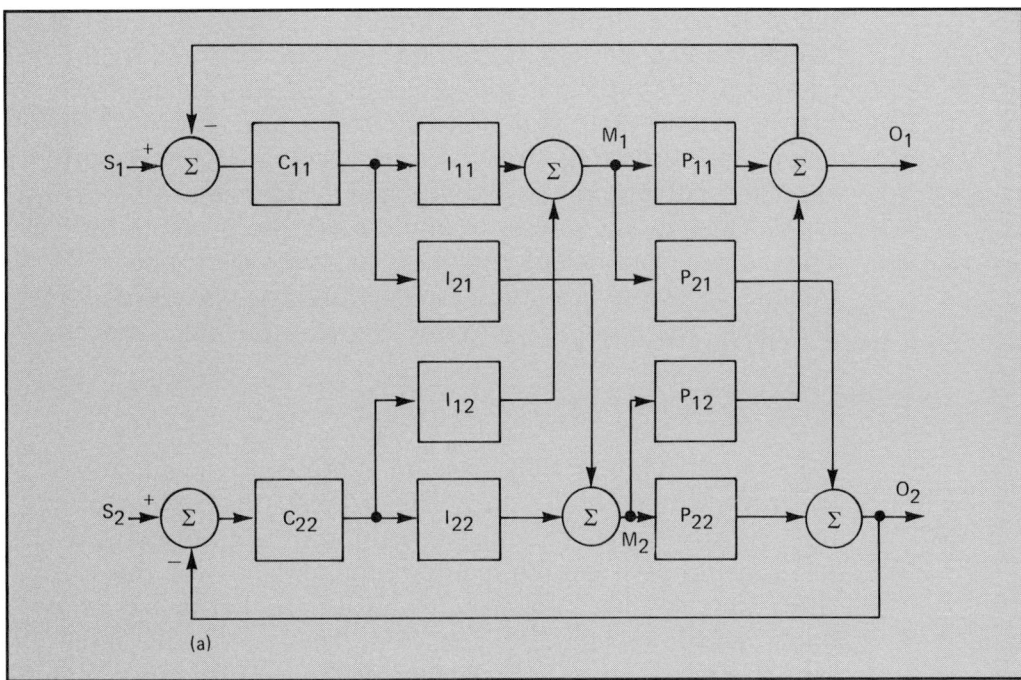

Fig. 7-7. (a) Block Diagram Showing Decoupling Networks for an Interacting Process[4]

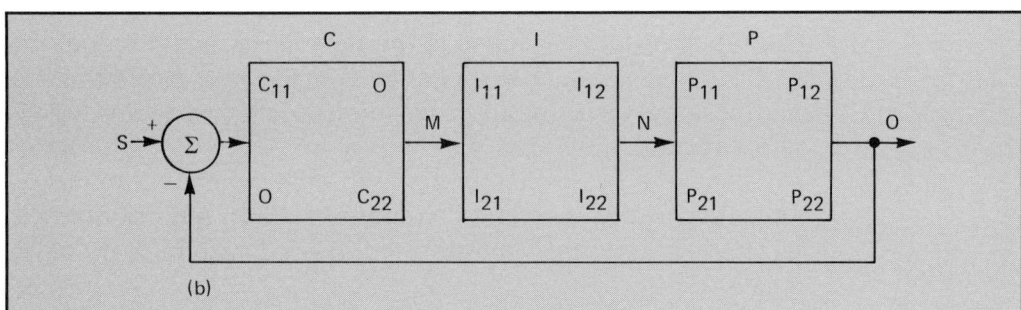

Fig. 7-7. (b) Matrix Block Diagram of Interacting System with Decoupling[4]

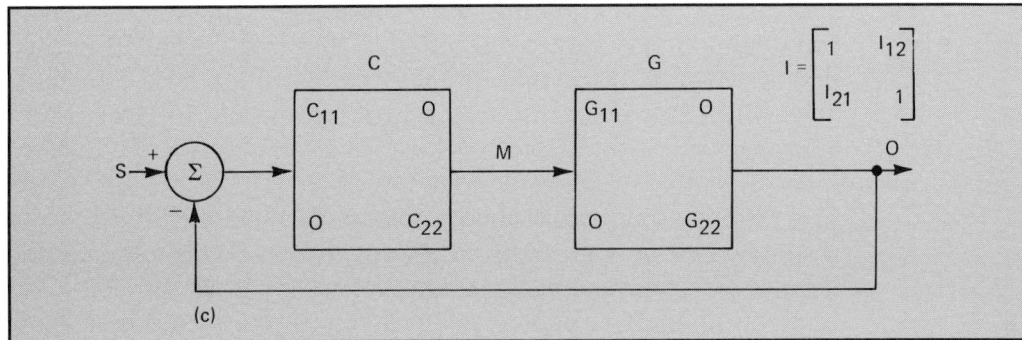

Fig. 7-7. (c) Resultant Block Diagram for Non-Interacting Control[4]

interacting control loops. However, the complexity (cost) of the design calculations for the decoupling networks increases exponentially with the rank of the matrix. But, reference to Fig. 7-2 shows that "real-world" systems have many control loops, whereupon, in addition to the two flow control systems shown, each compressor would have a surge control, and typically many compressors would serve the pressure header. Thus, a means of calculating the expected severity of interaction before designing the decoupling networks is very desirable. The relative gain array provides one such means.

7-4. Relative Gain Array[2]

A very useful interaction measure is the relative gain array (RGA) developed by Bristol.[6] The array is a matrix made up of relative gain terms. The relative gain term is a normalized relationship between the manipulated variable and the controlled variable in any one control loop that is part of a multivariable system. The gain term itself is defined as the ratio of the uncontrolled response of the controlled variable to the controlled response of the controlled variable. The response terms are described as follows:

- The *uncontrolled* response is the new steady-state value of a controlled variable after a change is made in a particular manipulated variable; all other *manipulated* variables are held constant.
- The *controlled* response is the new steady-state value of a controlled variable after a change is made in a particular manipulated variable; all other *controlled* variables are held constant.[7]

Some or all of the other manipulated variables must change in order to hold all of the other controlled variables constant. The changes in the manipulated variables result in a steady-state value of the controlled variable that is different from the one that results if the manipulated variables are held constant. The gain terms refer only to the process and not to the control systems.

The ratio of the uncontrolled response to the controlled response is thus a measure of the response of a simple isolated loop to the response of the same loop as a part of an interacting, multivariable system. Because the individual responses are process gain terms, their ratio is called the *relative gain term*.

A mathematical model is required to calculate the relative gain terms. Figure 7-8 is a block diagram representing the mathematical model of the pressure control-surge control on the single compressor shown in Fig. 7-1, and Fig. 7-9 is a block diagram showing the interaction between the two compressors shown in Fig. 7-2. Referring to Fig. 7-8, the uncontrolled responses are (1) the change in pressure (p) to pressure control valve movement (x_p) with the anti-surge valve (x_s) held constant and (2) the change in surge flow (w_s) to anti-surge valve movement (x_s) with pressure valve movement (x_p) zero. These gains are defined by partial derivatives:

$$\left.\frac{\partial p}{\partial x_p}\right|_{x_s} \qquad \left.\frac{\partial w_s}{\partial x_s}\right|_{x_p} \qquad (7\text{-}6)$$

Likewise, the controlled variables are defined:

$$\left.\frac{\partial p}{\partial x_p}\right|_{w_s} \qquad \left.\frac{\partial w_s}{\partial x_s}\right|_{p} \qquad (7\text{-}7)$$

And the relative gain matrix becomes

$$\begin{array}{c c} & \begin{array}{cc} x_p & x_s \end{array} \\ \begin{array}{c} p \\ \\ w_s \end{array} & \left[\begin{array}{cc} \dfrac{\left.\frac{\partial p}{\partial x_p}\right|_{x_s}}{\left.\frac{\partial p}{\partial x_p}\right|_{w_s}} & \dfrac{\left.\frac{\partial p}{\partial x_s}\right|_{x_p}}{\left.\frac{\partial p}{\partial x_s}\right|_{w_s}} \\ \\ \dfrac{\left.\frac{\partial w_s}{\partial x_p}\right|_{x_s}}{\left.\frac{\partial w_s}{\partial x_p}\right|_{p}} & \dfrac{\left.\frac{\partial w_s}{\partial x_s}\right|_{x_p}}{\left.\frac{\partial w_s}{\partial x_s}\right|_{p}} \end{array} \right] \end{array}$$

The relative gain terms in the matrix are commonly represented by the Greek lambda (λ) for simplicity of notation:

$$\begin{array}{c c} & \begin{array}{cc} x_p & x_s \end{array} \\ \begin{array}{c} p \\ w_s \end{array} & \begin{array}{cc} \lambda_{p,x_p} & \lambda_{p,x_s} \\ \lambda_{w_s,x_p} & \lambda_{w_s,x_s} \end{array} \end{array} \qquad (7\text{-}9)$$

Thus far, in any $\lambda_{i,j}$ the first subscript refers to the controlled variable while the second refers to the manipulated variable.

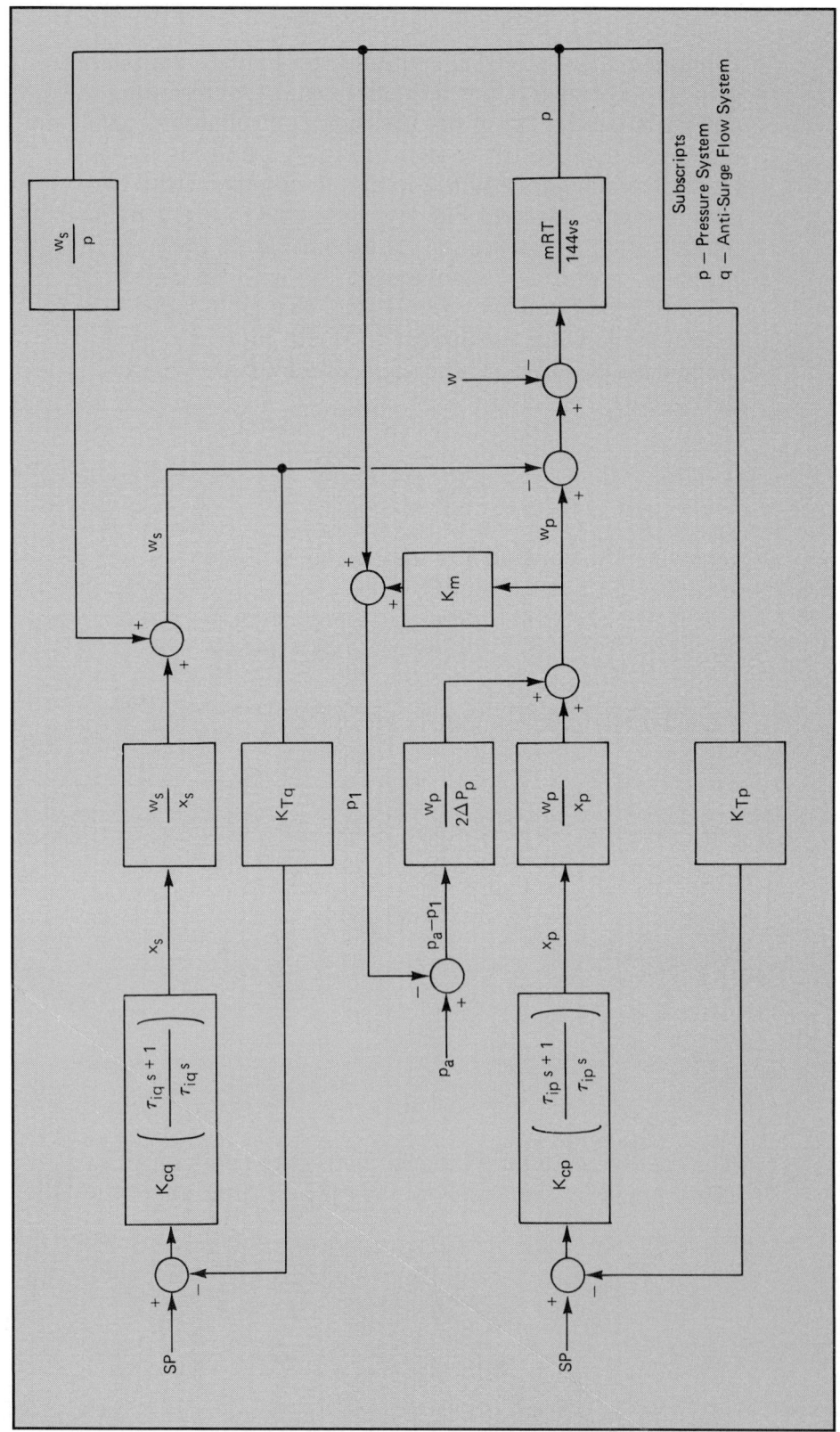

Fig. 7-8. Block Diagram of Constant Speed Centrifugal Compressor with Capacity and Surge Controls as Shown in Fig. 7-1

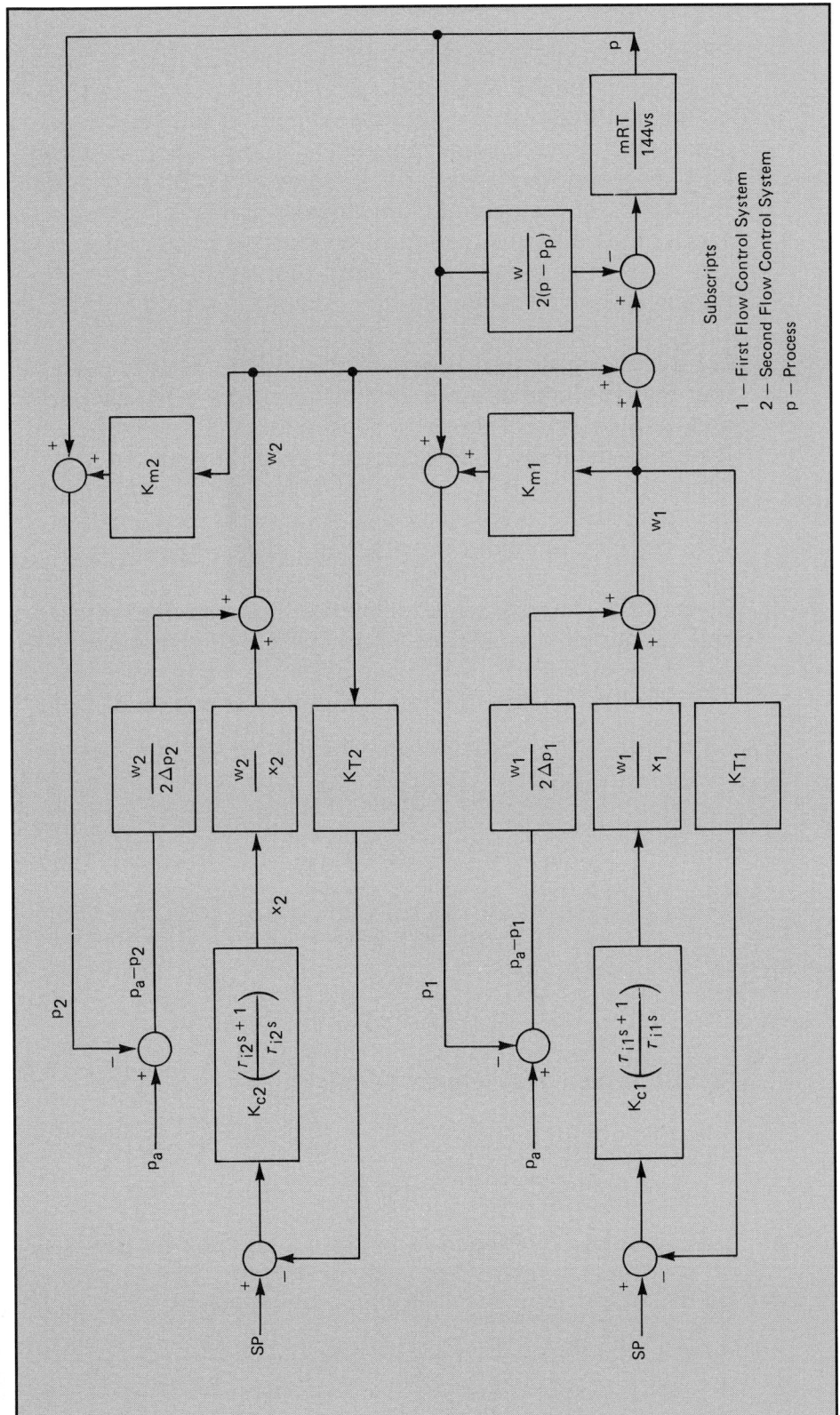

Fig. 7-9. Block Diagram of Two Parallel Centrifugal Compressors with Flow Control Feeding a Pressure Header as Shown in Fig. 7-2

Then the pairing of p with x_p can be stated as λ_{p,x_p}, interpreted as representing the ratio of the effect of x_p on p under two different situations. Since x_s is held constant in calculating the numerator, the numerator represents an open-loop situation for the second loop. Similarly, because w_s is constant in the calculation of the denominator, the denominator represents a special closed-loop situation in which perfect control is achieved in the second loop. Thus λ represents a dimensionless ratio of open- to closed-loop sensitivity.[8,9]

For a 2 × 2 system, it is necessary to calculate only one of the four matrix elements since Bristol[6] has shown that each row and each column of the array must sum to 1.0. A useful feature of the array is the pairing of manipulated and controlled variables. By using the largest positive number for variable pairing, one is choosing the control configuration with the least amount of interaction between the loops.

Consider a measure of the interaction between the pressure and anti-surge flow control loops shown in Fig. 7-1 and 7-7.

Example 7-1. Consider the following process equations from Fig. 7-8:[10]

$$C_{sp} = w_p - w_s - w \tag{7-10}$$

$$w_s = \frac{w_s}{p}\bigg|\, p + \frac{w_s}{x_s}\bigg|\, x_s \tag{7-11}$$

$$p_1 = K_m w_p + p \tag{7-12}$$

$$w_p = -\frac{w_p}{2(p_a - p_1)}\bigg|\, p_1 + \frac{w_p}{x_p}\bigg|\, x_p \tag{7-13}$$

Substitute eq. (7-12) into eq. (7-13):

$$w_p\left[1 + \frac{w_p K_m}{2(p_a - p_1)}\right] = -\frac{w_p}{2(p_a - p_1)}\bigg|\, p + \frac{w_p}{x_p}\bigg|\, x_p \tag{7-14}$$

Substitute eqs. (7-11) and (7-14) into (7-10) and let $\Delta w = 0$:

$$\left[C_s + \frac{\dfrac{w_p}{2(p_a - p_1)}}{1 + \dfrac{K_m w_p}{2(p_a - p_1)}} + \frac{w_s}{p}\right] p = \frac{\dfrac{w_p}{x_p}}{1 + \dfrac{K_m w_p}{2(p_a - p_1)}} x_p \tag{7-15}$$

and

$$\left.\frac{\partial p}{\partial x_p}\right|_{x_s} = \frac{\dfrac{w_p}{x_p}}{\left[\dfrac{\dfrac{w_p}{2(p_a - p_1)}}{1 + \dfrac{K_m w_p}{2(p_a - p_1)}} + \dfrac{w_s}{p}\right][\tau_1 s + 1]} \qquad (7\text{-}16)$$

Similarly, substitute eq. (7-14) into eq. (7-10), and let $\Delta w = 0$:

$$C_{sp} = -\frac{\dfrac{w_p}{2(p_a - p_1)}}{1 + \dfrac{K_m w_p}{2(p_a - p_1)}} p + \frac{\dfrac{w_p}{x_p}}{1 + \dfrac{K_m w_p}{2(p_a - p_1)}} x_p - w_s \qquad (7\text{-}17)$$

$$(\tau_2 s + 1)p = \frac{2(p_a - p_1)}{x_p} x_p - \frac{1 + \dfrac{K_m w_p}{2(p_a - p_1)}}{\dfrac{w_p}{2(p_a - p_1)}} w_s \qquad (7\text{-}18)$$

Then, for steady-state conditions:

$$\left.\frac{\partial p}{\partial x_p}\right|_{w_s} = \frac{2(p_a - p_1)}{x_p} \qquad (7\text{-}19)$$

And the relative gain array term becomes

$$\lambda_{p,x_p} = \frac{1}{1 + \dfrac{w_s}{p}\left(\dfrac{2(p_a - p_1)}{w_p} + K_m\right)} \qquad (7\text{-}20)$$

Equation (7-20) shows the relative gain term to have a value of one when the anti-surge valve is closed ($w_s = 0$). Thus the relative gain matrix diagonal terms have values of one, and the off-diagonal terms have values of zero. This configuration denotes non-interaction, which is obviously true with the anti-surge valve closed. The relative gain term decreases as anti-surge flow increases, showing an increased interaction in that circumstance.

Table 7-1 shows values of relative gain term λ_{p,x_p} for a number of different centrifugal compressors made by different manufacturers for different services. The gain terms have been calculated using eq. (7-20) with the anti-surge flow (w_s) 50% of the compressor throughput flow (w_p). In each case, the relative gain term is large enough that the matrix should have a strong diagonal, and little interaction would be expected.

Fehervari[11] has developed an RGA for a compressor with a surge control system similar to that shown in Fig. 6-12. He concludes that the off-diagonal elements are always significantly smaller than one and that the multivariable system is inherently stable. He also takes issue with designs that would reverse the controls in special applications, pointing out that controller detuning will be required to compensate for the unfavorable RGA.

The relative gain term for the multivariable system shown by Figs. 7-2 and 7-9 can be determined by the techniques shown in Example 7-1.

$$\frac{\left.\dfrac{\partial w_1}{\partial x_1}\right|_{x_2}}{\left.\dfrac{\partial w_1}{\partial x_1}\right|_{w_2}} = \lambda_{w_1,x_1} =$$

$$\frac{1 + \dfrac{w_2 K_{m2}}{\Delta p_2} + \dfrac{2(p-p_p)}{w}\dfrac{w_2}{2\Delta p_2}}{1 + \dfrac{w_2}{2\Delta p_2}\left[\dfrac{K_{m2} + \dfrac{w_1 K_{m1} K_{m2}}{2\Delta p_1} + \dfrac{2(p-p_p)}{w}\dfrac{w_1 K_{m2}}{\Delta p_1} + \dfrac{2(p-p_p)}{w} + \dfrac{2(p-p_p)}{w}\dfrac{w_1 K_{m1}}{\Delta p_1}}{1 + \dfrac{w_1 K_{m1}}{\Delta p_1} + \dfrac{2(p-p_p)}{w}\dfrac{w_1}{2\Delta p_1}}\right]}$$

(7-12)

Equation (7-21) is made up of performance curve gains (K_m) and inlet and discharge flow resistances. The value of the relative gain term approaches one as the flow through the "second" compressor (w_2) approaches zero. This confirms that interaction disappears as the second system is taken out of service.

The many possible control configurations on a compressor result in different analytical expressions for the relative gain terms and different values for those terms. Experience has

Compressor	Joy (Figure 5-9)		Dresser Clark (Figure 6-2)	Mitsui (Figure 6-14c)	Atlas Copco (Figure 6-21)	
Guide Vane Angle	−10°	60°			0°/0°	16°/16°
K_m psi, lb/min	−0.025	−0.086	−0.184	−0.023	−0.034	−0.014
*SCL flow, lb/min	853	742	566	1248	483	1140
$p_a - p_1$, psi	0.2	0.2	0.7	0	0	0
p, psia	82	67.5	234.4	315	34.5	34.5
λ_{p,x_p}	0.88	0.67	0.82	0.96	0.81	0.81

* SCL—surge control line

Table 7-1

shown that terms can have values outside the range $0 \leq \lambda \leq 1$, specifically terms greater than one, which necessitate other terms that are negative. A relative gain term between zero and one means that the effect of the second loop changes the measurement signal in the same direction. McMillan[1] calls this cooperation. A relative gain term greater than one means that the net effect of the second loop is to change the measurement signal in the opposite direction. This is called conflict. Interaction dominates in the conflict condition. Thus, if the first control loop is stable when the second is in the manual mode, it can quite possibly become unstable when the second loop is placed in automatic. In any case, it will be greatly underdamped.

Equation (7-21) shows that the value of the relative gain terms is a function of the flow resistances. These resistances are determined by the slope of the load curve and the performance curve, as shown by Fig. 7-10. Using the concept of cooperation and conflict, McMillan[1] has classified different pairings of variables on the basis of the steepness of the curves, or of the comparable resistances, in Table 7-2.

Pairing of Variables	Load Curve Steeper	Performance Curve Steeper
Pressure & speed	Cooperation	Conflict
Pressure & vane	Cooperation	Conflict
Pressure & valve	Cooperation	Conflict
Flow & speed	Conflict	Cooperation
Flow & vane	Conflict	Cooperation
Flow & valve	Conflict	Cooperation

Table 7-2. Effect of Relative Slopes and Pairing on Interaction[1]

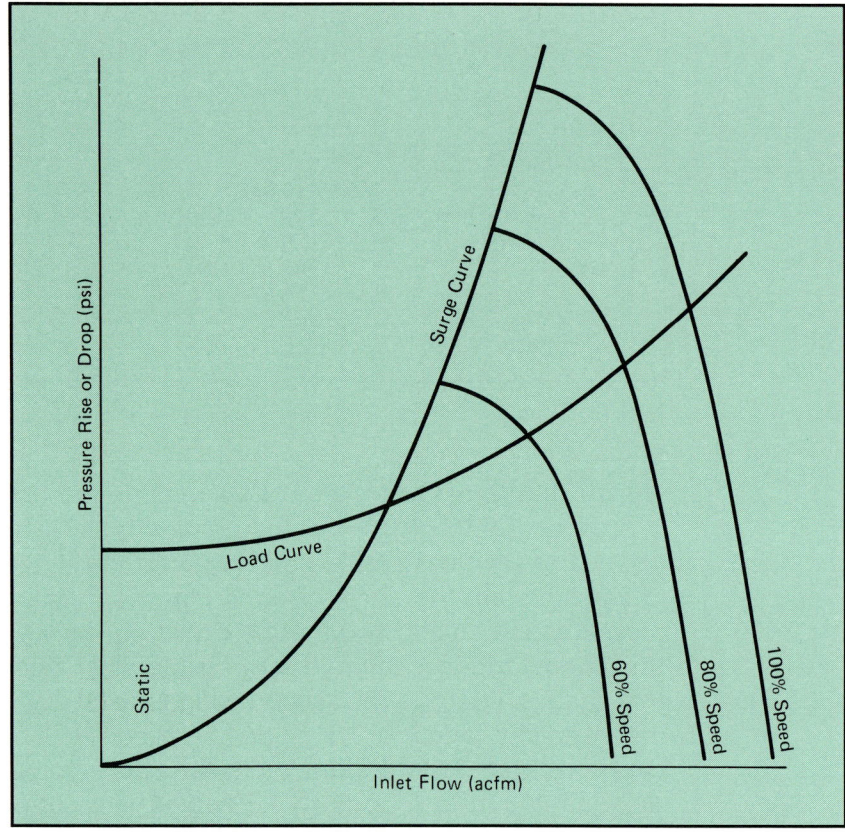

Fig. 7-10. Compressor Map for Axial Compressor and a System with a High Static Load[1]

7-5. Dynamic Interaction

The relative gain array has been successfully used[12,13,14] to predict interaction, to assess the severity of the interaction, and to provide guidance in minimizing interaction. However, it is a linear, steady-state mathematical expression and, as such, becomes an inaccurate representation when changes in production rates cause changes in gains, or when the interaction leads to dynamic instability. An example is shown in the relative gain term shown in eq. (7-20) for Fig. 7-1 and in eq. (7-21) for Fig. 7-2. In each case, the relative gain analysis shows little tendency toward interaction, yet interaction is a persistent problem in both multivariable systems. In each case, the interaction is a form of dynamic instability, and the RGA analysis must be supplemented by a strategy that "ignores the theory and resorts to a sound engineering approach."[15]

Modifications have been made to the RGA to accommodate system dynamics.[8] Consider the relationships shown in Fig. 7-7a.

$$O_1 = M_1 P_{11}(s) + M_2 P_{12}(s) \qquad (7\text{-}22)$$
$$O_2 = M_1 P_{21}(s) + M_2 P_{22}(s)$$

Then a dynamic lambda (λ) can be defined from the final value theorem:[10]

$$\lambda_{O_1,M_1} = \frac{\left.\dfrac{\partial O_1}{\partial M_1}\right|_{s,M_2=0}}{\left.\dfrac{\partial O_1}{\partial M_1}\right|_{s,O_2}} \qquad (7\text{-}23)$$

A dynamic measures of interaction can then be evaluated:

$$\lambda_{O_1,M_1} = \frac{1}{1 + \dfrac{P_{12}(s)P_{21}(s)}{P_{22}(s)P_{11}(s)}} \qquad (7\text{-}24)$$

This measure of interaction compares to those developed by others.[9,15]

Dynamic interaction will typically result from either (1) the system gain as seen by one controller being higher because of the manipulation of the second control valve[16] and/or (2) both control loops having essentially the same dynamic response. The latter condition applies to most compressor control configurations, where the only dynamics will be in the volumes of process vessels included in the compressor anti-surge recycle piping. Both Figs. 7-1 and 7-2 show control configurations where all control systems have essentially equal dynamic responses. Consider the characteristic equation of the multivariable control system shown in Fig. 7-9:[10]

$$1 + K_{cq}\left(\frac{\tau_{iq}s + 1}{\tau_{iq}s}\right)\frac{w_s}{x_s}K_{Tq} + K_{cp}\left(\frac{\tau_{ip}s + 1}{\tau_{ip}s}\right)\frac{2\Delta p_p}{x_p}\frac{K_{Tp}}{\tau s + 1}$$

$$+ \left[K_{cq}\left(\frac{\tau_{iq}s + 1}{\tau_{iq}s}\right)\frac{w_s}{x_s}K_{Tq}\right]\left[K_{cp}\left(\frac{\tau_{ip}s + 1}{\tau_{ip}s}\right)\frac{2\Delta p_p}{x_p}\frac{K_{Tp}}{\tau s + 1}\right]$$

$$\tau = \left(\frac{2\Delta p_p}{w_p} + K_m\right)C \quad (7\text{-}25)$$

Here the equation has been simplified by the omission of negligible terms. The resistance term in the time constant (τ) of the discharge piping and equipment volume is quite small (see Table 7-1). Thus, unless the discharge volume (capacitance) is extremely large, the time constant (τ) will be so small as to be negligible.

With the discharge volume dynamics negligible, optimum controller tuning will result in identical bracketed terms. Thus, the summed terms result in an open-loop gain twice that of the stable gain of either system tuned individually. The product terms, since both are well-tuned and have the same natural frequency, have a dynamic gain several times that of either system because of the logarithmic nature of the frequency response relationships.[10]

The only way out of the dynamic interaction situation is by a decoupling network, such as shown in Fig. 7-7a, or detuning one controller. Decoupling is seldom used on compressors because of the question of the validity of the data, formerly discussed. The effect of detuning can be observed by letting one of the controller gains equal zero in eq. (7-25); the equation immediately reduced to the response equation for the other controller alone. Thus, placing a very wide proportional band and a very slow reset on the pressure controller will cause it not to respond to changes in anti-surge flow, and interaction will be negligible. The price to be paid, of course, is poor performance by the pressure control system.

Suchanti et al.[17] have tried to interject a bit of order into the trial and error procedure of tuning interacting controllers. Their first option is an interactive procedure that suggests that controllers in each loop be adjusted repeatedly in sequence until satisfactory convergence is obtained. This procedure cannot be applied to the surge controller because, as pointed out in Unit 6, the possibility of the compressor going into surge makes experimental tuning of the surge controller very risky. Their second option is the use of a performance index (PI) in a simulated system, and they demonstrate useful results with that technique.

The most elegant analytical tool for predicting and analyzing interaction is the Inverse Nyquist Array.[17] The array has been found to be a useful tool for control loop design and for comparison of different control strategies. It has also been found to be useful for the design of decouplers and the tuning of controllers. However, like the more familiar Nyquist plot for assessing stability, the plotting of the graph is extremely labor-intensive. For all practical purposes, the Inverse Nyquist Array is useful for design only if it is available as a stored program on a computer.

7-6. Interaction Compensation

While sophisticated techniques are available for the design of decoupling networks, the reduction of interaction in compressor control systems is usually accomplished by heuristic compensators because of the difficulty of obtaining precise calibration data. Figures 7-11 and 7-12 show two typical circuits added to minimize interaction.[18,19]

Figure 7-11 shows the simplest possible system, which is based on the concept that the anti-surge valve must open quickly to protect the compressor but can close as slowly as desired. In opening the anti-surge valve, controller output must decrease. The output of the low-pass filter will decrease less quickly.[10] Thus, the decreasing controller output signal will be transmitted directly to the control valve by the low

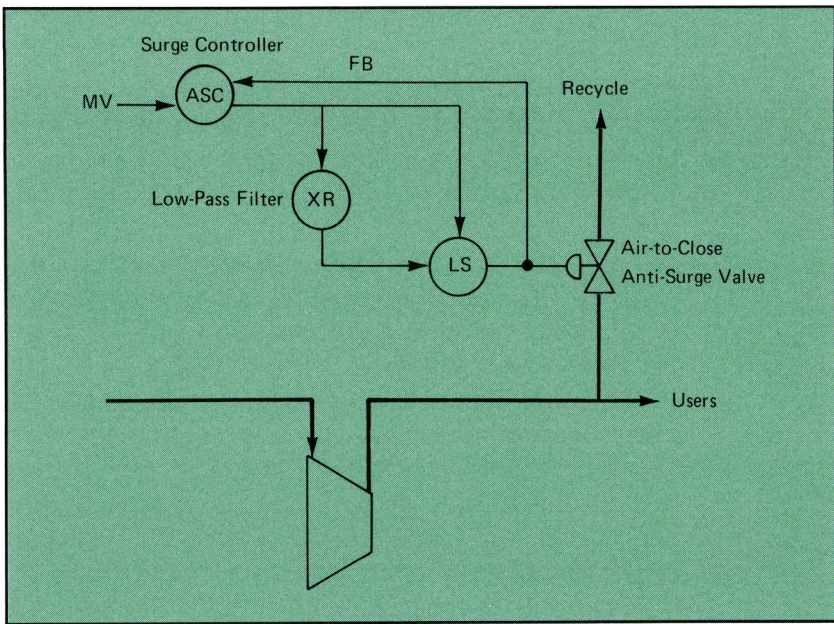

Fig. 7-11. Circuit to Produce a Nonsymmetrical Surge Control System Response to Reduce Interaction

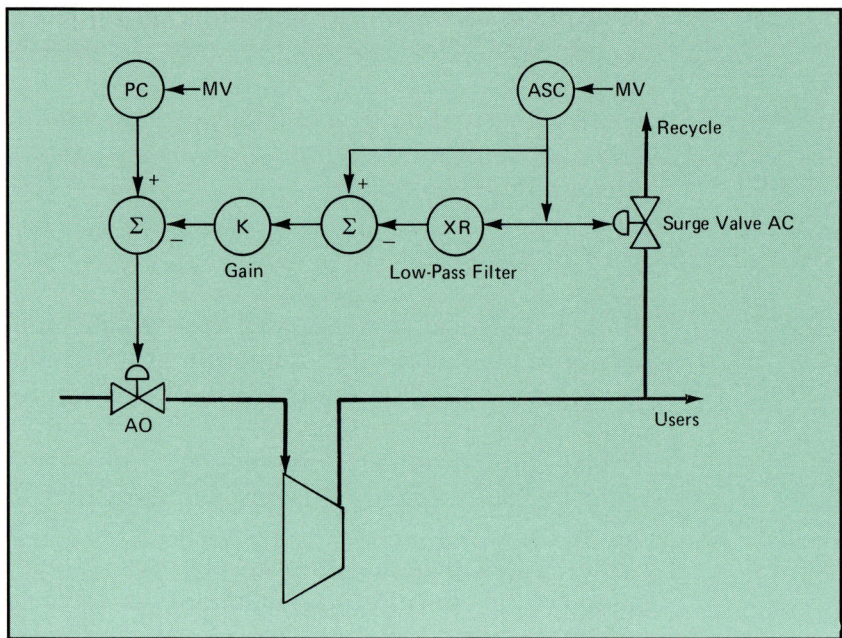

Fig. 7-12. Interaction Compensator for Reducing Pressure and Surge Control System Interaction[18]

selector, and the valve will open as quickly as directed by the controller. However, upon an increasing controller output signal, the output of the low-pass filter will lag behind the controller output, will be lower than the controller output at the low selector relay, and will regulate the valve position. Thus the valve will close more slowly, the rate of closure being adjusted by the time constant setting of the filter.

Interaction is usually observed on the recorder chart or the CRT as a sine wave. This pattern results from first one system adjusting, then a second system readjusting in response to the first. However, the surge control system shown in Fig. 7-11 cannot respond in a sine wave since the forced slow closing results in an asymmetrical sawtooth wave. This response tends to break up to interactive cycle, thus minimizing the problem without detuning either controller.

Figure 7-12 uses an impulse-totalizing relay to decouple the pressure and surge control systems. The output of the pressure controller goes to the inlet valve through the totalizing relay. The output of the relay is equal to the output from the pressure controller minus an impulse[10] from the surge controller. Both inputs to the low-pass filter summing junction are equal when the surge controller output is constant, and a zero signal is transmitted from the surge control system to the pressure control system. On a decrease in user flow demand, discharge pressure increases and pressure controller output decreases to close the inlet valve. Concurrently, the surge controller output decreases (to open the anti-surge valve), resulting in a signal to the totalizing relay that tends to open the inlet valve. The two signals to the inlet valve tend to cancel, and the valve does not move. The gain adjustment (K) on the impulse relay is used to match the characteristics of the two control valves, while the low-pass filter adjustment is used to tune the dynamic response of the pressure control loop.

The stationary position of the inlet valve is the correct control action because the valve is already at its minimum opening and should not be permitted to close. With time, the plus and minus inputs to the low-pass filter summer will become equal, and the output of the impulse relay into the pressure control system will become zero. During this time, the pressure controller output gradually changes to maintain constant discharge pressure.

The low-pass filter can be a pneumatic, electronic, or digital device to accommodate the type of control system. Its use permits the design of an interaction compensation network that does not depend on the parameters of the compressor.

Exercises

1. Why is there no interaction in the system shown in Fig. 7-1 when the compressor is operating at the design point?

2. Why does anti-surge flow increase in Fig. 7-3 when pressure control system set point is increased?

3. What is the relationship between output (o) and manipulated variable (m) in Fig. 7-6?

4. Why would a value of one for the term $\lambda_{p,xp}$ be expected when the anti-surge valve is closed?

5. Calculate the relative gain term from eq. (7-21) for the following conditions:

$$K_{m1} = 0.184 \qquad K_{m2} = 0.184$$

$$\Delta p_1 = 0.5 \text{ psi} \qquad \Delta p_1 = 0.5 \text{ psi}$$

$$w_1 = w_2 = 600 \text{ lb/min} \qquad w = w_1 + w_2$$

$$p = 300 \text{ psi} \qquad p_p = 287 \text{ psi}$$

6. Suppose in Exercise 5 that $\Delta p_1 = \Delta p_2 = 2$ psi. What is the relative gain term? What can be concluded from this result?

7. Why do the RGA terms change value with changing production level?

References

1. McMillan, G. K., *Centrifugal and Axial Compressor Control*, Instrument Society of America, Research Triangle Park, NC, 1983.
2. McAvoy, T. J., *Interaction Analysis: Principles and Applications*, Instrument Society of America, Research Triangle Park, NC, 1983.
3. Lloyd, S. G., "Basic Concepts of Multivariable Control," *Instrumentation Technology*, December, 1973, p. 31.
4. Zalkind, C. S., "Practical Approach to Non-Interacting Control, Part I," *Instruments and Control Systems*, March, 1967, p. 89.
5. Lefkowitz, I., Class Notes, Case Western Reserve University, 1958.
6. Bristol, E. H., "On a New Measure of Interaction for Multivariable Process Control," *IEEE Transactions on Automatic Control*, AC-11, No. 1, 1966, p. 133–134.
7. Nisenfeld, A. E., "How Do You Design Control Systems?" *CHEMPID Newsletter*, Instrument Society of America, Research Triangle Park, NC, 1970.
8. Witcher, M. F., and T. J. McAvoy, "Interacting Control Systems: Steady State and Dynamic Measurements of Interaction," *ISA Transactions*, Vol. 16, No. 3, p. 35.
9. Nisenfeld, A. E., and H. M. Schultz, "Interaction Analysis Applied to Control System Design," *Instrumentation Technology*, April, 1971, p. 52.
10. Moore, R. L., *The Dynamic Analysis of Automatic Process Control*, Instrument Society of America, Research Triangle Park, NC, 1985.
11. Fehervari, W., "Assymetric Algorithm Tightens Compressor Surge Control," *Control Engineering*, October, 1977, p. 63.
12. Wang, J. C., "Compute Relative Gain Matrices for Better Distillation Control," *InTech*, March, 1980, p. 40.
13. Woolverton, P. F., "How to use Relative Gain Analysis in Systems with Integrating Variables," *InTech*, September, 1980, p. 63.
14. Shinsky, F. G., "Interaction Between Control Loops—Park I: Positive Coupling," *Subject: Control Loop Interaction*, Foxboro Co. Publication 413-3.
15. Tyreus, B. D., "Variable Pairing in Process Control: A Nonproblem?" DuPont Co. Publication, 1981.
16. Shinsky, F. G., "Interaction Between Control Loops—Part III: Dynamic Coupling," *Subject: Control Loop Interaction*, Foxboro Co. Publication 413-3.

17. Suchanti, N., K. Chun, and C. D. Fornier, "Tuning Controllers for Interacting Processes," *Instruments and Control Systems*, April, 1973, p. 47.
18. Warnock, J. D., "Methods of Control of Centrifugal and Reciprocating Compressors," Moore Products Co., Spring House, PA.
19. Warnock, J. D., "Typical Compressor Control Configurations," *Centrifugal Compressor Operation and Control*, Instrument Society of America, Research Triangle Park, NC, 1976.

Unit 8:
Coordinated Compressor Control

UNIT 8

Coordinated Compressor Control

Previous units have shown compressor surge control to be a challenge. The analytical data on which anti-surge systems are calibrated is, at best, fuzzy. Uncertainty dictates wide surge control margins to reduce the possibility of damage to the compressor. However, the wide margins of safety result in interlocks shutting the compressor down and anti-surge valves remaining open, both long before necessary to protect the machine.

Special compressor control systems that permit minimum surge control margins are described in this unit. Essentially, surge control margins are reduced on the basis of confidence generated by a surge measurement feedback. The disadvantage of the special control systems is that they are different, from a maintenance and operational viewpoint, from the other control systems in a plant or factory.

Learning Objectives — When you have completed this unit, you should:

A. Be familiar with control systems designed specifically for centrifugal compressor control.

B. Be aware of the concepts incorporated into custom control systems.

C. Appreciate the advantages and disadvantages of custom control systems.

D. Know the source of supply of custom control systems. (Note: Most of the concepts and equipment described in this unit are patented by Compressor Controls Corporation, Inc., Des Moines, Iowa).

8-1. Introduction[1]

Surge in turbocompressors is costly in mechanical damage and in process disturbances and downtime. Since the effects of surge are too costly to ignore, anti-surge protection is necessary on all turbocompressors.

The consequences of surge are severe. Table 8-1 shows typical costs of repairing compressor damage resulting from surge. Repair costs are only a part of the total cost of surge damage to a compressor in an operating process. Table 8-2 shows the cost of the unavailability of the compressor, while Table 8-3 shows the costs of starting it up again. The compressor will almost always be the largest energy user in the process. Operating costs are shown in Table 8-4. Even a small percentage saving in such large expenditures by keeping the anti-surge valve closed when it doesn't have to be open is significant.

The set point of the surge controller is a line on the compressor map parallel to the surge limit line, as shown by eq. (6-17) and Fig. 6-13. Equation (6-17) can be rewritten as follows:

$$\Delta p_o = K \Delta p_c + b_1 \qquad (8\text{-}1)$$

The surge control line defined by eq. (8-1) is shown on the linearized compressor map in Fig. 8-1. The first objective of the surge control system is to protect the compressor from surge. Protection is assured by large values of b_1. But the area of operation of the compressor, shown in Fig. 8-2, must be maximized for efficient compressor operation. Efficient

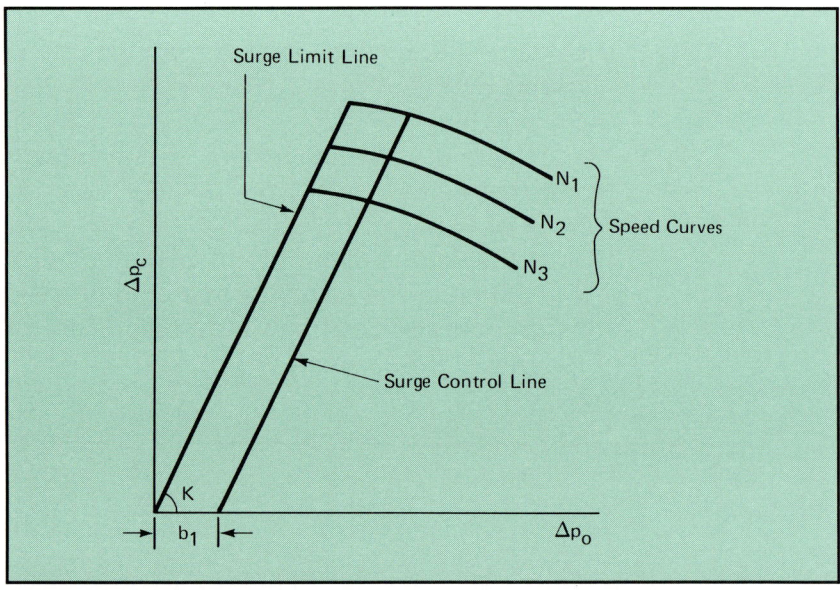

Fig. 8-1. Linearized Compressor Map for Variable Speed Compressor Showing Surge Limit Line (SLL) and Surge Control Line (SCL)[1]

	3,000-hp Process Gas Compressor	20,000-hp Axial Air Blower
Replace:		
Seals	$ 20,000	$ 50,000
Bearings	$ 10,000	$ 25,000
Rotor Assembly	$200,000	$750,000

Table 8-1. Repair Costs[13]

- 30,000-bpd Cat Cracker. $5,000 per hour, lost sales plus fixed expenses. The biggest units are 4 times this size!
- North Sea Platform, 100 mmscfd. $12,500 per hour lost sales plus fixed expenses.
- Consequences of downtime: lost profit, lost customer goodwill, repair costs, wrath of top managements, heads may roll!

Table 8-2. Cost of Downtime[13]

	100-mmscfd North Sea Platform	30,000-bpd Cat Cracker
Start-up cost per day	$375,000	$1,150,000

- Includes lost sales plus operating expenses
- Most compressor control systems design problems are discovered during commissioning.

Table 8-3. Commissioning Costs[13]

Cost to operate one compressor per year

1,000 hp	$327,000
4,000 hp	$1,307,000
40,000 hp	$13,100,000

Assumes power at $.05 per kilowatt hour for $327 per horsepower per year.

Table 8-4. Operating Costs[13]

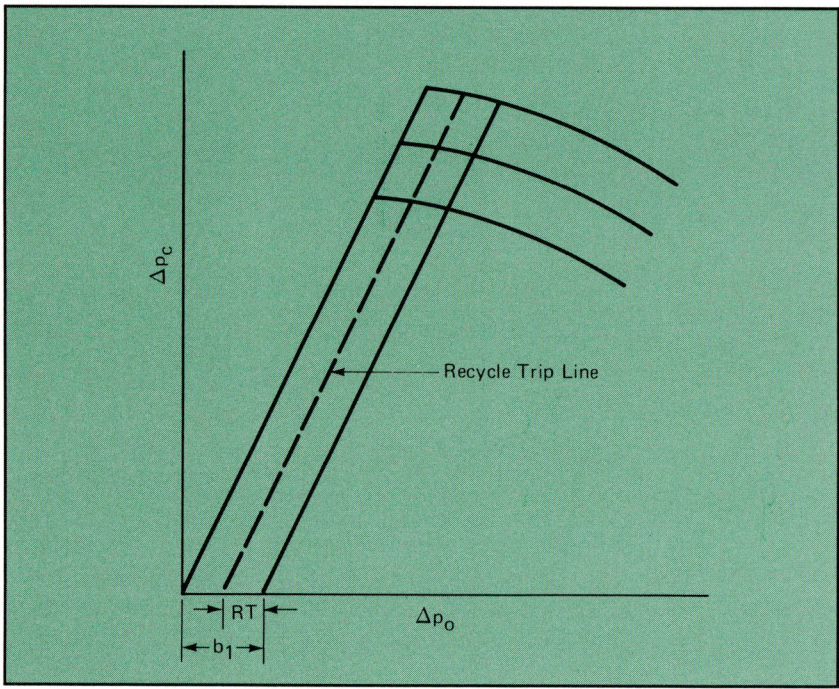

Fig. 8-2. Linearized Compressor Map per Fig. 8-1 with Addition of Recycle Trip Open-Loop Control Line[1]

operation is attained only with the anti-surge valve closed. From this perspective, small values of b_1 are desirable.

Surge control system reliability is enhanced by combining an open-loop control circuit with the closed-loop surge control system described in Unit 6. The open-loop control adds a recycle trip line, shown in Fig. 8-2, to the linearized compressor map. In essence, the recycle trip is an overshoot limit. Small or slow distrubances are handled by the surge controller. Large or fast disturbances will reach the recycle trip line and activate the open-loop control algorithm.

A part of the uncertainty in the calibration of the surge control system is the movement of the surge limit line, either by changes in the state of the gas or the condition of the compressor. That uncertainty can be eliminated by detecting the surge flow reversal and having the control system add an additional bias (b) to the surge control line. Further surge is prevented by the larger surge control margin. The operator is informed of the new bias for his evaluation.

Interaction with the capacity control system is detrimental to the response of the surge control system. Decoupling circuits from the surge controller to the capacity controller minimize the interaction and improve the response of the surge control system.

8-2. Incipient Surge[2]

Figure 8-3 shows the variation in the differential pressure across the compressor (Δp_c) and the differential pressure across the flow element (Δp_o) as the compressor goes into surge. Notice that the time span of the entire chart portion is only 30 seconds. The events on the chart are happening very quickly, and very high-performance or high frequency response instruments are required to sense and track the events shown. Figure 3-12 compares the response of the typical petrochemical plant pressure transmitter (Bourdon tube plus linkage) and the newer diffused silicon electronic transmitters. Surge often goes undetected with the slower typical process transmitters. Figure 8-3 has been obtained with fast transmitters and a high-speed recorder.

Figure 8-3 shows the flow signal (differential pressure across the flow element) to be more sensitive to surge than the pressure signal (differential pressure across the compressor). This dictates using the flow signal for a feedback measurement of surge and is to be expected since the performance curve on the compressor map is flat at the surge point.

The flow graph on Fig. 8-3 shows pre-surge oscillations before surge actually occurs at 14 seconds. The pre-surge oscillations, called incipient surge, are useful as a warning of the closeness of impending surge when testing the compressor. However, incipient surge is difficult to implement as a component of the surge control system because of the problem of distinguishing it from the flow measurement noise.

As surge begins, flow drops precipitously—typically from the design condition to a reverse flow in 0.05 second, regardless of pressure or compressor speed.

Incipient surge can also result from a mismatch between the performance curves of the different stages of a multistage compressor, as shown in Fig. 8-4. Assuming no sidestreams or condensate separation, the same mass flow (w) will pass through each stage during steady-state operation. Operation on load line I will not cause surge. As process resistance increases, the process is characterized by load line II. Here,

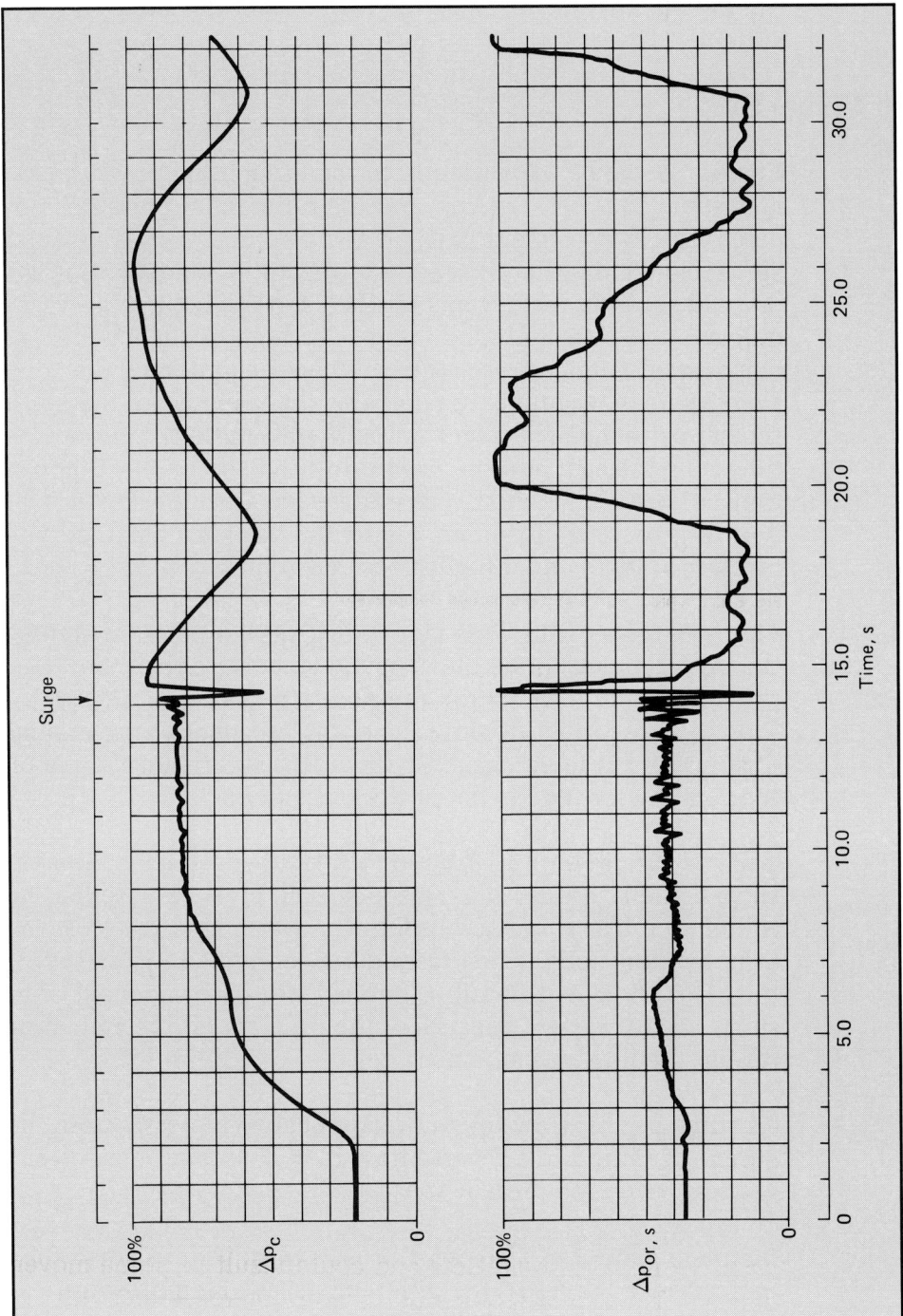

Fig. 8-3. Recordings of Differential Pressure across the Compressor (ΔP_c) and Differential Pressure across the Flow Element (ΔP_o) as Surge Begins[2]

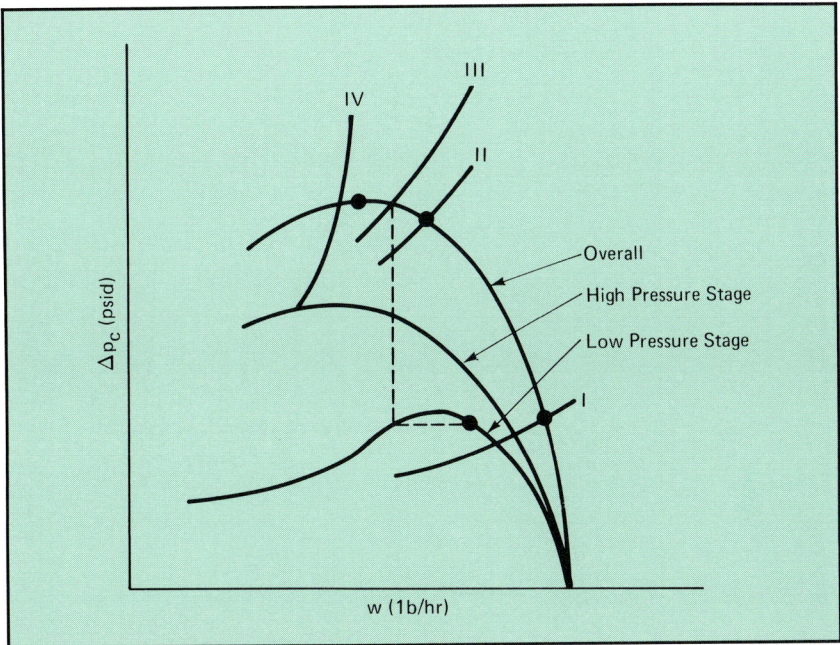

Fig. 8-4. Incipient Surge in a Multistage Compressor[13]

the high-pressure stage is well away from surge, but the low-pressure (first) stage is at the peak of its performance curve and is tending to surge. Incipient surge would be observed at this operating point. Load line III would cause the low-pressure stage to surge, and load line IV operation would cause the whole compressor to surge.

8-3. Experimental Performance Testing[3,4,5]

The operating envelope of a centrifugal compressor is shown as Fig. 8-5. A typical compressor can operate safely at any point within the envelope but has serious problems if it strays outside the envelope.

The operating envelope is bounded on the left by the surge control line. Operation to the left of this line incurs the danger of surge. The right side of the envelope is bounded by the line of maximum motor power. Operation to the right of this line provides no additional flow and could result in prime mover damage. The top of the operating envelope represents full opening of the inlet valve (or maximum compressor speed). This is a physical constraint that the compressor cannot cross. Likewise, the lower limit represents the load line of minimum process resistance.

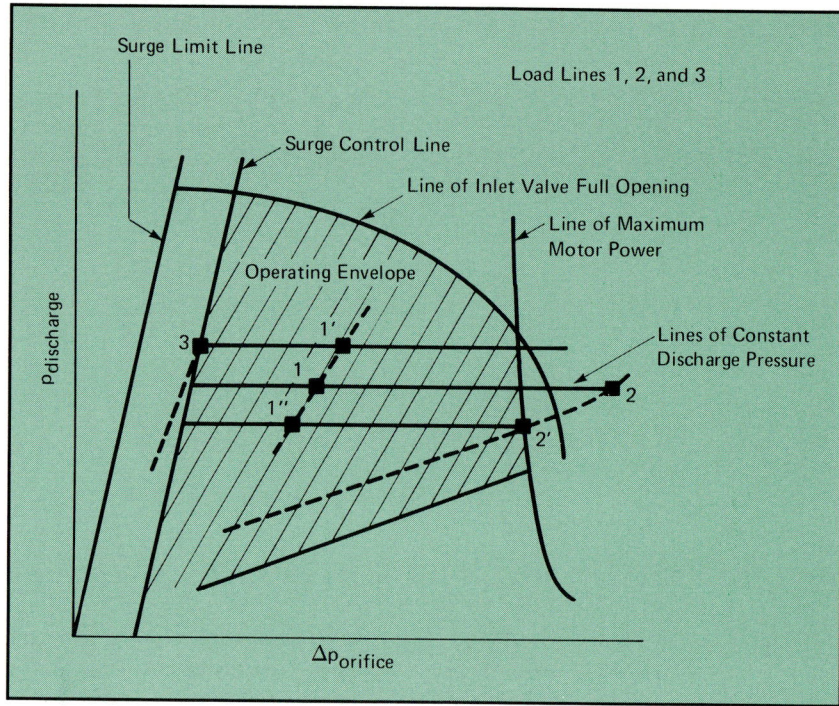

Fig. 8-5. Operating Envelope of a Centrifugal Compressor[4]

The operating envelope is traditionally composed of curves provided by the compressor manufacturers at the time of purchase to predict the operating ranges of their compressors. Factory tests are performed to verify the specified design point. The tests are performed with gas mixtures representing the process gas, and the results of the tests are not entirely accurate. Surge tests are generally not performed at the factory, and surge limits (the SLL) are estimated. With usage, the operating limits change because of wear, corrosion, and dirt accumulation. For these reasons, the original estimated performance curves are not sufficiently accurate to optimize either short range or long range decisions, and their use results in inefficient compressor operation. Thus, periodic experimental performance testing can result in significant operating economies.

Testing in the operating region of the surge limit line must be done with extreme care, because of the potential damage to the compressor, and is probably best done under the supervision of a representative of the manufacturer. Incipient surge should be used as a warning of impending surge, and thus high-performance measuring and recording instruments should be in use to display that phenomenon. A surge control system of the quality of that shown in Fig. 8-2 (and described later in this unit) should be in service. This system can stop surge within the first half cycle,[2] thereby offering protection against damage to the compressor should it inadvertently be forced into the surge area.

Figures 8-6a and b show the results of surge tests on two operating compressors.[5] Figure 8-6a is the compressor map for

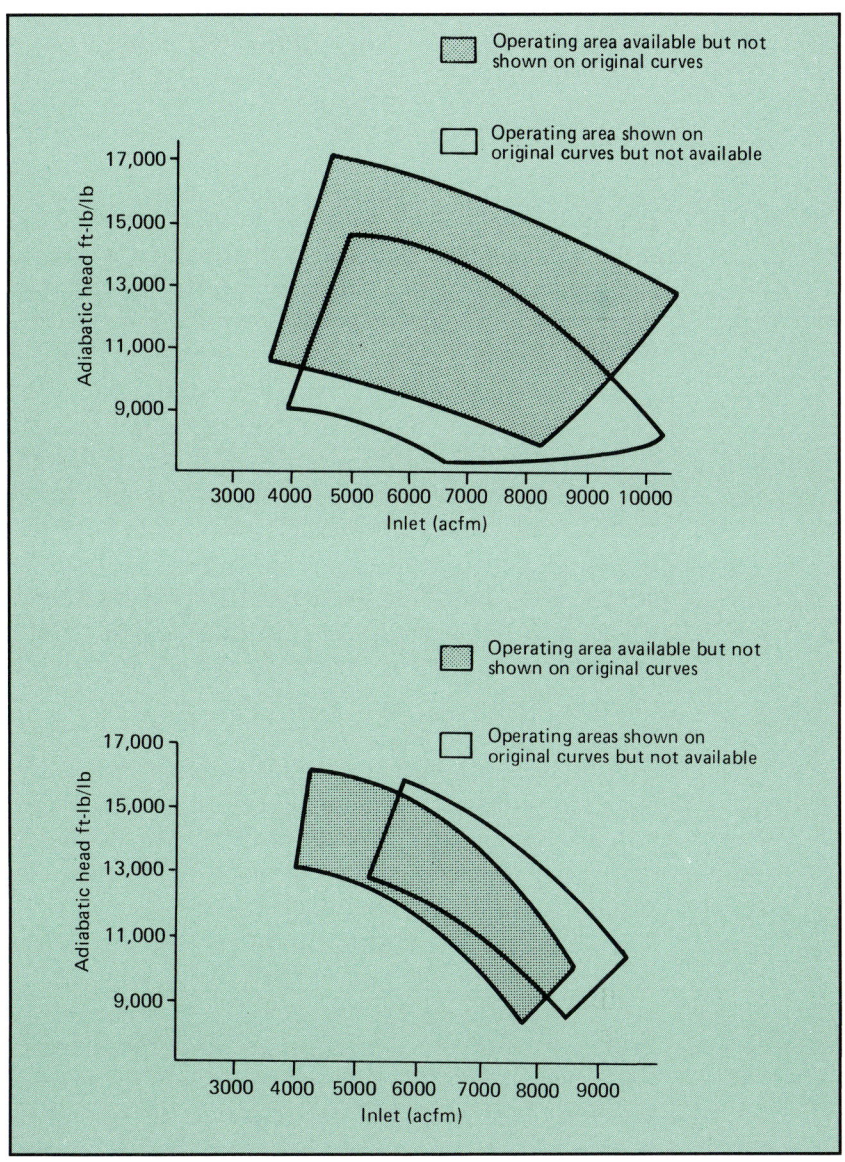

Fig. 8-6. Comparison of Operating Envelopes from Manufacturer and from Test Data[5]

a 19,800-hp, single-case, industrial-type gas turbine-driven centrifugal compressor. The performance test includes two experimental surge points, one at high rpm, and one at low rpm. The compressor was found to be capable of developing significantly higher heads and operating without surge at significantly lower flow rates than was formerly assumed. Also, the performance curves are somewhat flatter than the original manufacturer's curves.

Figure 8-6b is the compressor map for a 12,500-hp gas turbine-driven centrifugal compressor. The differences between the manufacturer's curves and the experimental curves are quite dramatic. The flow range of the compressor has increased by 15 to 19 percent.

8-4. Integrated Surge Control[4]

Previous discussions of surge control systems have described the fuzziness of the data on which the systems are designed; the movement of the surge line with the varying thermodynamic and mechanical parameters of the system; and the resulting necessary conservativeness of the offset of the surge control line from the surge limit line. Attention has also been given to the speed of opening of the anti-surge valve. However, a system that conforms to these design guidelines can still be inadequate if the flow through the compressor can be reduced very, very quickly. With this very fast disturbance, the surge controller will possibly not be able to get the anti-surge valve open fast enough; the compressor will go into surge, and the controller will quite possibly not be able to bring the compressor out of surge because of interaction with other controls.

The integrated control system is concerned with maximizing the area within the operating envelope shown in Fig. 8-5. The compressor design point is point 1. The capacity (pressure) controller will bring the operating point back to point 1 should a disturbance cause it to vary to point 1' or point 1". The surge control system will keep the anti-surge valve closed while constant discharge pressure is being maintained.

If the resistance of the compressor discharge network decreases (a valve is opened, or a new user is added), the pressure controller will try to move the operating point to new position 2. But the motor is overloaded at point 2. Before overload can occur, the load-limit (overload) controller will sense the demand for excess power and will override the pressure controller and throttle back the speed or inlet valve position. The operating point will settle back to point 2' on the compressor map. Although the discharge pressure is now lower than desired, it is the highest discharge pressure that the compressor can deliver under the existing conditions. Motor overload is prevented, and the compressor is 100 percent loaded.

If the resistance of the discharge network increases and the operating point reaches point 3, the surge controller starts opening the anti-surge valve, reducing the resistance and preventing surge. As previously discussed, the opening of the anti-surge valve upsets the pressure control system, leading to interaction. Decoupling networks in the form of compensating algorithms reduce the interaction. If the anti-surge valve is opening to prevent surge, decoupling consists of causing the pressure controller to open the inlet value further (or increase the speed) to prevent surge.

The potential for energy savings by the use of integrated surge control is shown graphically by Fig. 8-7. The graph shows air consumption over a 24-hour period. The air consumption is highly variable. With three-shift operation in the plant and numerous machines, instruments, processes and tools as loads, consumption peaks during the afternoon in the first shift. Air consumption drops dramatically during shift changes and breaks.

Unit 6 shows the development of the linearized compressor map, shown as Fig. 8-1. The surge limit line (SLL) is defined on these coordinates as a straight line coming out of the origin. It is defined by

$$\Delta p_o = K \Delta p_c \qquad (8\text{-}2)$$

The slope of the surge limit line, K, is determined from the compressor map supplied by the manufacturer, or, preferably, from the results of experimental testing.

A second parameter is the distance b_1 between the surge limit line and the surge control line in Fig. 8-1. Mathematically, b_1 is a bias in the defining equation, as shown in eq. (8-1), and it is known as the "surge margin" in designing the surge control

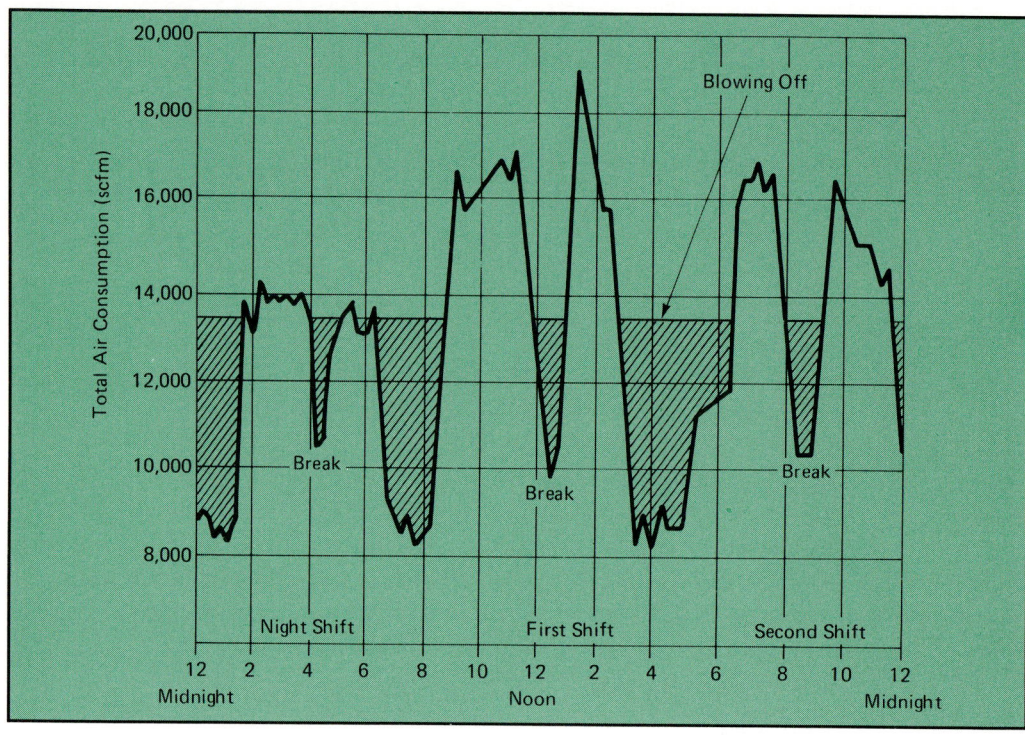

Fig. 8-7. Air Consumption in a Typical Industrial Plant[4]

system. The value of the surge margin is an indication of the quality of the anti-surge protection system. Small values of b_1 allow the compressor to be operated close to the surge limit line without recycle flow. This design is desirable since it reduces the energy consumed during low-flow operation as compared to a surge control system with a larger surge margin.

Operating close to the surge limit line in steady state with a small surge margin requires that the surge controller be tuned to a very fast response to open the anti-surge valve and prevent surge. However, stability considerations limit the speed of response to which the controller can be tuned. Thus, a very rapid decrease in throughput flow, such as that resulting from a valve interlocking shut to protect the process, would cause the compressor to surge before the controller could react.

The *open-loop override control algorithm* is added to the surge controller to increase the speed of response of the control system. The override is shown as the recycle trip line (RTL) on Fig. 8-2. Its objective is to prevent surge due to large or fast disturbances.

Unit 8: Coordinated Compressor Control

The open-loop override control algorithm operates when the compressor operating point is to the left of the surge control line and on the recycle trip line (RTL). The RTL is a line parallel to the surge limit line and the surge control line (Fig. 8-2). The RTL line is defined by the equation

$$\Delta p_o = K\Delta p_c + b_1 - RT \qquad (8\text{-}3)$$

The open-loop algorithm is activated when the operating point is moving to the left (decreasing flow) and reaches the RTL. The output of the RTL and PI control algorithms are added to form the output of the surge controller.

The RTL algorithm opens the anti-surge valve in a series of steps with magnitude C1 and a time interval between steps C2, as shown in Fig. 8-8. Opening the anti-surge valve in stepwise fashion continues as long as the operating point stays to the left of the RTL. When the operating point moves to the right of the RTL, indicating that the anti-surge valve is opening and flow is increasing, the RTL algorithm output decreases to zero exponentially with a time constant τ.

The fact that the RTL algorithm was activated indicates that

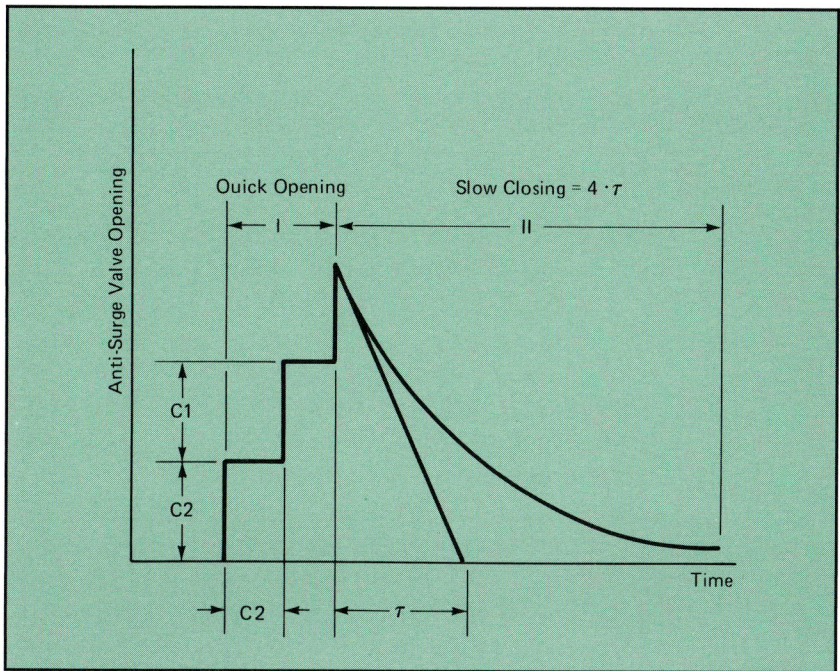

Fig. 8-8. Open-Loop Control Response[4]

the strength of the disturbance was too great for the best response of the PI control algorithm—flow was decreasing rapidly, and the compressor was moving toward surge. The RTL algorithm thus provides a strong response to quick, dangerous disturbances. Of course, for lesser disturbances, the PI algorithm would turn the flow around before the RTL was reached, and normal controller response would be observed.

An *adaptive surge detection* algorithm protects the compressor if b_1 or C1 are set too small, or if C2 was set too long. In this case the operating point would move to the left past not only the surge control line but also the recycle trip line. A really severe disturbance would move the operating point past the surge limit line, and the compressor would be in surge.

The surge detection line, shown in Fig. 8-9, is another line parallel to the surge limit line and located in the linearized compressor map to the left of it. The distance between the surge limit line and the surge detector line, shown on Fig. 8-9 as distance d_2, must be determined for each individual compressor. The equation for the surge detection line is

$$\Delta p_o = K \Delta p_c - d_2 \qquad (8\text{-}4)$$

Activation of the surge detection line algorithm increases the surge margin in accordance with the following relationship

$$b = b_1 + (n)(b_2) \qquad (8\text{-}5)$$

where:

n = number of surge cycles
b_1 = orginial surge margin
b_2 = surge detection bias increment

Each surge cycle increases the value of b and moves both the recycle trip line and the surge control line to the right, away from the surge limit line. The greater surge margin incurs the cost of decreased operating efficiency. For this reason, the additional b_2 bias can be reset to zero by a push button after the cause of the surge has been identified and eliminated.

Figure 8-10 shows the effect of the open-loop override control algorithm, or the recycle trip line. The recorder plots show the response of the compressor to the sudden closing of an inlet

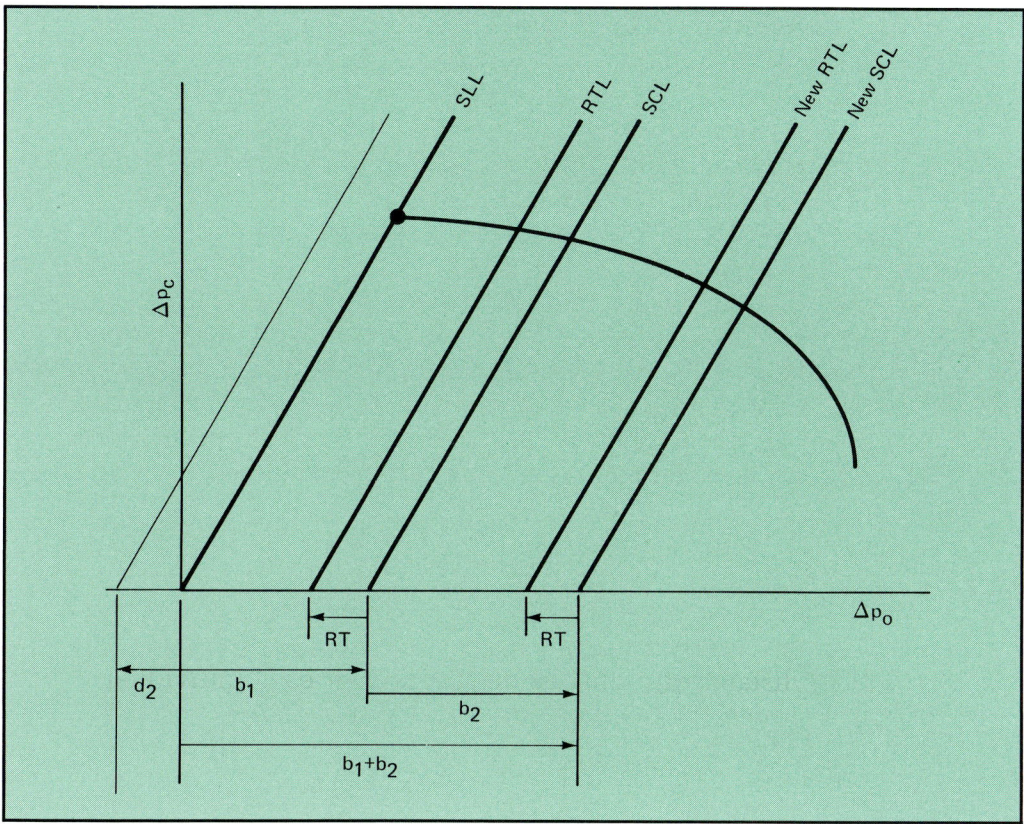

Fig. 8-9. Adaptive Surge Detection Response[13]

valve. The PI control algorithm responded as the flow decreased past the surge control line, opened the anti-surge valve to some 10% of stroke, and slowed the rate of flow decrease. However, the compressor still approached surge. Then the open-loop override called for complete valve opening. This produced a very fast reversal of the movement of the operating point. Surge was avoided, and the recycle trip circuit began to decay back to steady state with the time constant τ_1.

Factors that cause distortion or movement of the surge limit line have formerly been discussed. Those factors include changing inlet temperature or pressure, varying molecular weight, changing specific heat ratio, and shifting compressibility of the gas. Of these, one of the most significant is varying molecular weight because it is normally accompanied by a change in the specific heat ratio, $k = C_p/C_v$. Figure 8-11 shows the linearized compressor map plotted with

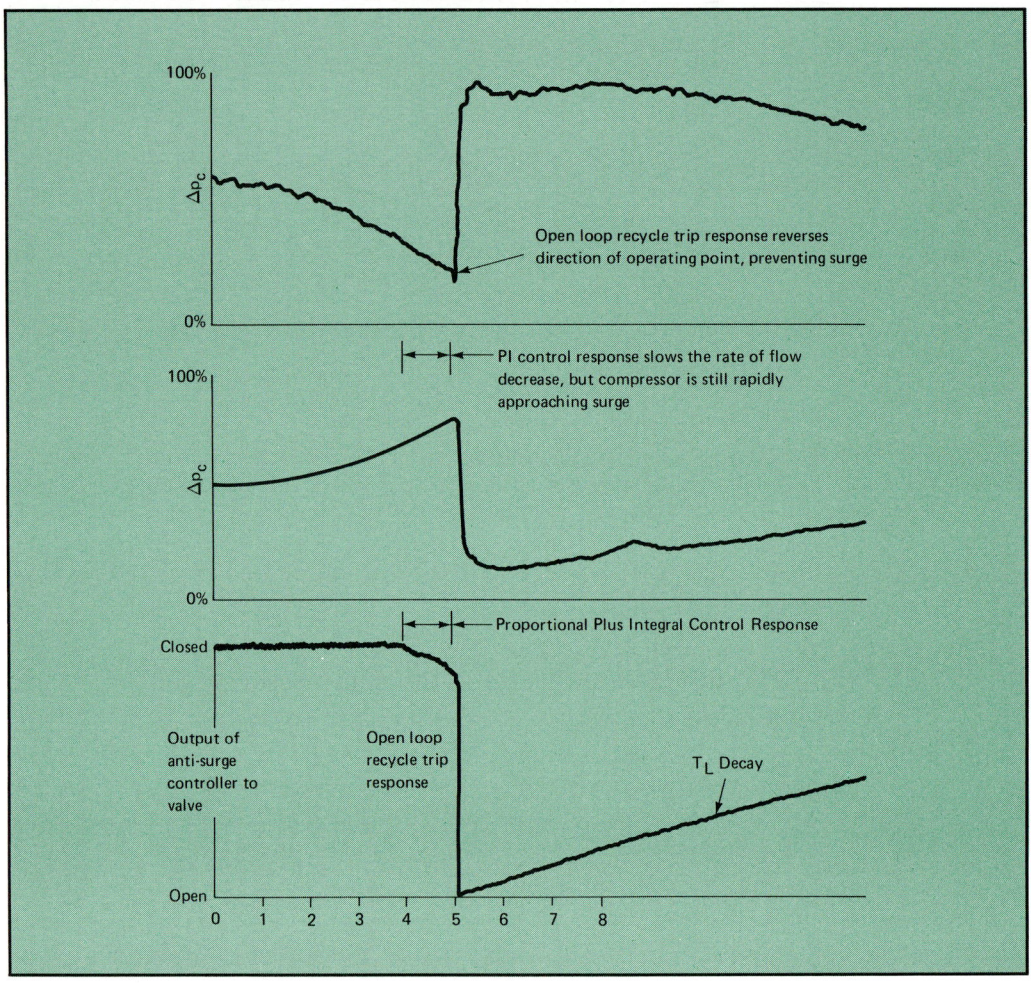

Fig. 8-10. Test Results of Integrated Surge Controller Response[1]

different coordinates, and the surge limit line defined by:[6]

$$K \frac{\Delta p_c}{p_s s^y} = \frac{\Delta p_o}{p_s s^x} \qquad (8\text{-}6)$$

This representation results in the location of performance curves and their surge limits that do not depend on speed of rotation nor any inlet conditions except gas composition. Exponents x and y are obtained from the regression analysis described in Fig. 6-8. The safe operating line for a set of different gas compositions becomes the nonlinear pumping limit line (PLL), shown by Fig. 8-11. The PLL is the locus of surge points on the various performance curves, defined by the peak or flat point of each curve. Operating to the right of

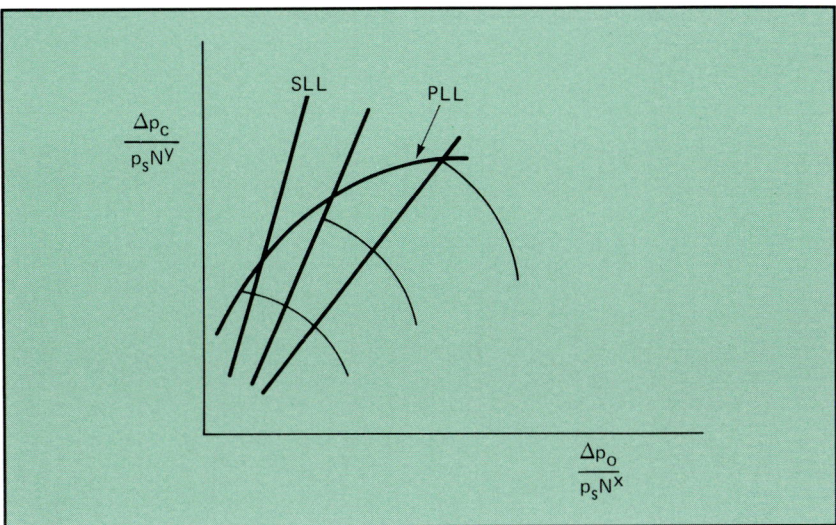

Fig. 8-11. Pump Limit Line for Compressors Operating with Gases of Variable Molecular Weight[6]

the PLL will prevent surge in the compressor as gas composition changes.

8-5. Integrated Surge Controller[7,8]

Figure 8-12 is a functional diagram of a commercially available integrated surge controller especially designed for dynamic compressor control. It has inputs that accommodate the foregoing equations. The input corresponding to Δp_c passes through the scaler (No. 1) and summer (No. 2) to transform it into eq. (8-1). The value of eq. (8-1) is compared to Δp_o by the PI controller (No. 3). If Δp_o is the greater signal, the controller's output becomes minimum (zero).

The output from the summer (No. 2) is decreased by the amount d_1 to handle strong disturbances in accordance with eq. (8-3). The addition is done by the summer in element 5. The comparator in element 5 checks the equality of eq. (8-3). When Δp_o is smaller, a signal is sent to the curve generator, and its output follows the shape shown—a rapid rise and an exponential decay. This signal is added to the output of the PI controller in the summer (No. 4).

Should the disturbance be so strong that flow decreases past the surge limit line despite the open-loop override, element 7 acts to increase the surge margin by a preselected bias b_2. This

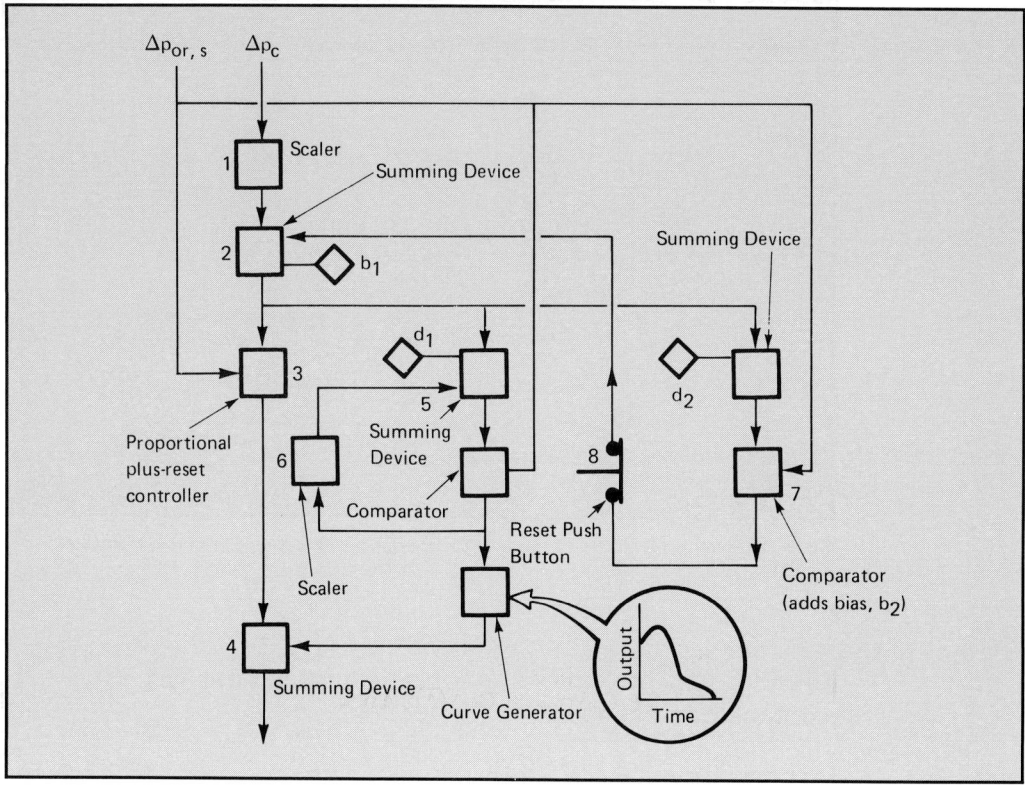

Fig. 8-12. Diagram of Integrated Surge Controller[2]

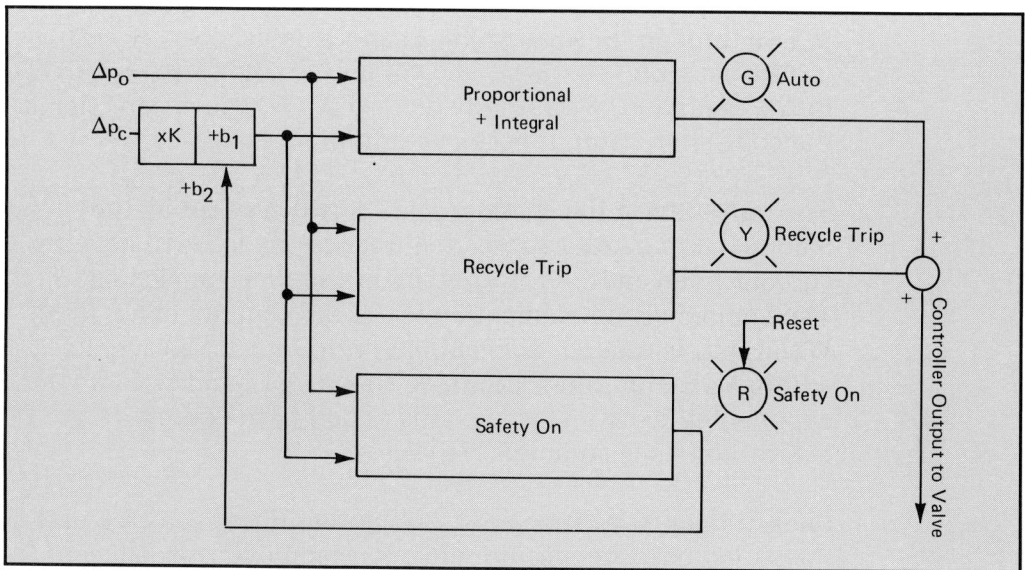

Fig. 8-13. Simplified Functional Diagram of the Integrated Surge Controller[8]

Mode	Formulation	Application
fA01	$\Delta p_o = K\Delta p_c$	Eq. 8-2
fA02	$\Delta p_o = K\Delta p_c f_3(\alpha)$	Guide vanes
fA03	$\Delta p_o f_1\left(\dfrac{p_d}{p_s}\right) = K\Delta p_c$	Flow measurement in discharge
fA04	$\Delta p_o f_1\left(\dfrac{p_s + \Delta p_c}{p_s}\right) = K\Delta p_c$	Flow measurement in discharge
fA05	$\Delta p_o f_1\left(\dfrac{p_d}{p_d - \Delta p_c}\right) = K\Delta p_c$	Flow measurement in discharge
fA06	$\Delta p_o f_1\left(\dfrac{p_s + \Delta p_c}{p_s}\right) = K\Delta p_c f_3(\alpha)$	FT in discharge & guide vanes
fA07	$\Delta p_o f_1\left(\dfrac{p_d}{p_d - \Delta p_c}\right) = K\Delta p_c f_3(\alpha)$	FT in discharge & guide vanes
fA08	$\Delta p_o = K\Delta p_c f_3(\alpha) f_4(T)$	Guide vanes & varying inlet temperature
fA09	$C_3 \Delta p_{o2} + C_4 \Delta p_{o1} f_1\left(\dfrac{p_{s2}}{p_{s1}}\right)$ $+ C_5 \sqrt{\Delta p_{o2} \Delta f_{o1} f_1\left(\dfrac{p_{s2}}{p_{s1}}\right)} = K\Delta p_c$	Side streams
fA10	$K\Delta p_c = f_1\left(\dfrac{p_{sm}}{p_s}\right) C_3 \Delta p_o$ $+ C_4 \Delta p_{om} \pm C_5 \sqrt{\Delta p_o \Delta p_{om}}$	Side streams
fA11	$\Delta p_o = K\Delta p_c f_3(p_s) + b, f_4(p_s)$	Varying inlet pressure

Table 8-5. Computational Modes of Integrated Surge Controller[8]

bias can be removed by manually resetting with the push button.

The functional diagram of the controller has been simplified in Fig. 8-13, which also shows the indicating lights on the front of the controller itself.

Figure 8-13 also shows the computational mode of the controller. Microelectronics permit a number of computational modes to be permanently programmed into the controller as subroutines, which are called according to menu to meet application requirements. The computational modes available with the controller as subroutines, along with their formulation and application, are listed in Table 8-5.

8-6. Integrated Capacity Control[9,10]

Capacity control, discussed in Unit 5, typically consists of the regulation of upstream pressure, downstream pressure, or throughput flow by the manipulation of compressor speed, inlet valve position, or guide vane position. Integrated capacity control is the combination of the capacity controller with the overload controller to prevent overloading the driver, with the surge controller to minimize interaction, and with other capacity controllers for load sharing.

Figure 8-14 shows a typical integrated compressor control system with discharge pressure being controlled by manipulation of an inlet control valve. Integration is accomplished by transmitting the motor load signal to the pressure controller and having the pressure controller communicate with the surge controller. Figure 8-15 shows a functional diagram of the capacity (performance) controller. Circuit A in Fig. 8-15 regulates the main controlled variable (discharge pressure in Fig. 8-15) using a PID algorithm plus an

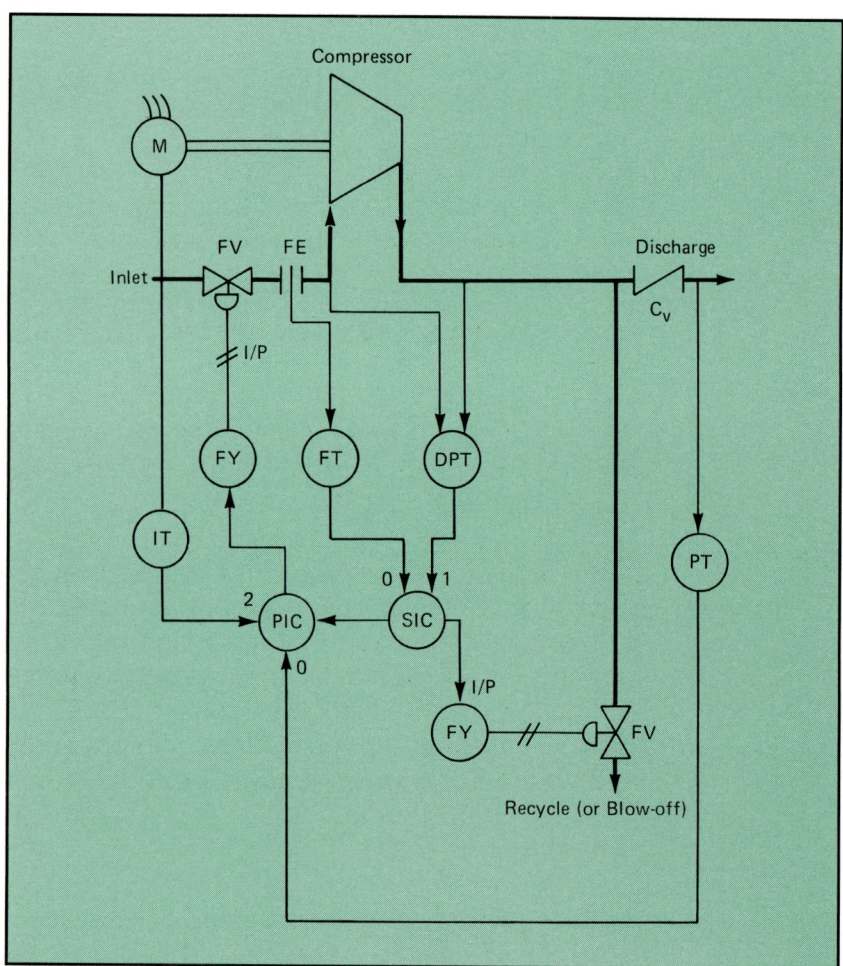

Fig. 8-14. Typical Integrated Compressor Control System[10]

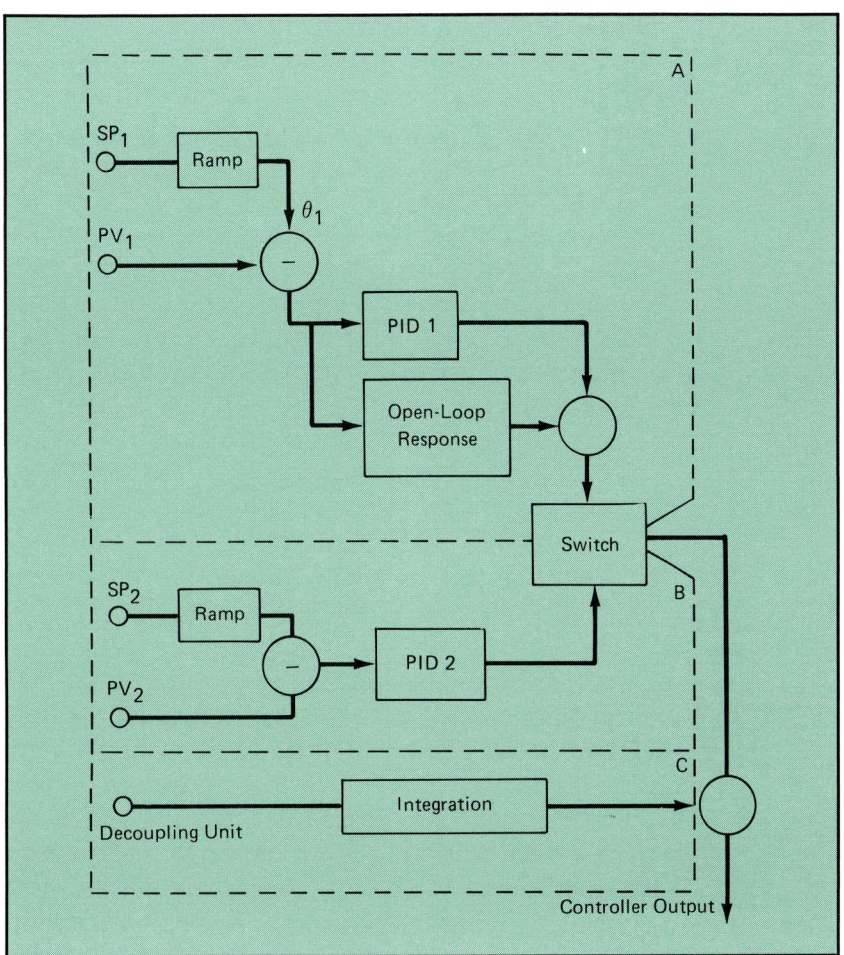

Fig. 8-15. Functional Diagram of Capacity Controller[10]

open-loop response. The open-loop response is initiated by the controller error exceeding the magnitude d_3. The circuit A output is incremented each d_2 time period by an increment d_1. Parameters d_1, d_2, d_3 are constants set on the controller's keyboard.

Circuit B in Fig. 8-15 controls the limiting variable PV_2 (motor load in Fig. 8-14) using a PID algorithm. As air becomes cooler and more dense, more power is consumed. Lower cooling water temperatures increase the efficiency of the compressor, also causing more power to be consumed. A 100°F change in temperature can cause as much as a 30 percent change in power demand.[4,11] Load-limiting control is implemented by Circuit B to prevent the compressor from overloading the motor during winter operation or when cooling water temperatures are low. When the motor power reaches its limit, the load controller algorithm in Circuit B will take control of the inlet valve. Pressure control is sacrificed to prevent overloading the motor, as shown in Fig. 8-5.

An adjustable ramp circuit is included on each set point to provide bumpless transfer from manual to automatic control.

Circuit C in Fig. 8-15 provides decoupling through an integrating algorithm to minimize interaction between the surge controller and the capacity controller. Circuit C implements the equation

$$O_c = (M_a)(PI) + (M_b)(RT) \tag{8-7}$$

M_a and M_b are coefficients set from the capacity controller keyboard, and PI and RT represent outputs from those respective circuits in the surge controller.

Circuit C provides decoupling by summing the outputs of the surge controller and the pressure (capacity) controller, as shown in Fig. 8-14. Referring to that diagram, suppose that user demand begins to decrease, pressure increases, and the controller begins to close inlet valve FV. This moves the compressor operating point closer to the surge limit line. As the load drops further, the operating point crosses the surge control line. Surge controller SIC in its independent flow loop beings to open anti-surge valve FV.

If load continues to drop and pressure to increase, the pressure controller will continue to close the inlet valve, reducing compressor flow, and driving the operating point closer to the surge limit line despite the opening anti-surge valve. Thus, the actions of the independent control loops are incompatible with each other. The incompatibility results in interaction. Circuit C minimizes the incompatibility by feeding the surge controller signal through the pressure controller to slow, or even reverse, the action of the inlet valve. Ultimately, discharge pressure will decrease because of the recycle flow through the anti-surge valve, which will reduce the discharge pressure.

8-7. Integrated Multistage Surge Control[12,13]

Figure 8-16 shows a two-stage (two-section) compressor with a surge control system on each stage. This configuration is

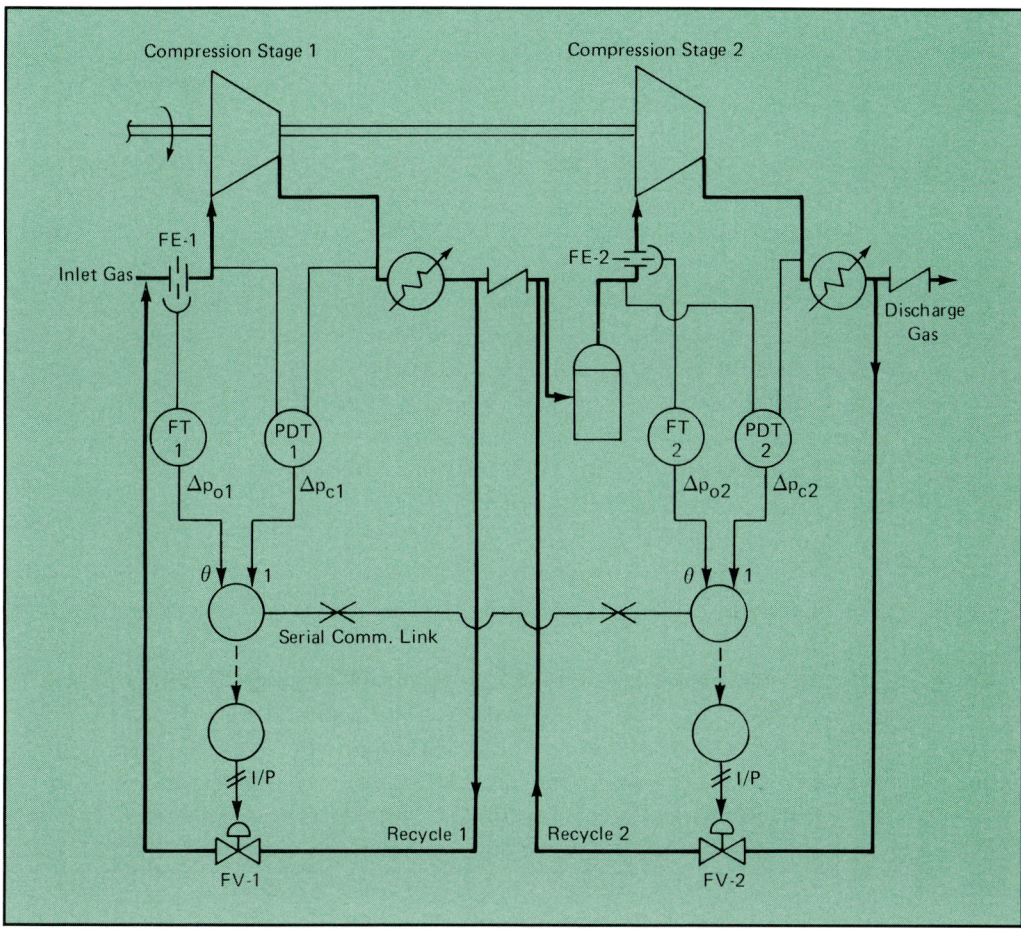

Fig. 8-16. Diagram of Surge Control Configuration for a Multistage Compressor[12]

similar to Fig. 6-16, where the two systems were justified by the unique compressor map for each stage of the compressor. Interaction takes place whenever one of the anti-surge valves opens to increase the anti-surge protection of its compressor stage. The operating point of the other stage can be driven toward the surge limit line by the transient change in flow and pressure.

Interaction is minimized by transmitting the changes in output of any one surge controller to the other surge controller (s), as shown in Fig. 8-16. The information transmitted is one signal proportional to the output of the PI section of the controller, and another signal from the recycle trip (RT) section. Each controller uses this information to protect its own stage from surge. The receiving controller scales the inputs in accordance with eq. (8-6) and adds them to its own output in transmitting a signal to the anti-surge valve. A diagram of the decoupling configuration is shown in Fig. 8-17.

Figure 8-18 shows an alternative surge control configuration using only one anti-surge control valve. This configuration removes the possibility of interaction among surge systems and is somewhat more economical. However, it is significantly less flexible than the system shown in Fig. 8-16.

Figure 8-19 shows the single anti-surge valve configuration using the integrated surge controllers. Anti-surge valve FV protects both stages. The primary controller protects the second stage by manipulating the anti-surge valve directly. The secondary controller protects the first stage by continuously updating the primary controller as to the operating point and the value of the recycle trip (RT) for the first stage. The primary controller then determines which stage has an operating point closer to its surge limit line and which has the greater value of recycle trip. With this information, it manipulates valve FV to protect both stages.

When two or more centrifugal compressors are operating in parallel, the greatest surge protection and efficiency results from all compressors operating equidistant from their respective surge control lines. The *S-Criterion* has been

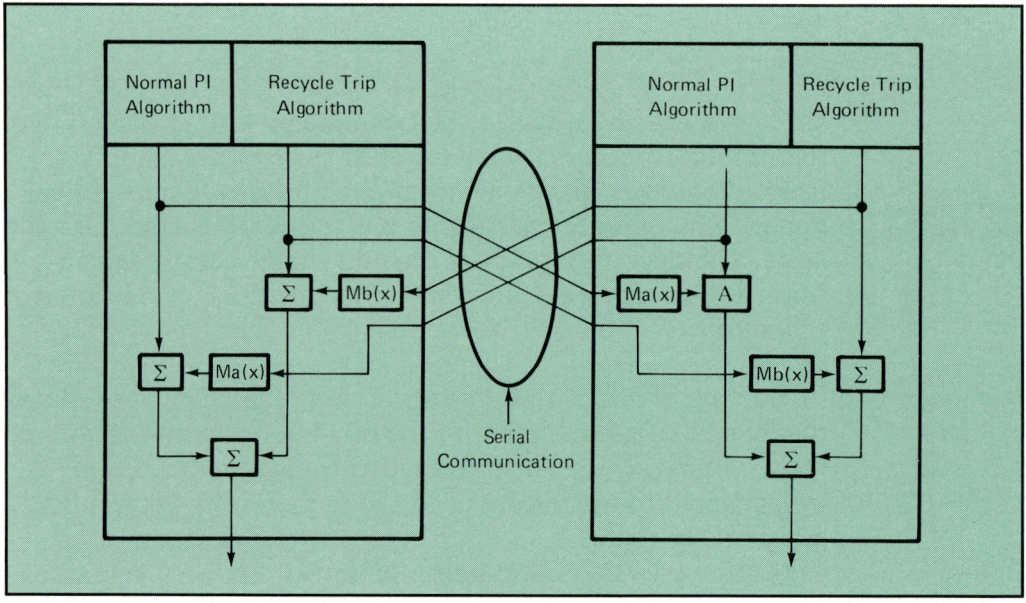

Fig. 8-17. Diagram of Decoupling Network for Multistage Compressors[13]

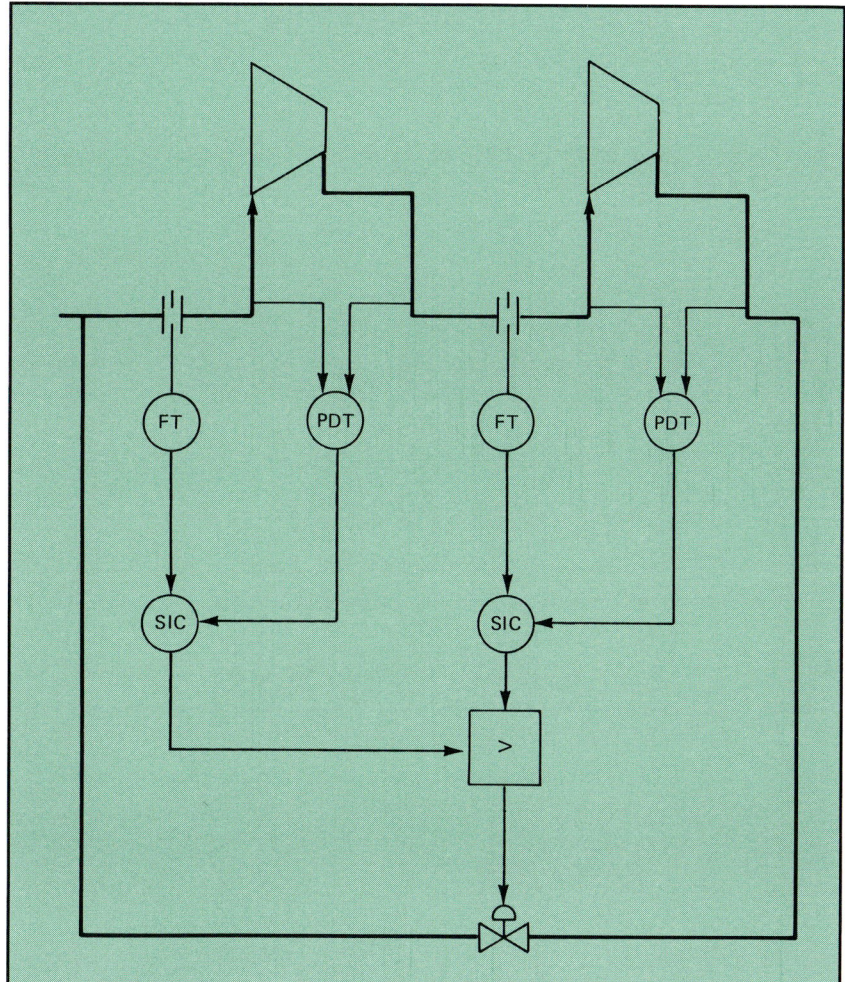

Fig. 8-18. Multistage Compressor with Single Anti-surge Valve[13]

developed as a measure of the angular distance between the operating point and the surge control line.[4,13] The S-Criterion is formulated from the equation of the surge control line as follows

$$S = \frac{K\Delta p_c + b}{\Delta p_o} \quad \text{or} \quad S = \frac{K\Delta p_c/p_s + b}{\Delta p_o/p_s} \quad (8\text{-}8)$$

If S is less than "one," the operating point is to the right of the surge control line and safely away from surge. The lower the value of S, the greater the distance from the surge control line. An S equal to "one" corresponds to an operating point located on the surge control line. Then a value of S greater

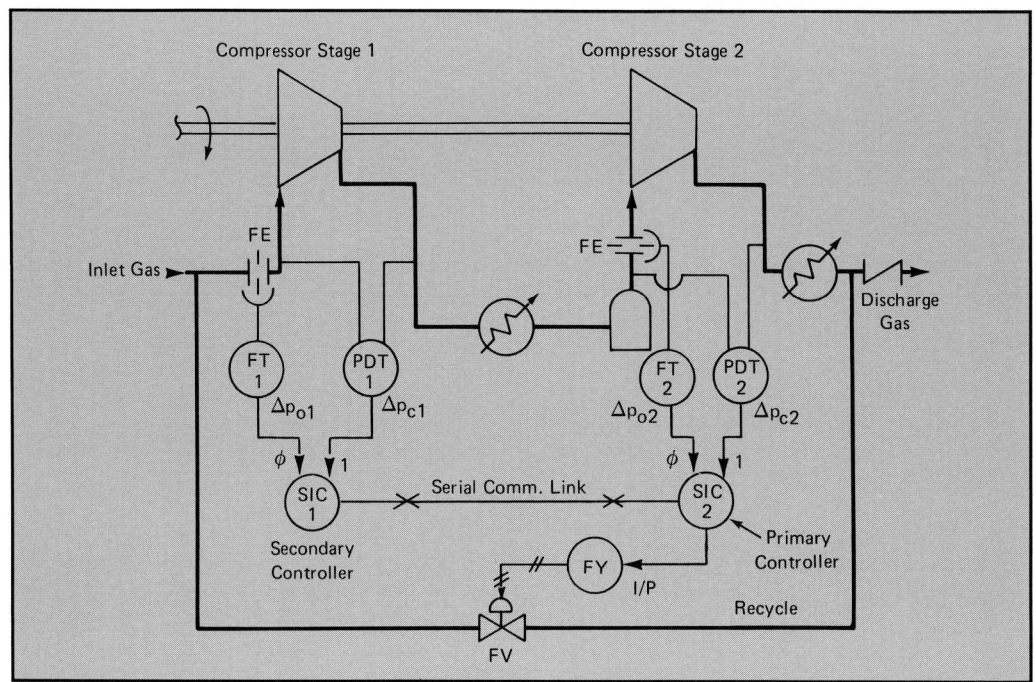

Fig. 8-19. Multistage Compressor with Single Anti-surge Valve Actuated by Integrated Surge Controllers[12]

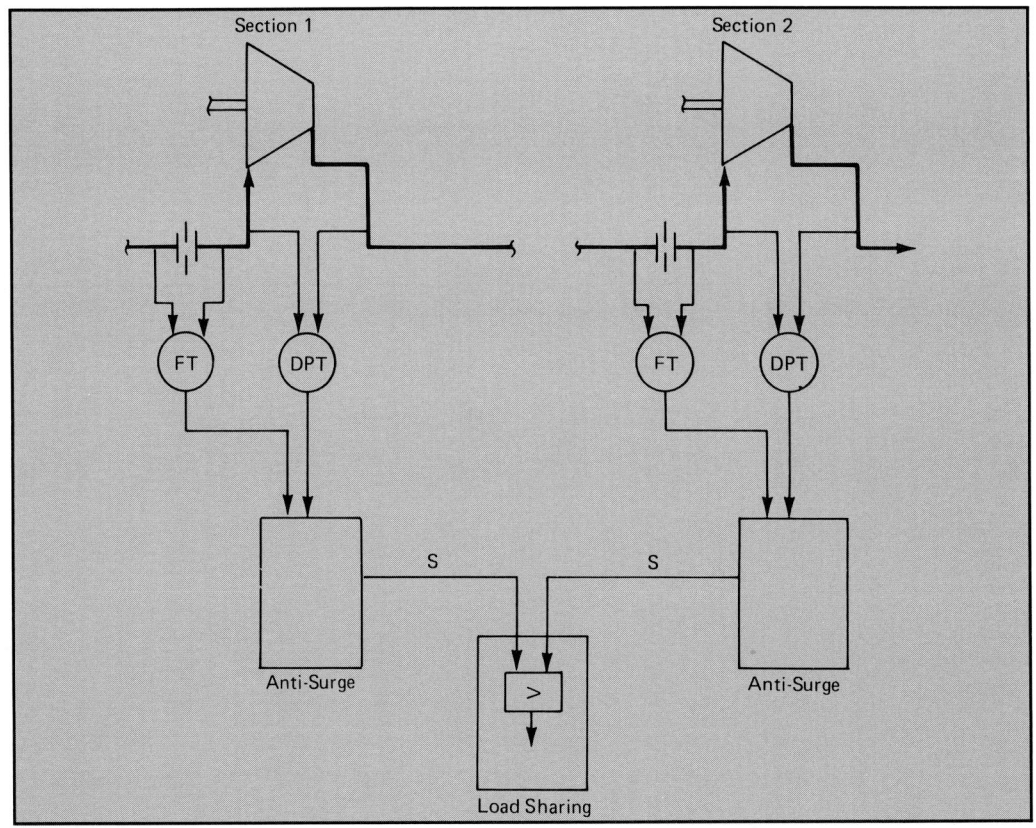

Fig. 8-20. Load Sharing Controller Used To Select the Highest Value of S for a Multistage Compressor[13]

than "one" corresponds to an operating point that has crossed the surge control line and is headed toward surge.

Criterion S is computed for each stage of a multistage compressor by each surge controller. The load sharing controller, shown in Fig. 8-20, selects the highest value of S for use as the compressor's S value in load sharing with other compressors.

8-8. Automatic Start-Up and Shutdown

The capacity controller contains automatic start-up and shutdown mode fB. Figures 8-21 and 8-22 show the time sequence of the automatic starting and stopping.

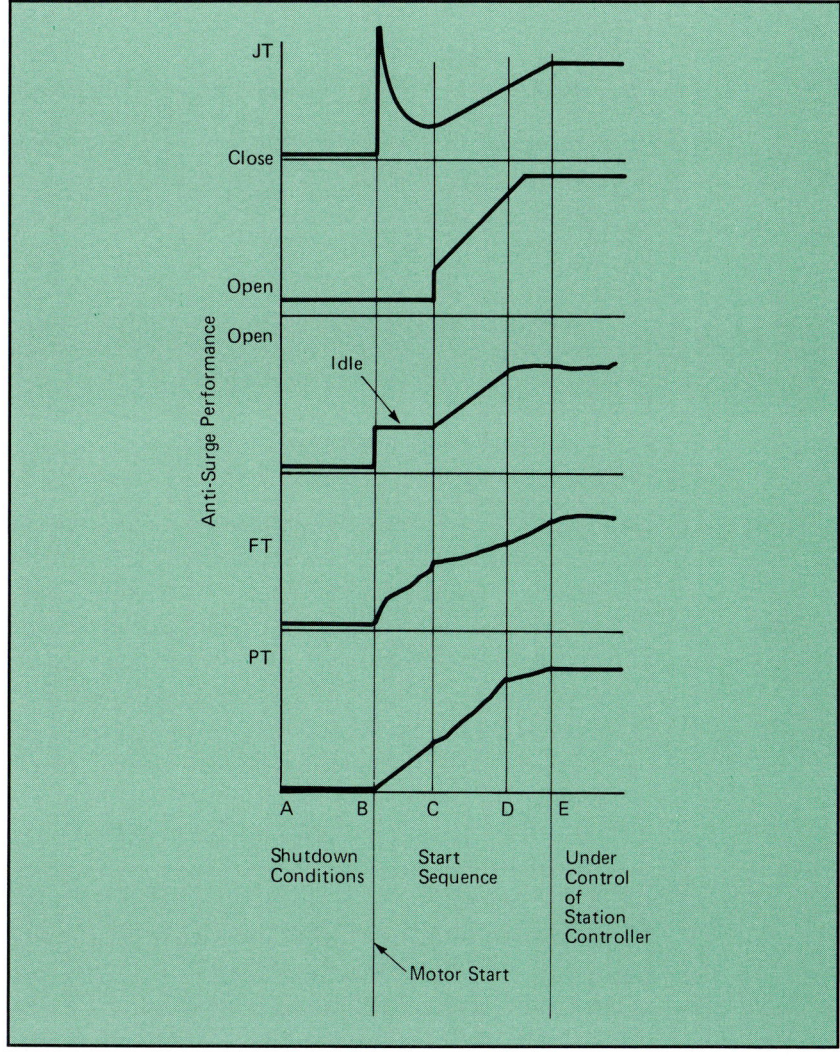

Fig. 8-21. Time Sequence of Automatic Start-up[4]

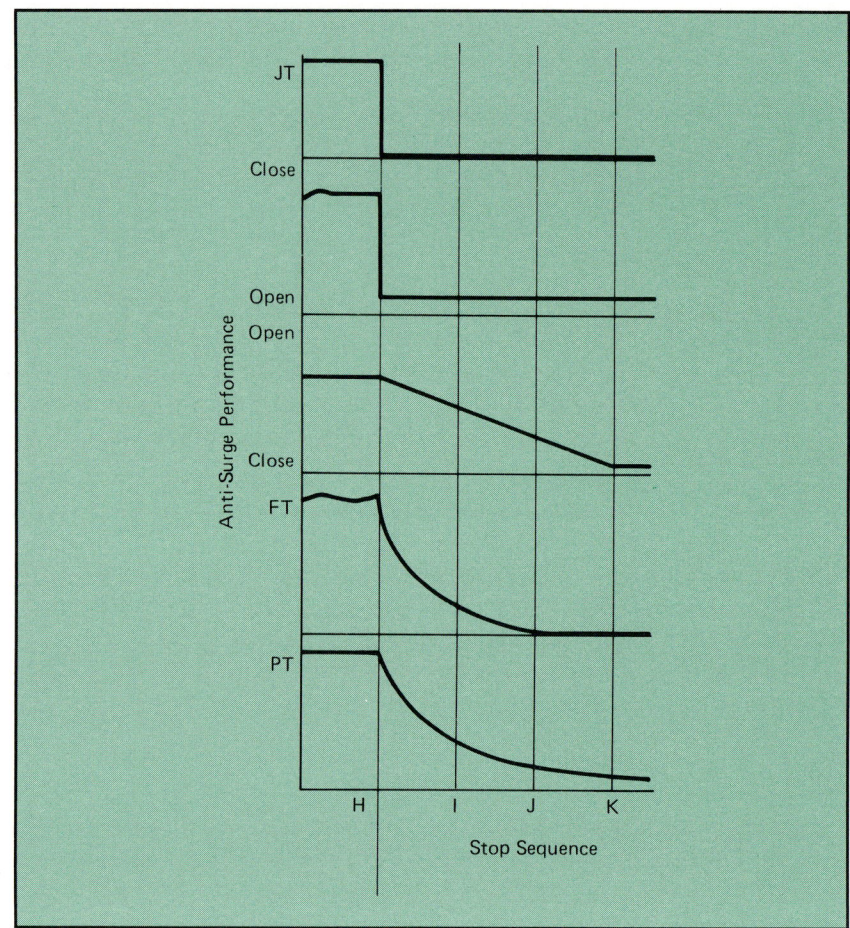

Fig. 8-22. Time Sequence of Automatic Shutdown[4]

Figure 8-21 illustrates the sequence of events necessary to bring a compressor on line from complete shutdown. Period AB in Fig. 8-21 shows the compressor shut down, with no current to the electrical motor, the anti-surge valve fully open, and the inlet valve closed. The motor start push button is pushed at time B. The inlet valve jumps to a nearly closed position where it is sufficiently open to avoid surge but not open far enough to overload the motor. The signal from power transmitter JT goes to a very high value since motor speed is low and the operating back emf has not been attained, as discussed in Unit 4.

Motor acceleration period BC must be minimized to avoid the high-temperature interlock. The control system must wait until time C to begin loading the compressor. At that point, the surge controller output goes to the "start" level, as shown. The capacity (performance) controller will ramp open the inlet valve at a preselected rate during period CD. Simultaneously, the surge controller is placed in the "automatic" mode and closes the anti-surge valve at a rate dictated by the controller tuning. The valve will assume a position that will maintain the operating point on the surge control line. The controller's reset time control mode is automatically set to a much smaller value during the period before the check valve is opened because of the smaller discharge volume.

As the inlet valve is ramped open during period CD, pressure rises. As pressure approaches the set point value to time D, the ramp rate of valve opening is reduced to prevent a pressure overshoot. As the compressor discharge pressure reaches system pressure, the check valve opens at E, the inlet valve opening ramp is aborted, and the compressor is under automatic control.

Similar automatic sequences take a compressor from on line to idling, from on line to shutdown, and from on line to a crash (emergency) shutdown.

Exercises

1. If typical process instruments are too slow to sense surge, how is surge detected?

2. What is the period of the surge cycles on Fig. 8-3? How does this compare to Fig. 3-11?

3. What type of disturbance is the recycle trip line (Fig. 8-2) added to protect against?

4. Why is the surge control system in Fig. 8-18 more economical than that in Fig. 8-16?

5. Give an example of the penalty paid for the inflexibility of the configuration in Fig. 8-18.

6. Refer to Fig. 8-22 and develop an automatic sequence for shutting down a compressor.

References

1. Hampel, J., "What Engineers Need to Know about Antisurge Protection," *Proceedings of Control Expo '84*, Chicago, Illinois, 1984.

2. Staroselsky, N., and L. Ladin, "Improved Surge Control for Centrifugal Compressors," *Chemical Engineering*, McGraw-Hill, New York, N.Y., May 21, 1979.
3. Anon., "A Field Test of the EAS Emergency Antisurge Control System," Compressor Controls Corp., Des Moines, Iowa, 1977.
4. Mirsky, S., and John Hampel, "Improving Efficiency and Reliability of Packaged Centrifugal Air Compressors Operating in Parallel," Technical Paper TP13, Compressor Controls Corp., Des Moines, Iowa, 1986.
5. Bath, C., and John Hampel, "Improved Antisurge Controls," *Pipeline and Gas Journal*, Harcourt Brace Jovanovich, Inc., 1986.
6. Staroselsky, N., "Antisurge Protection for Compressors Operating on Gases of Variable Molecular Weight," Application Note AN12, Compressor Controls Corp., Des Moines, Iowa, 1985.
7. Anon., "Series II Antisurge Controller," Bulletin PB12, Compressor Controls Corp., Des Moines, Iowa, December, 1984.
8. Anon., "Series II Antisurge Controller," Theory Manual MS18, Compressor Controls Corp., Des Moines, Iowa, 1985.
9. Anon., "Series II Performance Controller," Product Bulletin PB13, Compressor Controls Corp., Des Moines, Iowa, 1984.
10. Anon., "Series II Performance Controller," Theory Manual MS17, Compressor Controls Corp., Des Moines, Iowa, 1986.
11. Nolar, R. L., & N. H. Robson, "Compressor Load Limit Controls," *Compressed Air*, Ingersoll Rand Co., Phillipsburg, NJ, June, 1982, p. 20.
12. Anon., "Series II Antisurge Controller," Instruction Manual IM12, Compressor Controls Corp., Des Moines, Iowa, 1984.
13. Anon., "Saving Money While Controlling Centrifugal and Axial Compressors," Training Manual FB103, Compressor Controls Corp., Des Moines, Iowa, 1986.

Unit 9:
Multiple Compressor Systems

UNIT 9

Multiple Compressor Systems[1,2]

Multiple compressor systems are assembled with either series- or parallel-connected dynamic compressors. Compressors are connected in series to provide the higher pressures required in the petrochemical industries. Compressors in parallel provide more flow at the same pressure, and the configuration provides greater rangeability and reliability. The performances of the two configurations are shown in Fig. 9-1. The classical system of parallel compressors of all kinds (reciprocating, axial, and centrifugal) is the engine room that supplies compressed air to a plant site; generally, the room will have grown one compressor at a time as the plant has grown, resulting in a wide variety of machines.

The number of variables to be considered increases dramatically with the number of compressors in the configuration. The tendency to merely extrapolate the control strategy for a single compressor should be resisted. The control of a single compressor has two objectives: capacity and surge protection. Multiple compressor systems have two more objectives: balancing the load among dissimilar machines, and the safe start-up and shutdown of compressors as needed. Finally, controls that manage the compressor net at maximum energy efficiency are economically justified.

Learning Objectives — When you have completed this unit, you should:

A. Be familiar with the functions of various compressor networks.

B. Understand the objectives of multiple compressor control.

C. Be aware of the various control configurations applied to multiple compressors.

D. Appreciate the advantages and the complexities of controls that maximize efficiency.

9-1. General

Multiple centrifugal gas compressors operating in series and/or parallel are a major component of the chemical, steel, oil

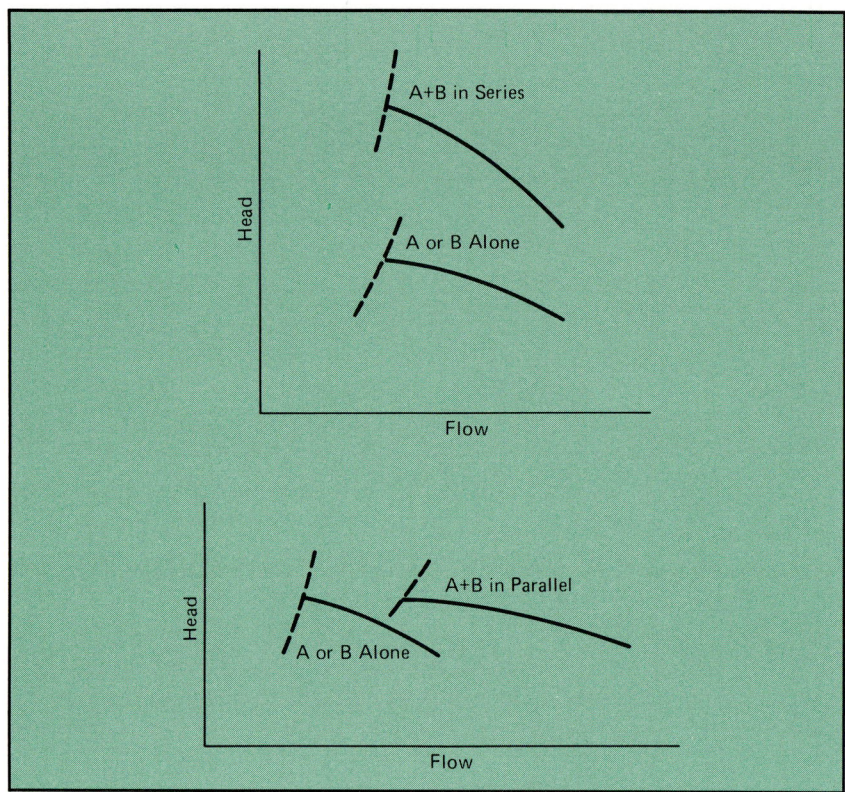

Fig. 9-1. Performance of Multiple Compressors in Series and in Parallel

refining, and natural gas production and transmission industries. The compression station must have enough capacity to supply the peak load and enough stages to meet the demand pressure. Loads can be expected to vary widely, even in plants that are base loaded. Further, the compression station will always be designed with surplus capacity to allow for scheduled and unscheduled downtime on the compressors. Operators will invariably call for more capacity (or energy) than is required in order to have an on-line reserve "to be safe." This tendency, along with the inherent limited turndown of the compressors and the limitations of some control configurations lead to excessive blow-off and wasted energy.[3,4]

Multiple systems are composed of either series- or parallel-connected dynamic compressors. Series compressors can have a common driver or independent drivers. Thus, similar control concepts apply to multistage (multisection) compressors, multistage compressors with side streams,[1] and independent compressors in series. The following approximate principle describes compressor stages in series:

> The polytropic heads, as given by the performance curve of each machine at a specified flow, are added to result in the total head across the series-connected units. Intermediate cooling will reduce the foregoing sum slightly.[2]

Series configurations are used to meet the very high pressure requirements of the chemical industry. A variety of control systems can be implemented for independently driven series-connected units. Figure 9-2 shows a typical series-connected

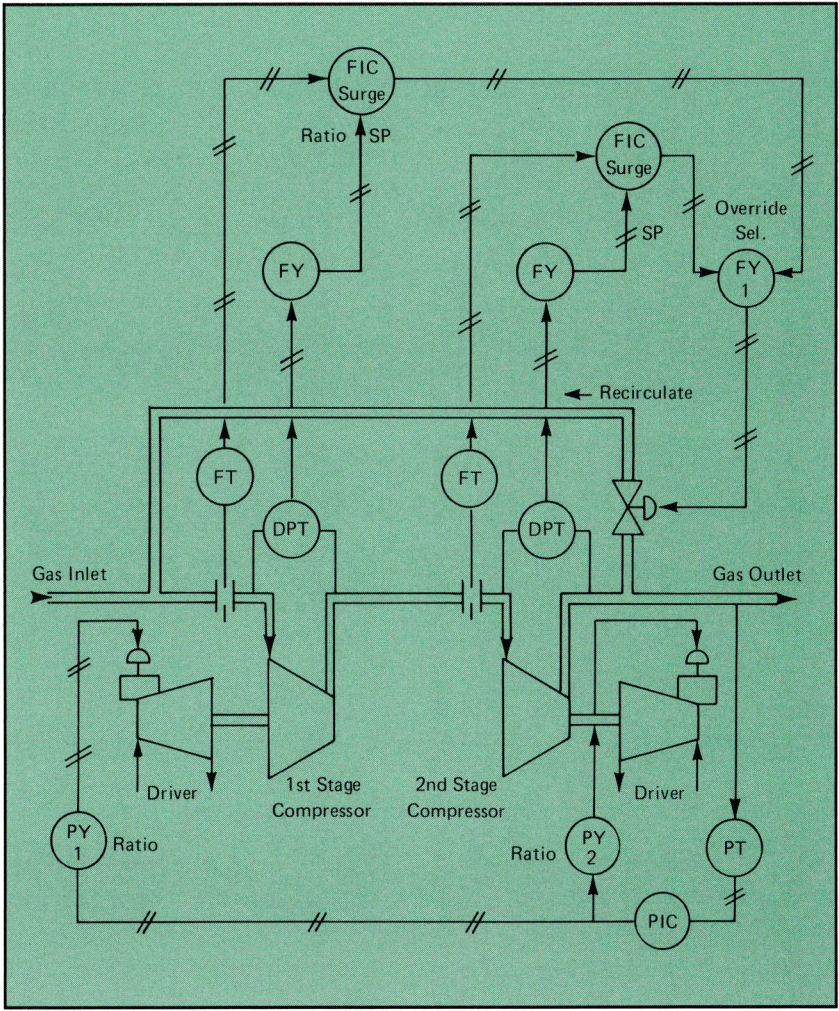

Fig. 9-2. Typical Control System for Series Compressor Configuration

configuration. Discharge pressure from the second compressor is regulated by manipulating the drivers' speeds. Relays PY1 and PY2 are adjusted to accommodate the differences in the performance curves of the two machines. Each compressor has a separate surge control system, but a common anti-surge valve is operated by either one of the surge controllers through low selector relay FY1.[2] This compares to the multiple-stage anti-surge systems shown in Figs. 8-18 and 8-19 but introduces the possibility of first-stage surge on start-up, as described in Fig. 6-17.

Tezekjian[5] offers the configuration in Fig. 9-3 for variable speed compressors in series. Inlet mass flow is controlled. The low-pressure compressor operates at constant speed, permitting discharge pressure rise and fall from the design point in accordance with the compressor's performance curve. The high-pressure compressor is then speed-controlled to maintain constant weight flow. Recycle around each compressor provides anti-surge protection but is omitted from the diagram for simplicity.

The control system for the base compressor and booster compressor shown in Fig. 9-4 was proposed by Sweet.[6]

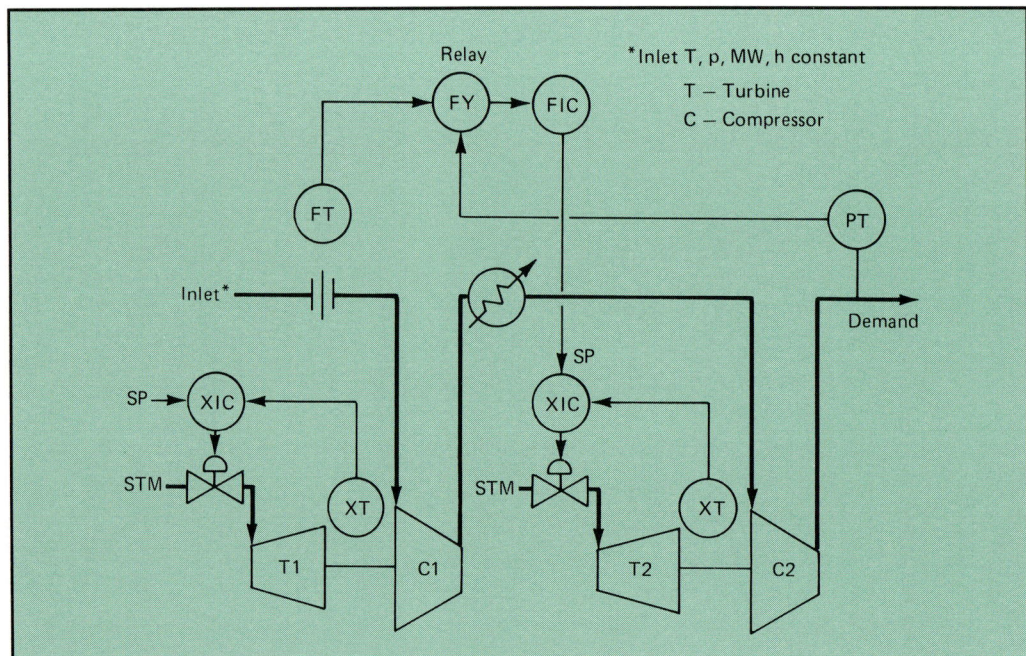

Fig. 9-3. Typical Control System for Series Compressor Configuration

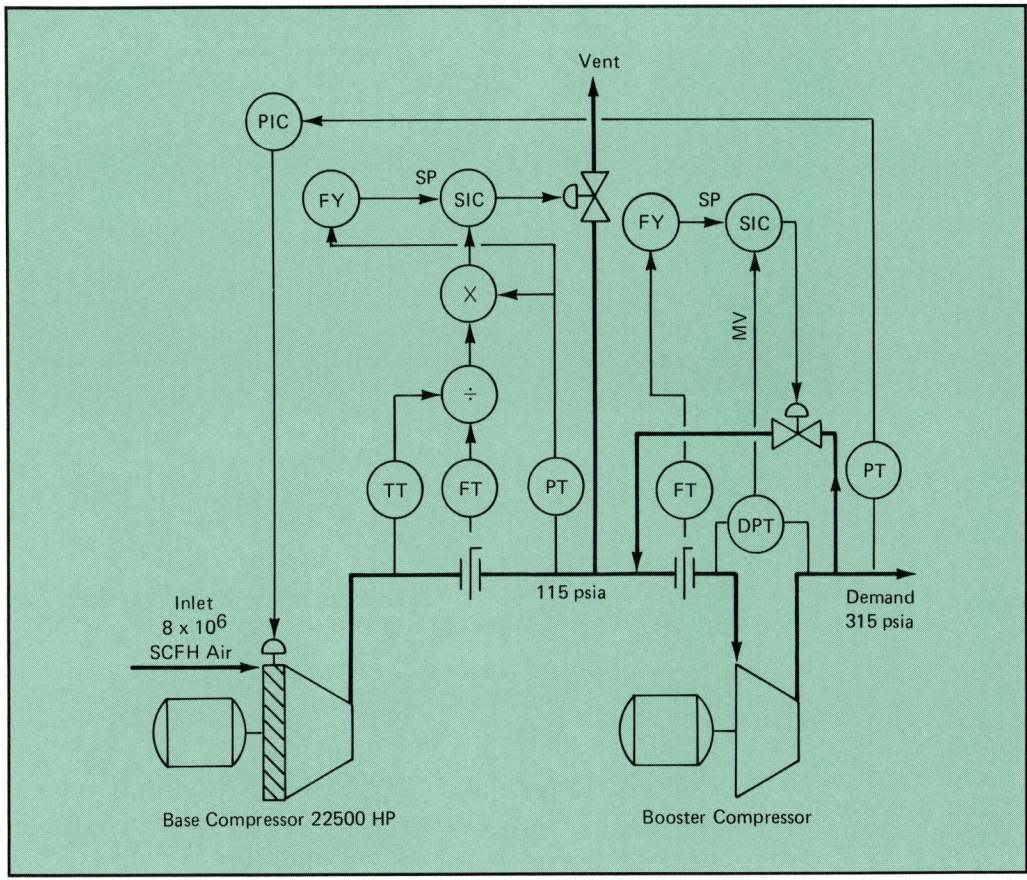

Fig. 9-4. Typical Control System for Air Compressors in Series

Capacity control is provided by discharge pressure control manipulating inlet guide vanes on the base compressor. The compressor map for the base compressor is in terms of mass flow and discharge pressure. Surge controller measurement is a compensated flow signal, and set point is a biased discharge pressure signal. The booster compressor has the linearized surge control system formerly discussed. Each surge controller has its set point determined by a computing relay, which is adjusted upon start-up of the compressors.

Parallel compressor configurations are encountered more often than series configurations. Air compressors are installed in parallel to provide improved reliability and ease of maintenance. The parallel configuration shown in Fig. 9-5, where each of the compressors is selected to carry 60% of the design load. In normal operation, only two of the compressors would be running at a time, the third being an "installed spare."[1] This gives rise to the general principle for machines

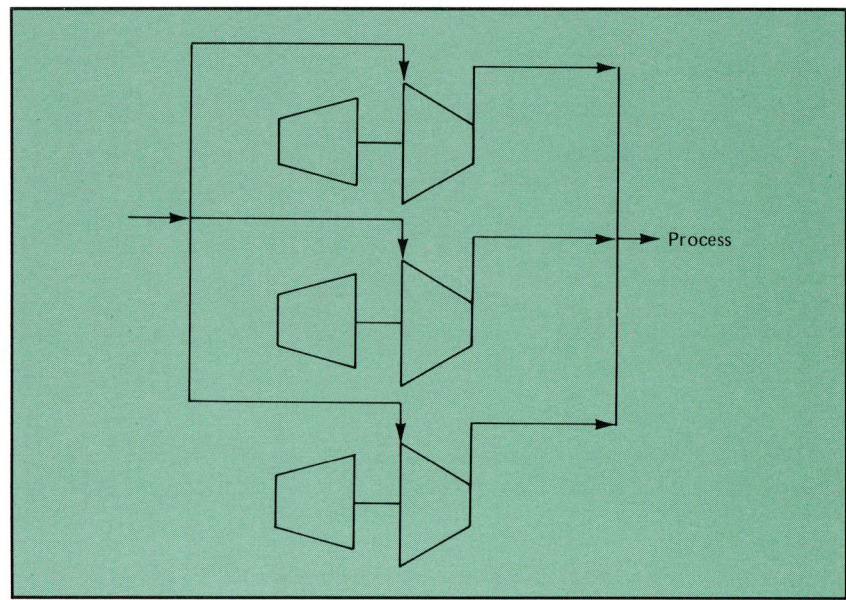

Fig. 9-5. Air Compressors Installed in Parallel to Assure Reliability

connected in parallel:

> The flows corresponding to a given discharge pressure on the performance curve of each individual machine add up to give the total flow through the parallel system at the discharge pressure.[2]

Figure 9-6 shows a parallel air compressor configuration commonly used for aeration tanks in sewage treatment plants. Obviously, the sum of the flow rates from each machine must be equal to the total flow rate. Pressure is not important in the configuration shown, but it is set by the mass balance in the discharge piping and, in turn, sets the individual flows through the individual performance curves. If the actual performance curves of the machines were identical, one third of the total flow rate would be supplied by each machine. Since the actual performance curves will invariably differ slightly, the load distribution among the compressors will not be even, and surge can quite possibly take place in one compressor before the other two. Figure 9-6 shows a surge control system for each compressor.

The difficulty in load balancing can be seen in Fig. 9-7, where the performance curve of one compressor is superimposed on the performance curve of a second compressor operating in parallel. Speed is constant. The surge points are different. Compressor A can pump flow B1 easily, but compressor B will actuate its surge control system at that flow. At discharge pressure p_{d2}, both machines cannot pump flow B1. Machine A must operate at a higher speed for equal load distribution.[1]

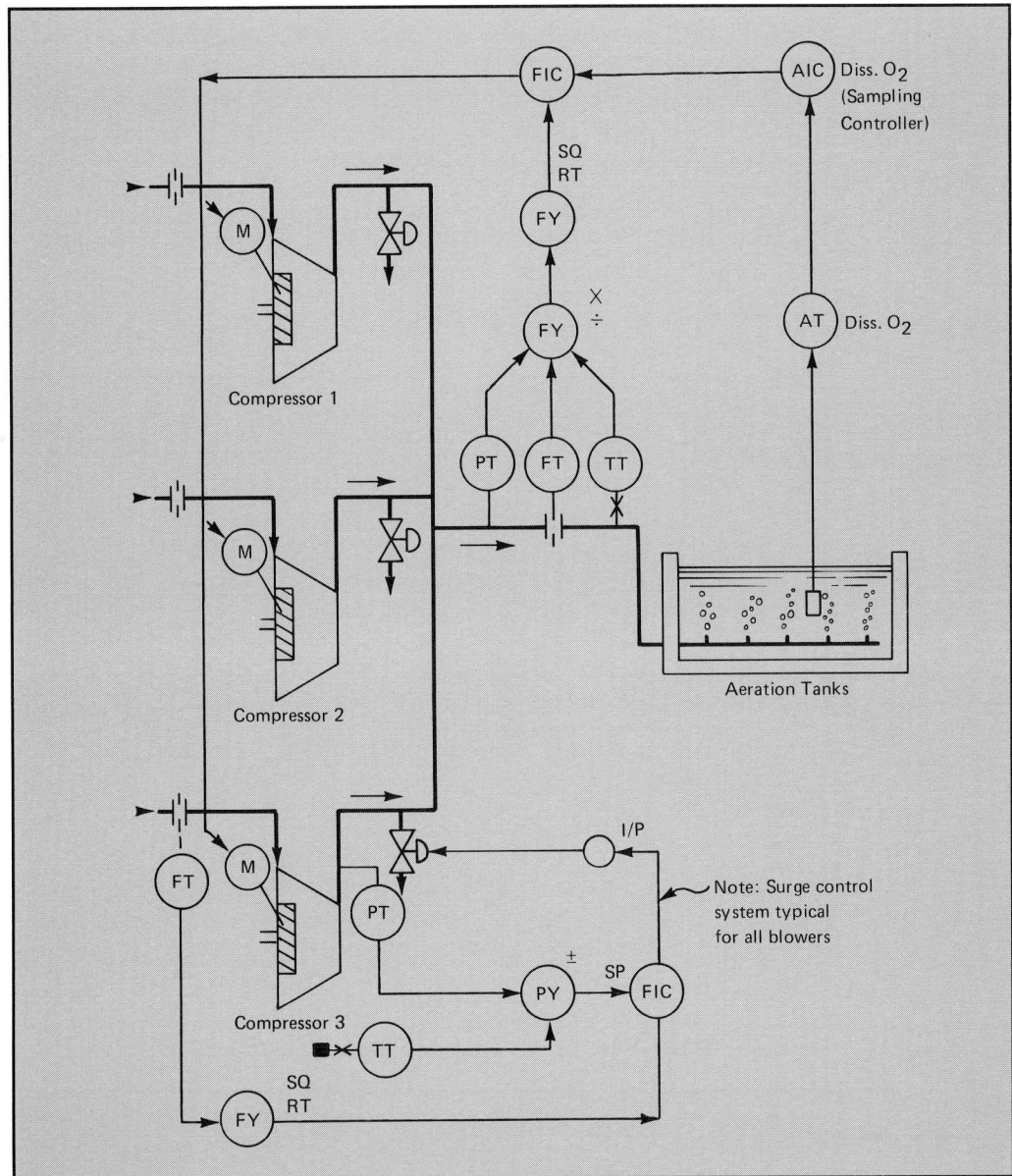

Fig. 9-6. Parallel Air Compressor Configuration for Dissolved Oxygen Control in Sewage Plant Aeration Process

Figure 9-8 shows one solution to the problem of differing performance curves. Each constant speed compressor has its individual flow control system. The set point of each flow control system sets the load on that compressor. Then the discharge header pressure controller is cascaded on the flow set points. The relays FY allow a trimming of the set points to better accommodate the performance curves.[5] Check valves have been installed in the discharge of each compressor to prevent backflow of fluid as a result of any slight imbalance in the characteristics of the two compressors. (Figure 4-2 shows process conditions that require check valve installation so far from the compressor that it is ineffective.) Surge control has been omitted from Fig. 9-8 for simplicity.

Figure 9-9 shows another typical control system for a parallel compressor configuration. The set points of the individual

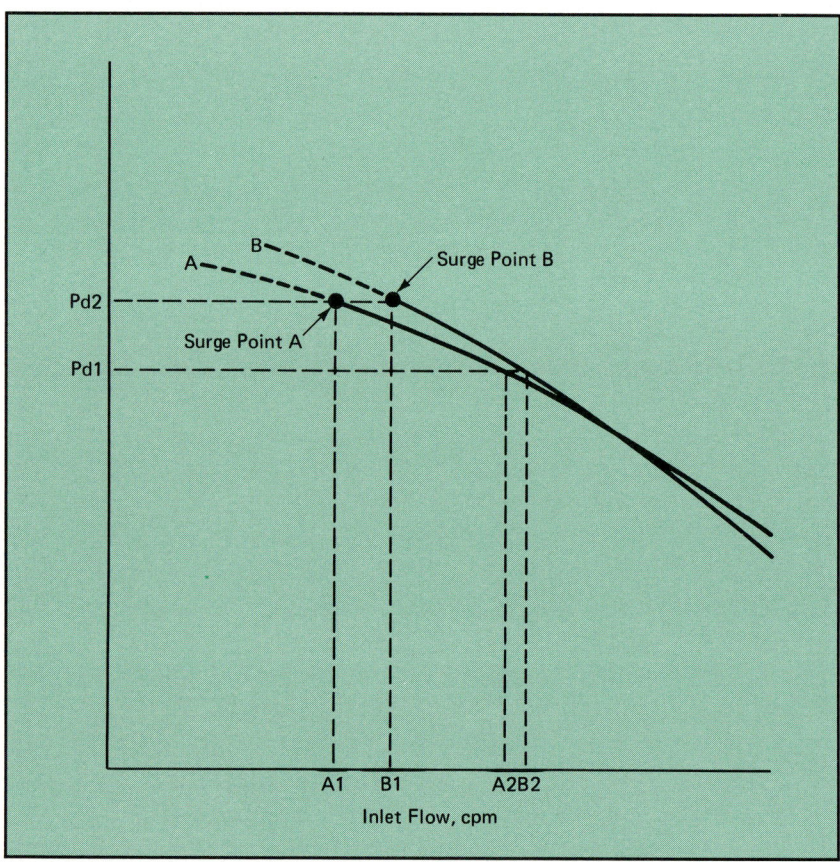

Fig. 9-7. Superimposed Performance Curves of Compressor A and Compressor B

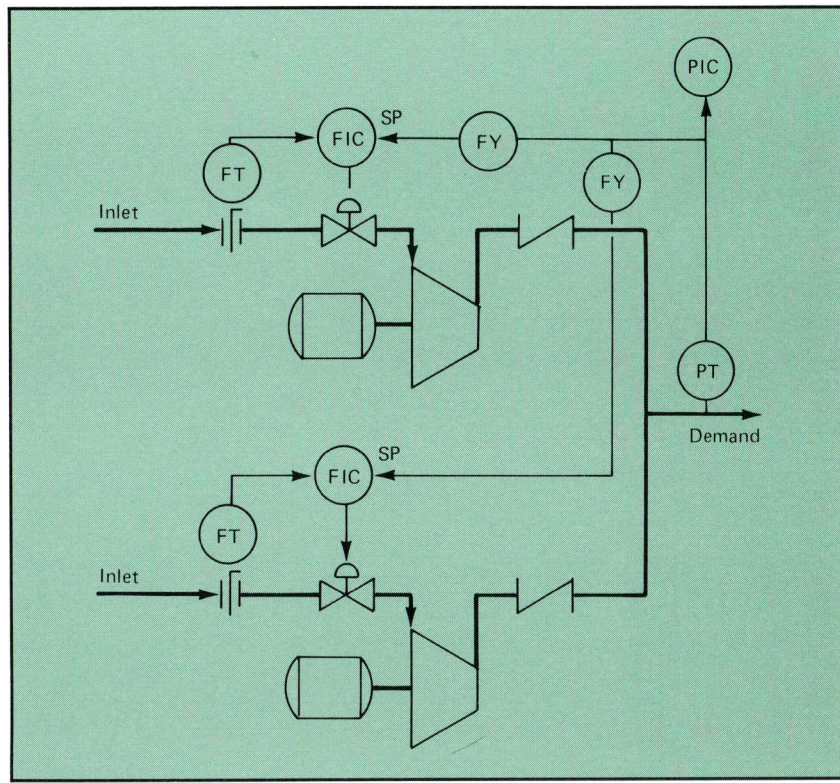

Fig. 9-8. Discharge Pressure Control of Parallel Compressors with Dissimilar Operating Characteristics

pressure controllers are tuned in steps. The compressor with the lowest pressure set point will be modulating its inlet valve to maintain header pressure at that set point. The other compressors will "see" a pressure lower than their set points and will run at full capacity. During the unloading of the compressor station, header pressure will rise. The first compressor (lowest set point) will "see" high pressure first and will close its inlet valve until the surge control line is reached, at which time the anti-surge valve will open, relieving the pressure. The set point of the pressure controller of the second compressor will then "see" a high pressure and unload itself, and so on. Again, the anti-surge system is not shown in Fig. 9-9 for simplicity.[7]

The parallel compressor configuration of Fig. 9-8 forces the compressors to operate in inefficient areas of the compressor map and entails a complex surge control system since the anti-surge system and the flow cascade system operate from

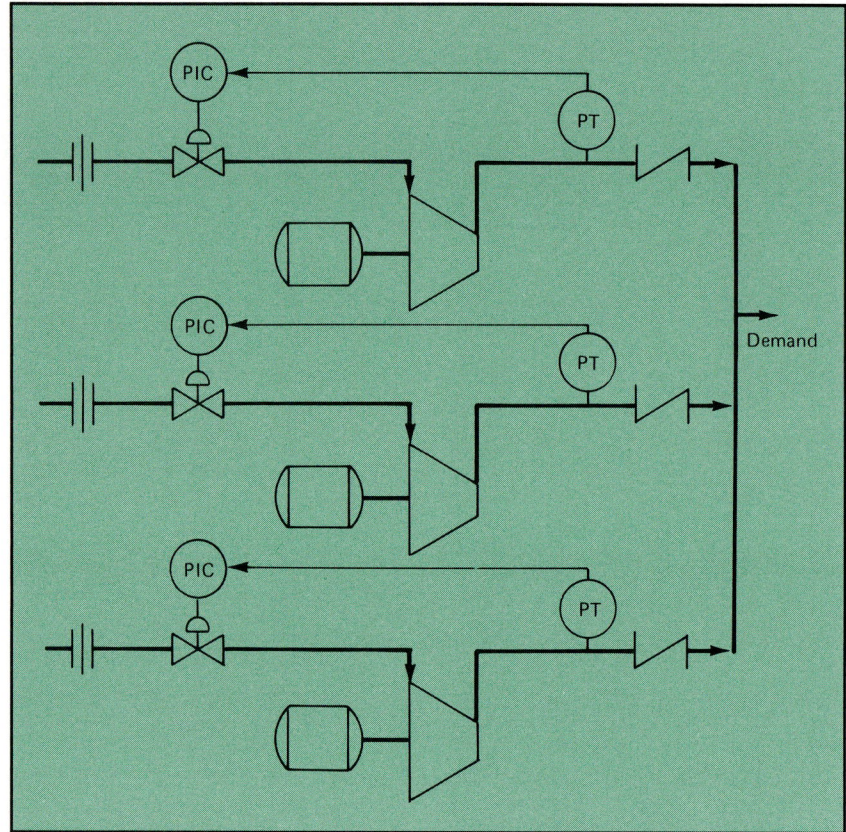

Fig. 9-9. Discharge Pressure Control of Parallel Compressors with Controller Set Points Adjusted in Steps

the same flow measurement. The configuration of Fig. 9-9 suffers from inefficient operation and limited rangeability.[7] More efficient control strategies for compressor stations can reduce energy costs by up to 25%.

9-2. Specific Power Consumption[4,8,9,10]

The additive feature of polytropic head in series-connected compressors has already been discussed. Polytropic efficiency is a measure of the energy required to compress the gas between the inlet and discharge ports of the compressor. A more useful parameter is the specific power consumption, which is a measure of the energy required to deliver the gas to the user. The difference between the two terms is the losses in intercoolers, valves, and discharge piping. "Specific power consumption," a somewhat wider definition than "polytropic efficiency," is defined as follows:

> *Specific power consumption* is the energy input to a compressor system required to maintain the controlled gas parameter and divided by the number of flow units of gas delivered to the discharge header.

Consider a compressor that requires 500 kW of power to deliver 1000 scfm of gas. Then,

$$\text{Absolute Specific Power Consumption} = \frac{500 \text{ kW}}{1000 \text{ scfm}} = 0.5 \frac{\text{kW}}{\text{scfm}} \quad (9\text{-}1)$$

A dimensionless measurement of specific power consumption is required to compare different compressors. The relative specific power consumption has units of power divided by maximum power and flow rate divided by maximum flow:

$$\text{Relative Specific Power Consumption} = \frac{\text{kW}}{\text{kW}_{max}} \div \frac{\text{scfm}}{\text{scfm}_{max}} \quad (9\text{-}2)$$

Since power consumption increases with energy consumption, specific power consumption is inversely proportional to energy efficiency.

Specific power consumption finds application in determining loading and unloading strategies for the compressor station. Compressors can be loaded and unloaded by speed changes, guide vane changes, throttling, and recirculation, as has already been discussed. The unloading of the compressors within a station can be simultaneous (as shown in Fig. 9-8), sequential (as shown in Fig. 9-9), or mixed. The algorithm controlling loading and unloading significantly influences the station efficiency.

Consider the compressor map shown as Fig. 9-10, which shows the performance curves for two identical centrifugal compressors. The process requires a constant compression ratio of 2.75. The relative specific power consumption is shown as Fig. 9-11. Operating points A, B, and C lie on a line of constant pressure ratio but varying throughput flow rates. Point A defines maximum compressor compression ratio and flow rate. Point B is the design point and has a better efficiency and lower specific power consumption. Point C lies on the surge control line, a point of lower efficiency and

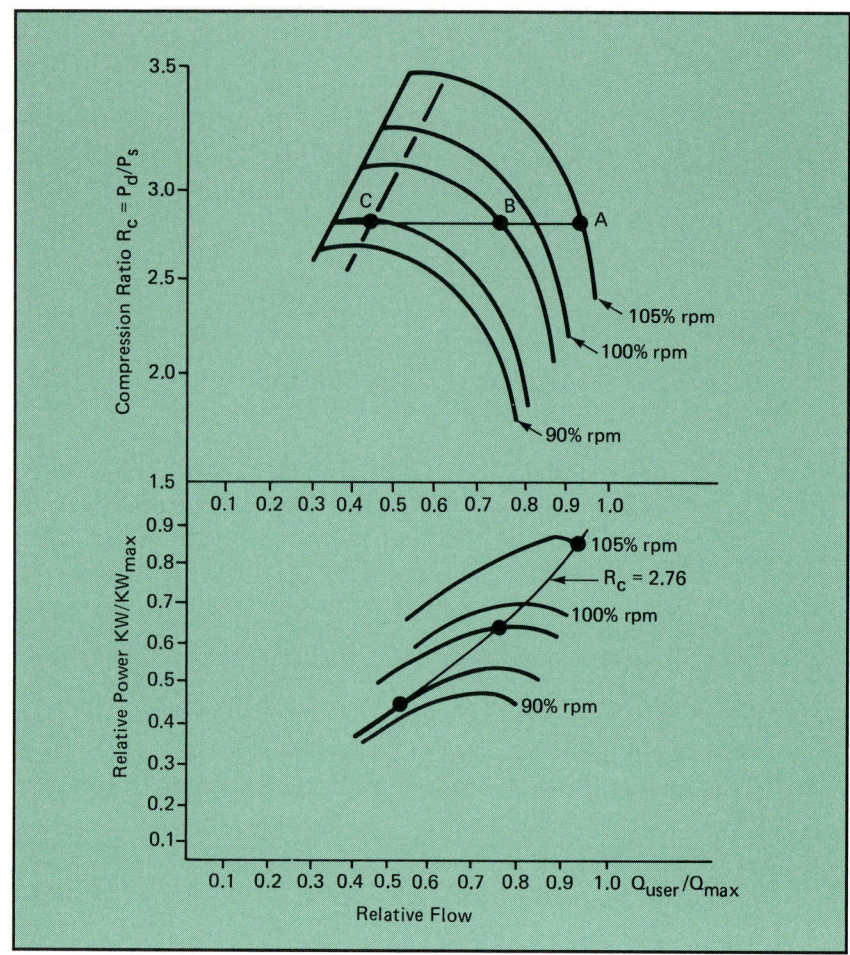

Fig. 9-10. Compressor Map for a Centrifugal Compressor

higher specific power consumption. Point C is the minimum safe flow for these compressors at the given speed. To reduce process demand flow further necessitates recirculation around the compressors to prevent surge damage. So flow through the compressor is constant at point C and below, and compressor power is constant. As a result of the recirculation, specific power consumption rises sharply to the left of point C. In general, then:

> The system efficiency of any compressor drops sharply once recirculation begins, and the cost of compressing a flow unit of gas rises sharply.[4]

The effectiveness of the surge control system is thus critical to efficient operation. The more effective the surge control

Unit 9: Multiple Compressor Systems

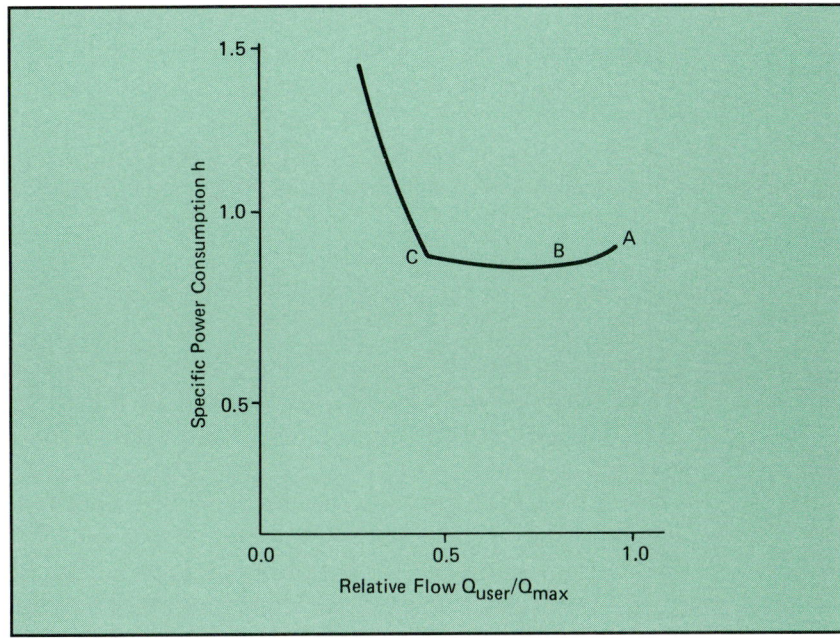

Fig. 9-11. Specific Power Consumption for a Single Compressor

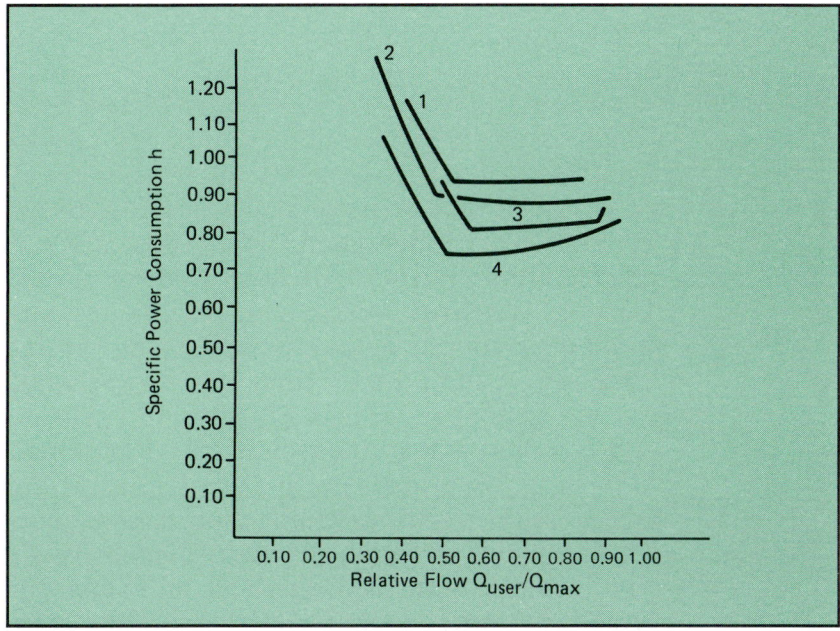

Fig. 9-12. Specific Power Consumption of Four Different Centrifugal Compressors

system is, the narrower the control margin can be. Narrowing the surge margin on one compressor is multiplied by many compressors in a compressor station, and the savings in operating costs rapidly become significant. Experience shows that an increase of 5% in control margin will decrease efficiency by 10% when operating in the region to the left of point C (Figs. 9-10 and 9-11).

In contrast to the foregoing, the change in polytropic efficiency across the compressor operating range is quite small. As shown by Fig. 9-11, specific power consumption rises dramatically to the left of point C, the surge control line. Figure 9-12 shows this same characteristic specific power consumption for four other compressors manufactured in America and Europe.[4] This leads to the conclusion:

> Increasing the safe operating range without recirculation is more important for compression efficiency than maintaining the compressors' operating points at the point of highest polytropic efficiency.

9-3. Multiple Compressors in Series[11,12]

Figures 9-2, 9-3, and 9-4 show some of the widely applied control systems on series compressor configurations. However, further analysis that considers intercooling and aftercooling yields more efficient strategies for controlling series compressors.[12]

Compressor stages (same drive shaft) and compressors (separate drivers) are installed in series to increase pressure ratio. A dilemma exists as to the best design of the anti-surge control valve configuration. One design calls for a complete surge control system for each compressor or compressor stage (section), as shown in Figs. 8-16 and 9-4. This configuration has the advantages of (1) fast response to protect each compressor, and (2) protection of the low-pressure compressor on start-up and shutdown. The difficulty with the configuration is that the opening of a low-pressure compressor anti-surge valve momentarily decreases flow to downstream compressors, possibly causing their anti-surge valves to open. An alternative design is shown by Figs. 8-18, 8-19, and 9-2. These control systems use a single anti-surge control valve around the entire compressor configuration. The single valve is actuated through a low-signal selector (anti-surge valves are always air-to-close) to choose the output of the surge controller that is closest to surge. This configuration has the advantages of (1) economy (a large low-noise anti-surge valve can cost $25,000) and (2) guaranteed absence of interaction. This configuration can provide somewhat slower surge protection for upstream stages or compressors because of the volume of coolers, separators, and piping in the system.[11]

The Fan Laws are typically extended for use with compressors in series, even those with intercoolers and separators that cool and condense liquids and remove them from the system. This assumption in the series application is questionable because:

A. The compressors are considered a single unit compressing polytropically. This assumption is true with no intercooling. However, intercooling results in a thermodynamic process that is neither polytropic or isothermal.
B. Separation means that some of the flowing fluid is removed between compressors, thus

$$m_2 = K_3 m_1 \qquad (9\text{-}3)$$

where K_3 = constant less than 1.

The removal of fluid within the compressor configuration can cause significant inaccuracies in the Fan Law relationships.

Even if all stages or all compressors in series have the same steady-state flow, large volumes in piping, intercoolers and separators cause significant differences in transient flow rates. Figure 9-13 shows a flow recording taken from a two-stage (section) compressor, where suction flow was increased. Second-stage flow increased slightly later than first-stage flow.

Two variable speed compressors in a series configuration are shown in Fig. 9-14. An aftercooler and separator between the compressors result in different temperature and flow in the second compressor. A strategy for improved efficiency and reliability, called optimum dividing of total pressure differential, is implemented by the control system shown.[12]

The total pressure differential across the compressor configuration is:

$$\Delta p_T = \Delta p_{C1} + \Delta p_{C2} - \Delta p \qquad (9\text{-}4)$$

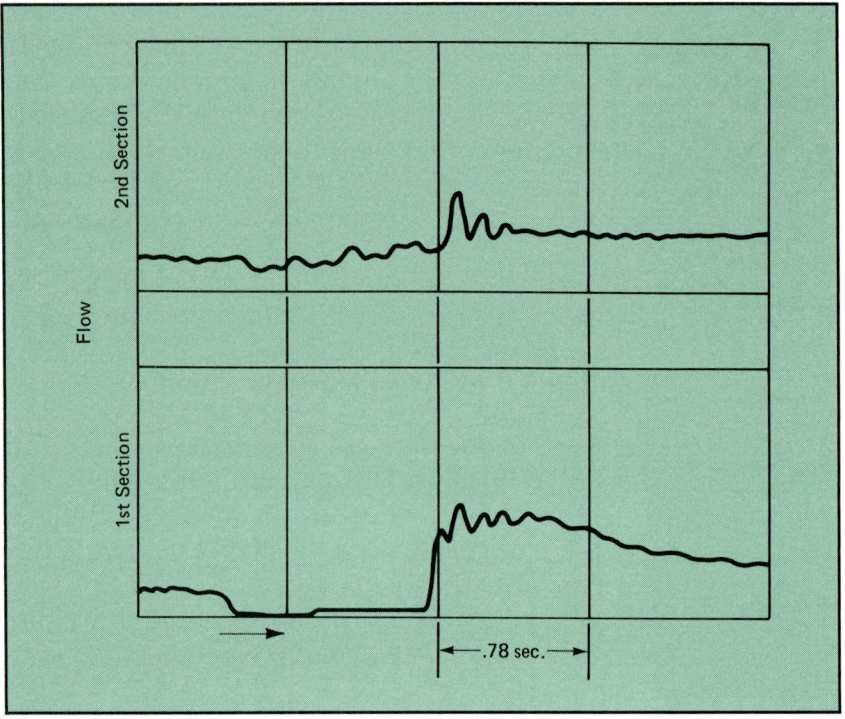

Fig. 9-13. Control Strategy for Total Compression Ratio Dividing for Two Compressors in Series

where:

Δp_T = total pressure increase across configuration
Δp = pressure loss in piping and equipment between compressors

Obviously, many different combinations of Δp_{C1} and Δp_{C2} can equal Δp_T for a given flow rate through the configuration. An arbitrary division of the total pressure differential will cause one compressor to operate at its surge control line while the other is operating far from the surge zone. One compressor will start to recirculate before the other reaches design capacity, possibly interfering with the ability to meet process demand.

The linearized compressor map results in the following expression for the surge line:

$$K_1 \Delta p_{Ci} = \frac{m^2 T_{si}}{P_{si}} \qquad (9\text{-}5)$$

or,

$$K_1 \frac{p_{si}}{T_{si}} \Delta p_{Ci} = m^2 \qquad (9\text{-}6)$$

where:

subscript s = suction
subscript i = compressor identification index

Then, eqs. (9-3), (9-4), and (9-6) can be combined into an equation that describes a load dividing that will avoid recirculation by one of the compressors in series:

$$K_1 \frac{p_{s1}}{T_{s1}} \Delta p_{C1} + b_{11} = \frac{K_2}{K_3} \frac{p_{s2}}{T_{s2}} \Delta p_{C2} + b_{12} \qquad (9\text{-}7)$$

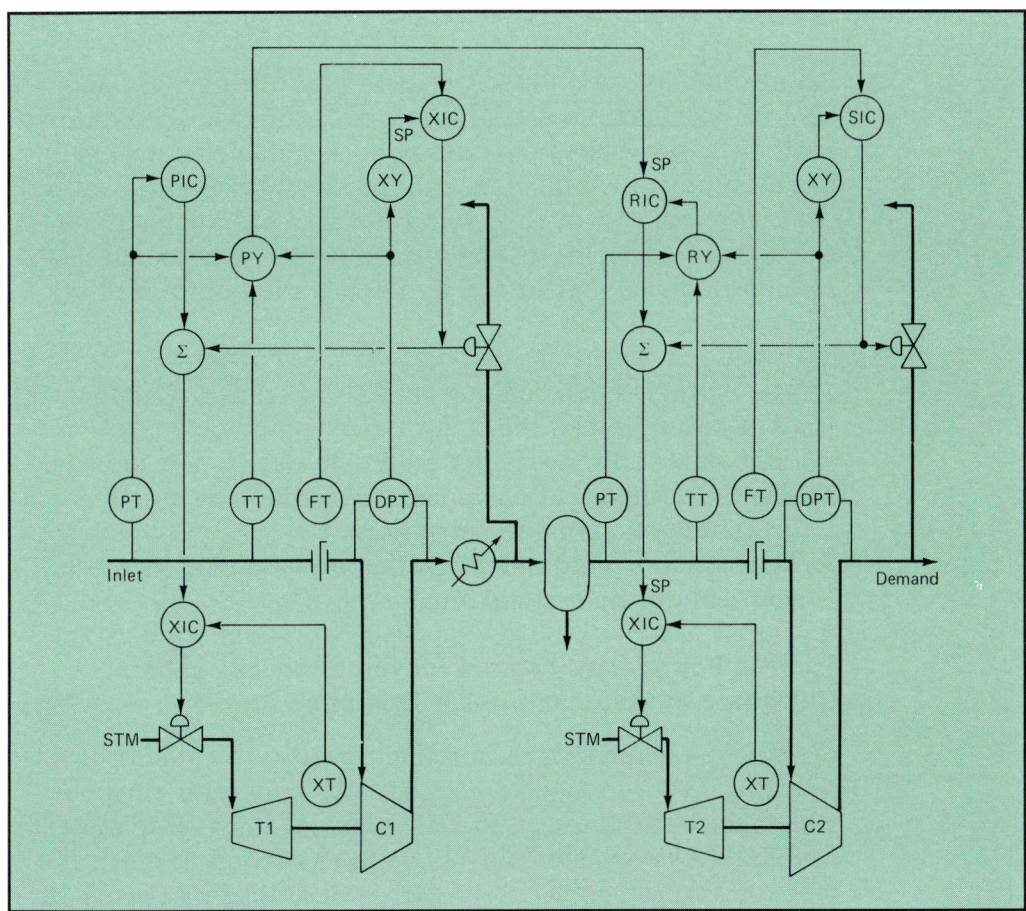

Fig. 9-14. Delay in Flow Change from First Stage to Second Stage of a Multistage Compressor

Figure 9-13 shows the implementation of eq. (9-7). Each compressor has the linearized surge control system and the circuit to reduce interaction that were formerly described. Inlet pressure of the low-pressure compressor is controlled by pressure control cascaded on turbine speed. The high-pressure compressor side of eq. (9-7) (index "2") is calculated by relay RY and is used as the measured variable for controller RIC. The set point of controller RIC is the low-pressure compressor side of eq. (9-7) as calculated by relay PY. Thus, controller RIC forces the equality. Gains K_1 and K_2 and biases b_{11} and b_{12} can be calculated from the compressor maps and given an experimental final adjustment during operation of the compressor. Microprocessor-based controllers are required by the complex calculations required by the control system shown on Fig. 9-14.

9-4. Multiple Compressors in Parallel[3,9,10,12]

Figures 9-6, 9-8, and 9-9 show some of the widely applied control systems for compressors in a parallel configuration. Each of these systems provides simple, reliable control and has been broadly applied on existing compressors. However, the increasing cost of energy and the ease of implementing calculation on microprocessor-based control equipment has lead to the development and implementation of three additional guidelines:[4]

1. The most effective compressor station operating strategy is to load and unload the compressors simultaneously. Efficiency and rangeability are improved by this strategy.
2. The most important component of station energy efficiency is the reduction of anti-surge flow.
3. The station should unload so that all compressors reach their control lines simultaneously.

Specific Power Consumption for Simultaneous Loading/Unloading in Multicompressor Operation

Consider a compressor station that operates two identical centrifugal compressors. A constant pressure ratio must be maintained, and throughput flow must be adjusted to meet the demand. Capacity is regulated by either suction or discharge pressure control. Each compressor is described by Figs. 9-10 and 9-11.

Three major alternative techniques can be used in changing station throughput flow:

- Load and unload compressors simultaneously.
- Load and unload compressors sequentially, first unloading the least efficient compressor.
- Combine the simultaneous and sequential load/unload strategies.

The solid line on Fig. 9-15 shows the specific power consumption for a simultaneous approach to the surge control lines of the two parallel compressors. The station compression ratio is maintained at all points on the graph.

Point A on Fig. 9-15 is the maximum capacity of the station, while Point B is the surge control lines of the combined compressors. With decreasing flow, the station efficiency first increases slightly, then decreases, reflecting the polytropic efficiency of Fig. 9-10. Both compressors approach their surge control lines simultaneously at Point B. Anti-surge flow begins between Points B and C as station output is adjusted to lower gas demand.

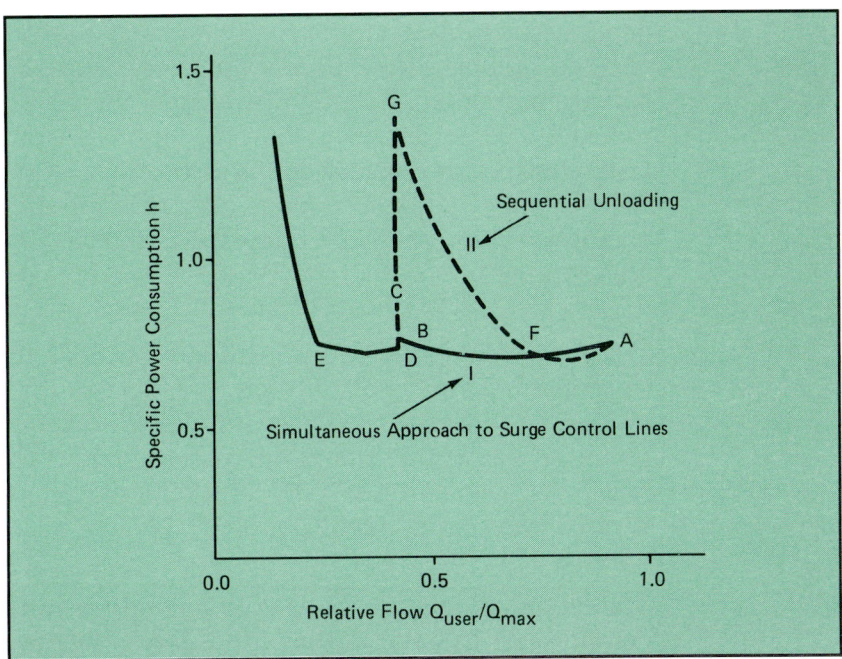

Fig. 9-15. Specific Power Consumption of Two Identical Compressors in a Parallel Configuration

At Point C, one compressor is shut down. The control system then loads the other compressor to maximum flow, Point D. Efficiency is improved by loading the machine. Then the station continues to be unloaded by decreasing flow until the remaining compressor reaches its surge control line, Point E. Further decrease in flow is compensated once again by anti-surge flow.

These observations result in compressor station operation with a range of from Point A to Point E on Fig. 9-15 at essentially good efficiency.

Specific Power Consumption for Sequential Loading/ Unloading for Multicompressor Operation

The dotted line on Fig. 9-15 shows the specific power consumption for the sequential loading and unloading of the compressor station. Between Points A and F, one compressor is running at design speed and throughput flow, while flow through the second compressor is decreasing to accommodate decreasing demand. The second compressor reaches its surge control line at Point F. However, the second compressor cannot be shut down at this point, since the base loaded compressor cannot supply the demand. To meet the demand without reducing the compression ratio, the second compressor cannot be shut down until the flow demand is reduced to Point G. At this point, station efficiency is very low. Shutting the second compressor down at Point G increases the station efficiency to Point D. With only one compressor running from Point D to Point E, Curves I and II are identical.

The curve for sequential unloading will invariably be higher in specific energy consumption than curve I for simultaneous unloading in Fig. 9-15. Thus, simultaneous unloading is to be preferred because of the improved efficiency. This conclusion is valid, however, only if all compressors reach their surge control lines simultaneously.

The foregoing conclusion is still generally valid for compressors in parallel that are different in size, power, and efficiency. The advantage may disappear for compressors whose efficiencies differ by more than 10%.[4]

The S-Criterion for Simultaneous Approach of Each Compressor to Its Surge Line

The S-criterion, defined by eq. (8-8), is a dimensionless number that expresses the relative distance from the operating point to the surge line of a compressor. Since the number is relative, the absolute value has no meaning. However, compressors having the same S number will all be operating equidistant from their respective surge lines. A control system that causes all of the S numbers to be equal will therefore insure that all compressors are approaching the surge control line simultaneously.

The S-Criterion, eq. (8-8), will be less than 1 when the operating point is safely away from surge and will equal 1 when the operating point is on the surge control line. The deviation "e" of the compressor operating point from the surge control line is:

$$e = S - 1 \qquad (9\text{-}8)$$

Figure 9-16 shows a parallel compressor configuration with suction pressure control, load sharing control based on eq. (9-8), and anti-surge control. The calculating module on each load-sharing controller computes the value:

$$S' = \beta_1(S - 1) + \beta_2 \qquad (9\text{-}9)$$

The parameters of eq. (9-9) allow adjustment for best efficiency. The rate of change of flow for the compressor is set by β_1, while equal values of β_2 guarantee that all compressors will reach their surge lines simultaneously.

The value of S' from eq. (9-9) serves as the measured variable for the load-sharing controller. The set points of all load-sharing controllers are the output of the pressure controller. This is a typical cascade control system. The load-sharing controller is typically tuned to a high gain (narrow proportional band), resulting in a cascade system that performs well. Load sharing has also been implemented by throttling inlet control valves to cause the amperage to the electrical motor drive on each of parallel compressors to be equal.[13]

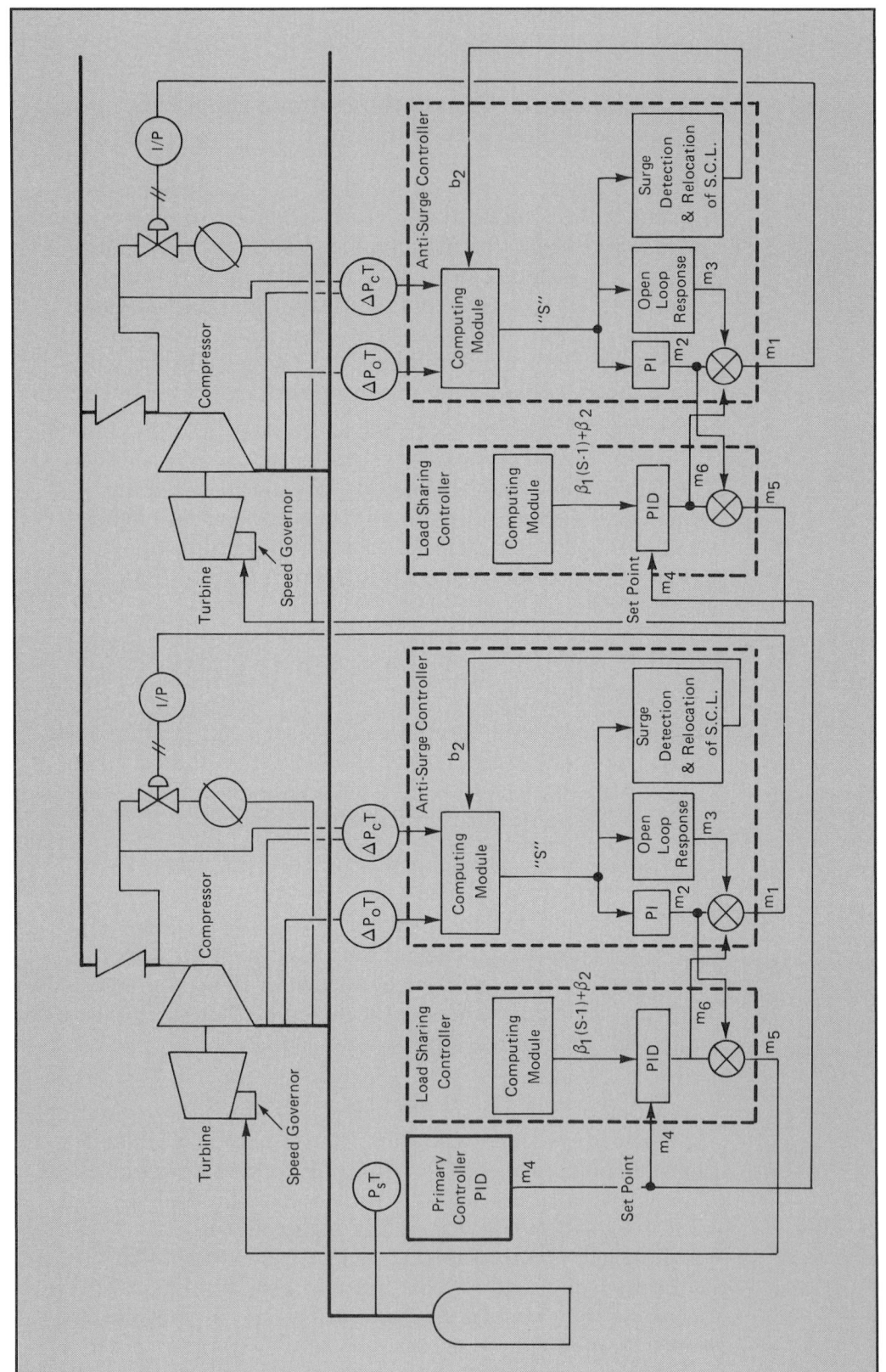

Fig. 9-16. Control System Using S-Criteria for Compressors in Parallel Configuration

Figures 3-14 and 3-15 show the user system load line on the compressor map. The discharge pressure that must be supplied by the compressor is determined by the demand flow of the users on the load line. Thus, as demand flow decreases, the pressure required to deliver that flow also decreases. Appreciable energy savings are available by reducing the discharge pressure with decreasing demand.[14] A serendipity is that the operating range (the distance between surge and stonewall lines) gets greater as the discharge pressure is reduced.

An example of the energy savings resulting from reduced discharge pressure for a compressor station of mixed compressors is shown in Figs. 9-17, 9-18, and 9-19. The parallel compressors shown in Fig. 9-17 are of different capacities. The output of centrifugal compressor 2 is 50 percent of centrifugal compressor 1, and the outputs of each of the reciprocating compressors are approximately 20 percent of that of compressor 1.

The compressor map for compressor 1 is shown in Fig. 9-18. The surge control line is "nn". Compressor 1 operates at point

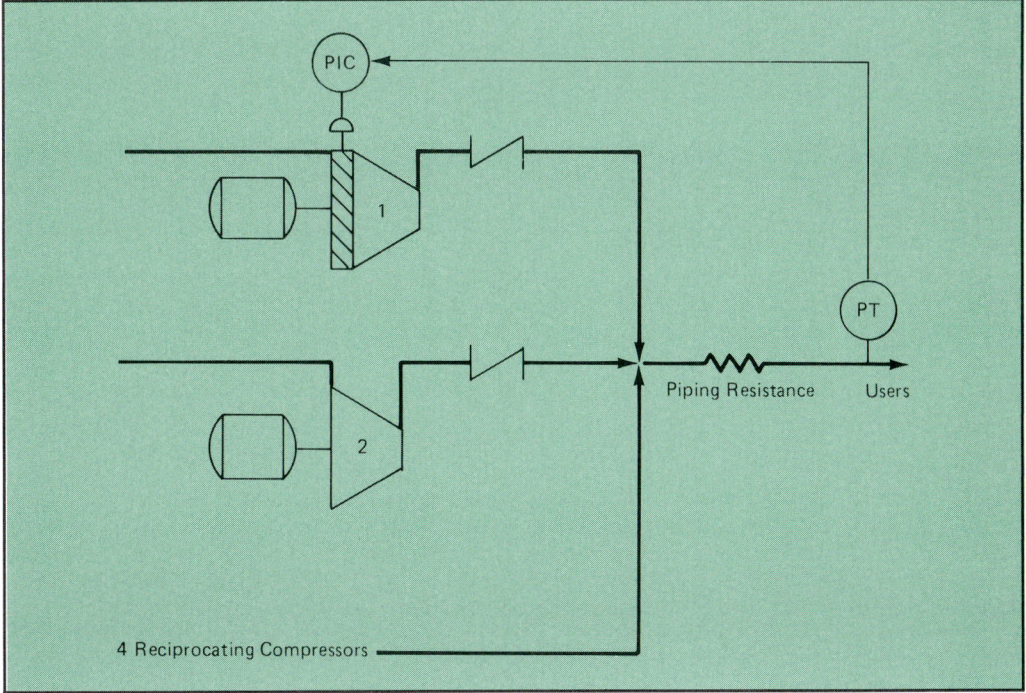

Fig. 9-17. Parallel Compressors Controlled by Pressure at Point of Use

"a" during station operation at design conditions with all compressors on line. Decreasing user demand causes the compressor 1 operating point to move to point "b," where one reciprocating compressor is shut down. Compensating for the decrease in station output by a magnitude corresponding to the output of the reciprocating compressor, the pressure control system increases the output of compressor 1 to point "c." Operation at point "c" results in compressor 1 operating at maximum efficiency. Further decrease in user demand results in the other three reciprocating compressors shutting down at points "d," "f," and "h" on the compressor map. After each shutdown, compressor 1 returns to high efficiency operating points "e," "g," and "i."

Further unloading at point "i" causes the operating point to move to point "k," where compressor 2 is shut down. Upon shutdown of compressor 2, the operating point of compressor 1 moves to "l." Further decrease in demand causes the operating point of compressor 1 to move along "lm," and the surge control system will maintain compressor flow at point

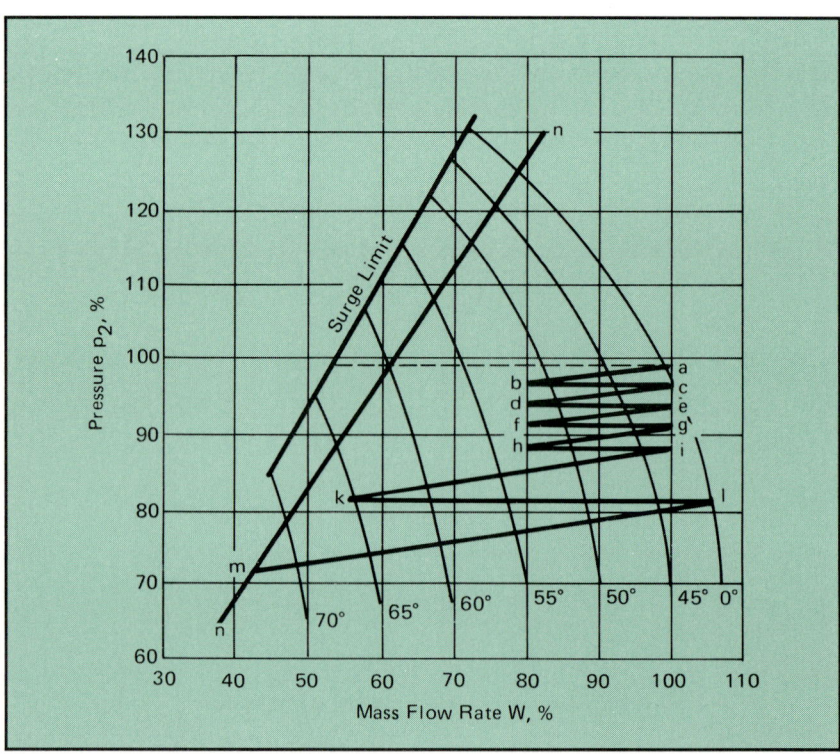

Fig. 9-18. Compressor Map for Compressor No. 1 in Compressor Station Shown in Figure 9-17

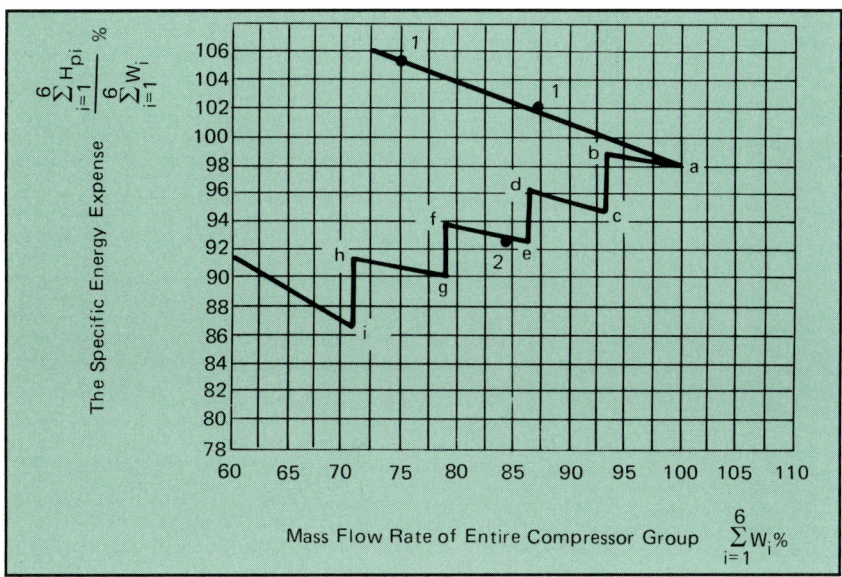

Fig. 9-19. Specific Energy Expense of Operating Compressor Station

"m." The surge control system is not shown for simplicity in Fig. 9-17.

Curve 2 in Fig. 9-19 shows the change in specific energy expense using the foregoing strategy. The energy expense of operating with constant compressor discharge pressure is shown as curve 1 of Fig. 9-19. Energy savings are significant.

Exercises

1. In Fig. 9-4, why is one anti-surge system a recycle, while the other is vented?

2. In Fig. 9-7, suppose that compressor B has its anti-surge valve open at flow B1. What has to be done to cause it to provide net forward flow B1?

3. Why is the check valve ineffective in preventing reverse flow and reverse compressor rotation in Fig. 4-2?

4. Sketch an anti-surge system for one of the compressors in Fig. 9-8.

5. Sketch a relationship similar to Fig. 9-15 for three parallel compressors as they are shut down sequentially to accommodate decreasing demand.

6. Why not shut down compressor 2 in Fig. 9-17 before shutting down the reciprocating compressors?

References

1. Nisenfeld, A. E., *Centrifugal Compressors, Principles of Operation and Control*, Instrument Society of America, Research Triangle Park, N.C., 1982.
2. Carlo-Stella, G., and W. S. Buzzard, "Control of Dynamic Compressors," Bulletin No. 91-53G-06, Fischer and Porter Co., Warminster, PA
3. Mirsky, S., and John Hampel, "Improving Efficiency and Reliability of Packaged Centrifugal Air Compressors Operating in Parallel," Technical Paper TP13, Compressor Controls Corp., Des Moines, Iowa, 1986.
4. Staroselsky, N., and L. Ladin, "More Effective Control for Centrifugal Gas Compressors Operating in Parallel," American Society of Mechanical Engineers Paper 86-GT-204, 1986.
5. Tezekjian, E. A., "How to Control Centrifugal Compressors," *Hydrocarbon Processing*, Gulf Publishing Co., Houston, TX, July, 1963.
6. Sweet, S. W., "A Look at Two Surge Control Systems," *ISA Transactions*, Vol. 14, No. 2, p. 172, 1975.
7. Staroselsky, N., and A. Rutshtein, "Control Strategy for Compressor Stations," *Centrifugal Compressor Operation and Control*, Instrument Society of America, Research Triangle Park, NC, 1976.
8. Staroselsky, N., and L. Ladin, "Improved Surge Control for Centrifugal Compressors," *Chemical Engineering*, McGraw-Hill, New York, N.Y., May 21, 1979.
9. Staroselsky, N., and L. Ladin, "Parallel Centrifugal Gas Compressors Can Be Controlled More Effectively," *Oil and Gas Journal*, Nov. 3, 1986, p. 79.
10. Staroselsky, N., and L. Ladin, "Compressors Simultaneously Approach Surge," *Oil and Gas Journal*, Nov. 10, 1986, p. 94.
11. McMillan, G. K., *Centrifugal and Axial Compressor Control*, Instrument Society of America, Research Triangle Park, NC, 1983.
12. Staroselsky, N., "Better Efficiency and Reliability for Dynamic Compressors Operating in Parallel and in Series," *ASME Proceedings of the Energy Technology Conference and Exhibition*, New Orleans, LA, 1980.
13. Elliott, H., "Retrofitting Compressor Controls," *InTech*, September, 1987, p. 29.
14. Rutshtein, A., and N. Staroselsky, "Some Considerations on Improving the Control Strategy for Dynamic Compressors," *ISA Transactions*, Vol. 16, No. 2, 1976.

Unit 10:
Optimization of Compressor Operation

UNIT 10

Optimization of Compressor Operation[1,2]

The transportation of gases and vapors is a major element in the operating cost of processing plants. The advanced control strategies described in Units 8 and 9 are directed toward increasing efficiency and can substantially lower the operating cost of the plant. Optimization causes the compressor or combination of compressors to be loaded in a particular way to satisfy demand at least cost or: optimization maximizes efficiency.

One study found that 140 billion kilowatt-hours of electricity is used each year to power electrical motor-driven compressors and compressor systems,[2] Thus, providing the power for compression is an enormous cost, and optimization configurations that provide only a few percentage points of savings can result in substantial savings. Even minor improvements in control strategy reduce operating costs significantly.

Generally, optimization will depend on minimizing anti-surge valve opening and on matching compressor output to the user demand. Multiple-compressor systems are optimized by the flow distribution with compressors in tandem.

Learning Objectives — When you have completed this unit, you should:

A. Be aware of compressor configurations and multiple-compressor networks that are amenable to optimization.

B. Be familiar with the techniques of optimization, including the application of linear programming and dynamic programming.

C. Understand the operating features of the control equipment used to implement optimization strategy.

D. Appreciate the value (savings divided by cost) of optimizing compressor operation.

10-1. General

Optimization is typically a supervisory control strategy that slowly drives set points from one steady state to another to cause the operating equipment to run at maximum efficiency. Maximum efficiency is the minimum cost of input power that meets the demand load requirements. This immediately suggests measuring the demand load and supplying only the power required to accommodate the load. The floating pressure set point capacity modulation control techniques,[2] which provide timely adjustment of compressor suction and discharge pressures, satisfy instantaneous refrigeration loads while minimizing instantaneous energy use and peak demands. This approach is in sharp contrast to the common compressor control practices of operating at the most stringent pressure and temperature requirements, even when these maximum requirements are seldom encountered. This is a dynamic technique; it provides real-time modulation of pressures as required to meet changing process needs.

Another dynamic technique is the use of feedforward-feedback with a digital self-tuning controller[3] to improve the time response of the control loop when inlet flow has changed.[4] Application of this control technique to a compressor demonstrates the usefulness of the adaptive structure to maintain stability at the process in the face of large disturbances in inlet flow rate.

One process technique for recovering power and increasing efficiency when high-pressure gas is available and must be reduced to a lower pressure is to add a waste gas (or vapor) turbine to the shaft of the compressor. This turbine, called an expander, is applicable in refrigeration systems and in some chemical processing systems. Monitoring the turboexpander system and taking corrective action will maintain efficiency as performance swings with changing demands.[5,6]

Compressors in cross-country pipeline compressor stations[7,8] typically use the gas in the pipeline as a fuel. They use large quantities of fuel, which is the saleable product of the company, and the use subtracts directly from profits. Relatively modest improvements in the control system can yield substantial savings.[9] Various optimization strategies have been applied to the compressor stations. Dynamic programming leads to the selection of a procedure that serves this particular set of circumstances.

Computer-based control systems permit expansion of optimization techniques to an entire power house, where the system specifies equipment loading to satisfy steam, electrical, compressed air, and refrigeration demands while minimizing expenditures for coal and purchased electricity. Obviously, the equipment is interactive. Starting a turbine-driven compressor increases the load on the boilers, which in turn increases load on the boiler feed pumps and the generators.[10,11] The system provides optimized on-line selection and loading of a variety of types of industrial power equipment.

10-2. Optimized Surge Control Line Adaption[1]

The factors that cause movement of the surge limit line (SLL) have been discussed. The surge margin must be increased to insure protection from surge. Unit 9 discusses specific power and the concept that optimum operation requires that anti-surge valve opening be minimized. Figure 8-13 shows a surge controller that recognizes shifts in the surge limit line and automatically adapts the surge control line (SCL) to the new condition. Figure 10-1 is an expanded view of the controller in Fig. 8-13. Blocks 7 to 11 of Fig. 10-1 are the recycle trip portion of the controller, which recognizes the approach of a surge condition that the feedback controller (PID) is unable to compensate and opens the anti-surge valve to protect the compressor.

The set point adaption blocks (1 to 6) recognize shifts in the surge limit line and respond by moving the surge control line to the right on the compressor map, or: they increase the secondary bias b_2 in Fig. 8-12. Block 6 is a comparator that determines the relationship of the operating point to the surge control line. Should the operating point cross to the left of the surge control line (h < SP), a timer (block 5) starts and is set for a duration of some 0.5 second. After waiting for the completion of time delay D_2, bias b is increased by Y%, and the additional bias b_1 remains in effect until the reset button is manually pushed.

The adaptive features shown in Fig. 10-1 permit a narrower surge margin, which minimizes anti-surge valve opening and optimizes the operation.

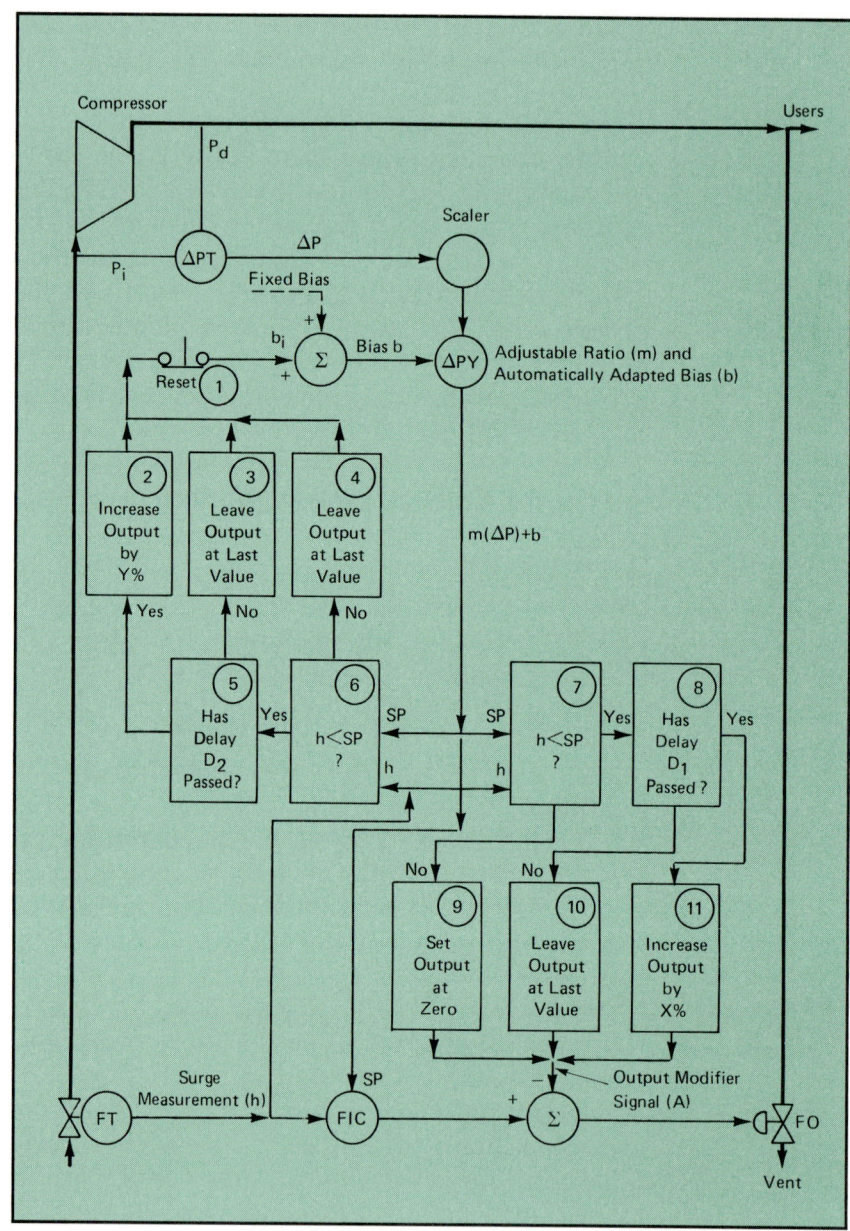

Fig. 10-1. Control System To Recognize Change in Surge Limit Line and To Adapt Surge Control Line (Patent No. 856,302 assigned to N. Staroselsky)

10-3. Compressor Optimization with Multiple Loads[1]

Many compressors serve a multitude of users, each one of which represents a variable demand. The compressor maintains the mass balance on the header by a pressure control system that regulates capacity, as shown in Fig. 10-2.

Each user control valve position is approximated by measuring the signal to that valve. The valve position measurements are transmitted to a high selector relay, which transmits the highest signal (widest open valve) to the position controller (VPC). The position controller has a set point of some 90% of full scale and is cascaded to the set point of the pressure controller (PIC). The control mode of the position controller is integral only (floating), and that mode is adjusted to respond very slowly to avoid interaction with the pressure controller.[3]

Energy optimization is attained by the position controller (VPC), which lowers the pressure set point when the widest open valve beings to close past 90% open. Thus less demand, as shown by the closing valve, requires less header pressure. In this way, the compressor provides the minimum energy

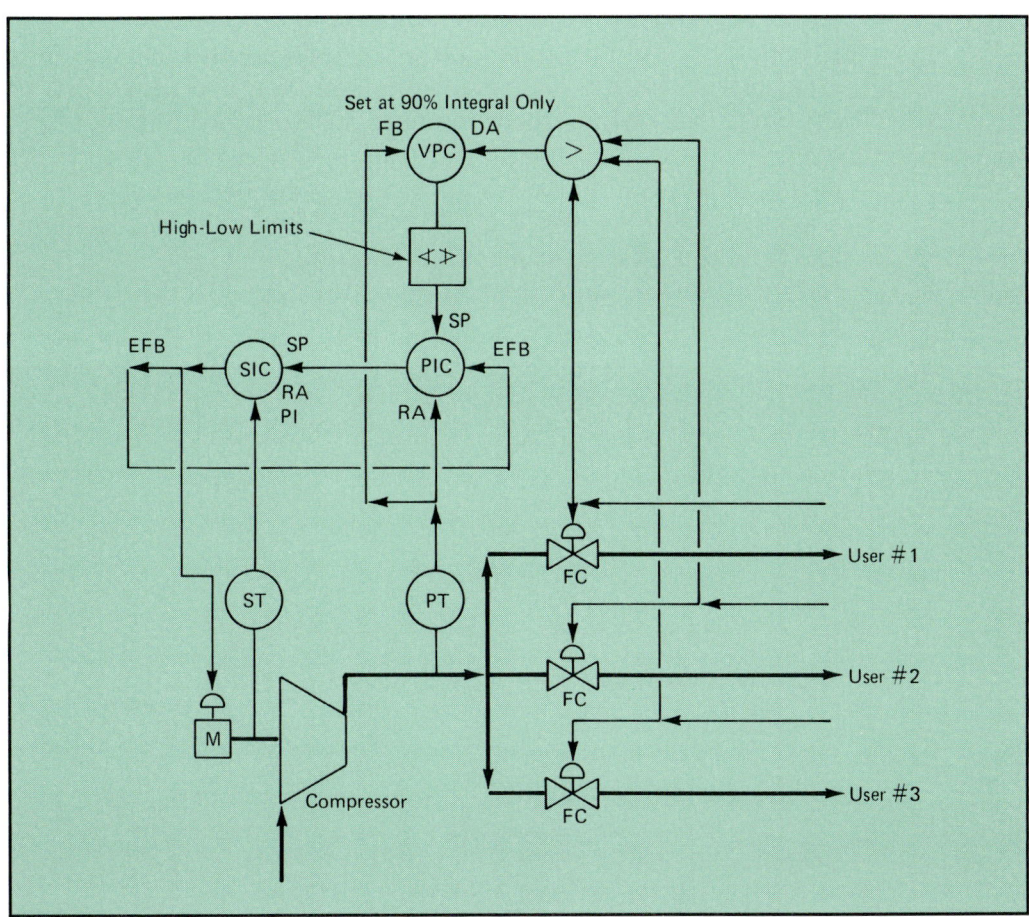

Fig. 10-2. Optimum Load-Following Control System To Meet Demand with Minimum Power Demand

required for the demand. By contrast, the system without the valve position cascade would provide constant header pressure high enough for any contingency, and the unneeded pressure (or power) is "thrown away" in pressure drop across the control valve.

Inlet pressure and motor overload overrides would be added to Fig. 10-2 if required.

10-4. Optimization by Improved Control Response[4]

The compressor map shows nonlinear relationships among the state variables. However, most commercially available controllers are based on the use of linear algorithms. Thus, process gains will change as demand on the compressor changes, and the controller gain (proportional band) will not be adjusted properly for the new condition. Similarly, process time constants are typically a function of volume over throughput, and they also change as demand on the compressor changes. The controller can be tuned near optimum only when operating near design conditions.

A digital self-tuning controller has been applied as the pressure controller (PIC) in Fig. 10-3. The compressor receives gas from four chemical reactors. The inlet flows are independent variables that create disturbances in the process.

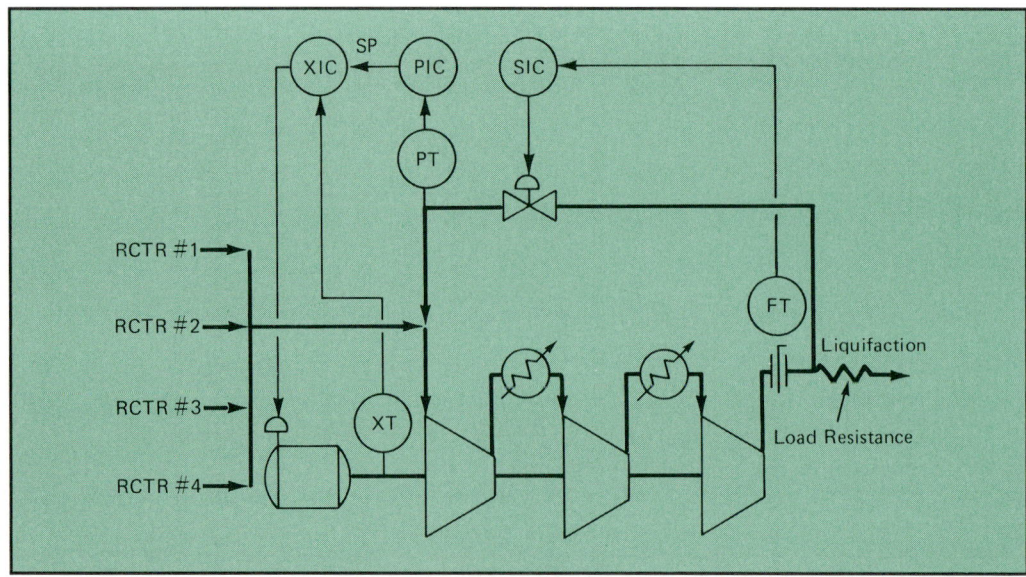

Fig. 10-3. Multistage Compressor with Self-Tuning Pressure Controller

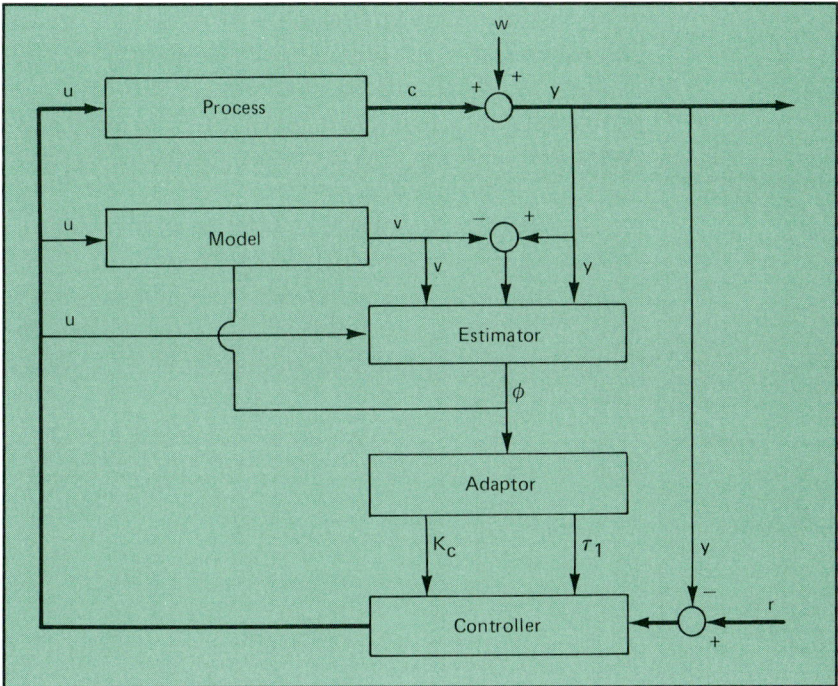

Fig. 10-4. Block Diagram of Adaptive Control System with Estimator

Different compressor throughput flows require different controller settings for optimum performance.[12] The adaptive pressure controller (PIC) is capable of recognizing the changing dynamics and adapting the controller parameters to meet these changes as needed for optimum control of the compressor.[4]

The instrumental variable estimation algorithm is utilized to predict the parameters of a first-order discrete time model. A block diagram of the adaptive control scheme is presented in Fig. 10-4. The algorithm is based on the classical least-squares method and has the ability to track parameter variations using a low-order linear discrete time model. It has the advantages of simplicity, a recursive form, and better performance than the least-squares technique in the presence of correlated noise.[3]

The estimator shown in Fig. 10-4 requires initial values to begin the estimation calculations. Start-up of the estimator is performed in the non-adaptive mode to allow tuning of the estimator without affecting the process. After obtaining the required constants, the estimator is placed in the adaptive mode where the controller parameters (proportional band and reset) are updated at the interval of the estimator sample time. The estimator sample time represents an integer multiple of the control sample time and constitutes the interval of time at which estimation is performed and the model parameters are updated.

The determination of the optimum controller constants for the pressure controller at different throughput flows is shown in Table 10-1. A single value for the reset rate was found to be near optimum for all cases. Table 10-1 shows that a larger controller gain is needed for higher flow rates and that the

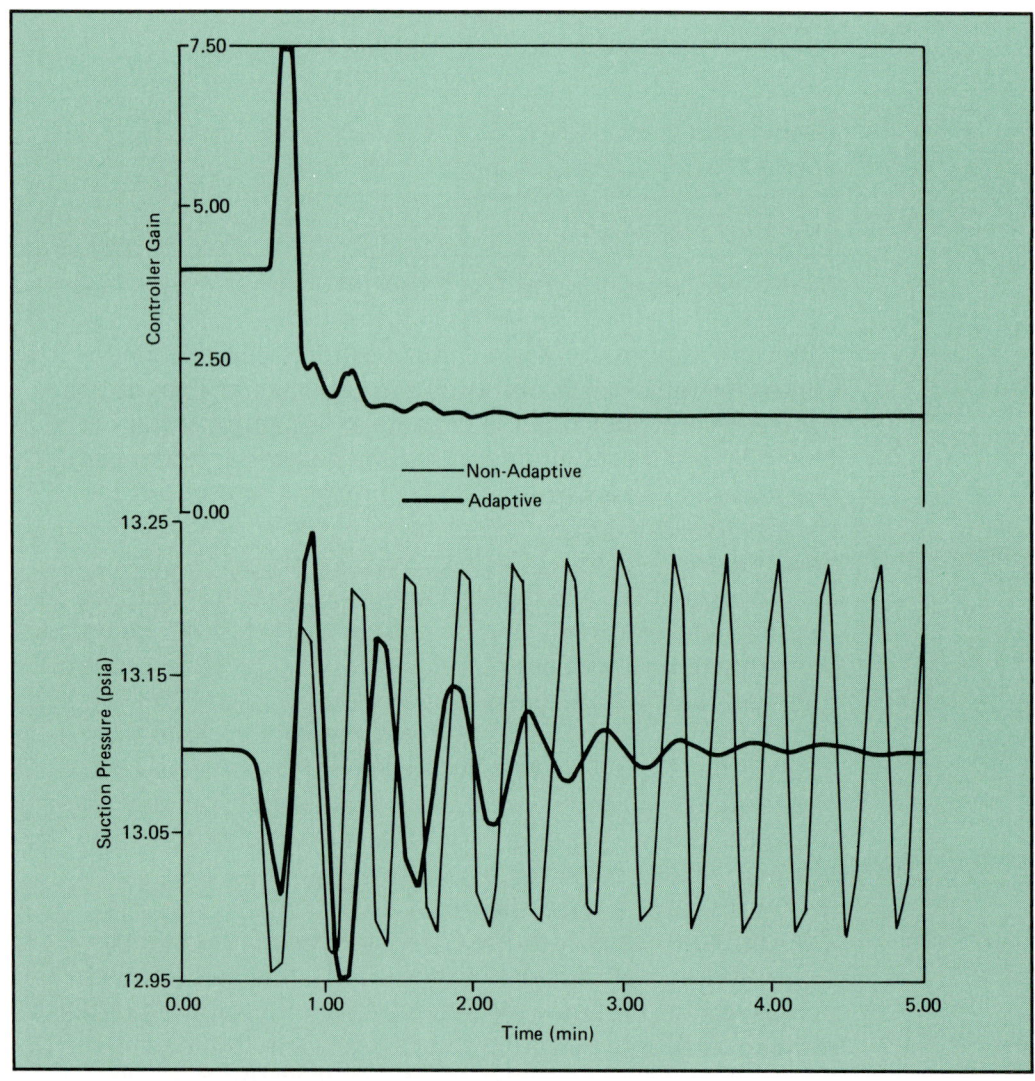

Fig. 10-5. Response of Inlet Pressure and Adaptive Controller Gain to a Change in Throughput Flow from 4 to 2 Reactors in Fig. 10-3

Change in Flow	IAE		Controller Gain	
	Initial*	Final		
# of Reactors	Non-Adaptive	Adaptive	K_c	K_c
4-3	0.035	0.032	4.0	3.09
4-2	0.774 +	0.128	4.0	1.60
4-1	0.527 +	0.163	4.0	1.73
3-4	0.030	0.028	3.0	3.22
2-4	0.093	0.082	1.6	1.85
1-4	0.146	0.104	1.7	1.94
4-1-2	0.769 +	0.150	4.0	1.72

* Same constants used for non-adaptive case.
+ Indicates oscillatory or unstable response.

Table 10-1. Comparative Results for Process To Disturbances[4]

optimized controller gain resulted in significantly smaller integrated absolute errors (IAE). Figure 10-5 shows the stability provided by changing the controller gain with throughput flow.

10-5. Optimization of Compressors with Turboexpanders[5,6]

Compressors are provided with process gas turbines on the same shaft as the driver and the compressor, where gas at high pressure must be reduced in pressure for use in the process. In this way, power is conserved. Expansion is a part of the process cycle in some operations, notably refrigeration, and the expander is commonly applied. The ammonia oxidation process (AOP) in the chemical field also has significant quantities of high-pressure gas that is used at lower pressures, and turboexpanders are traditionally supplied with the large compressors used in that process.

Figure 10-6 shows a typical application of a turboexpander to a compressor in a chemical plant. The speed of the compressor turboexpander is set by the synchronous electrical motor prime mover. Electrical power demand is reduced by the torque supplied by the turboexpander, which reduces motor torque as well as motor slip and motor current demand.

Turboexpander efficiency relates directly to the production of refrigeration, and to the horsepower available to the compressor, in a refrigeration system. It relates directly to horsepower savings in other systems. Efficiency depends on

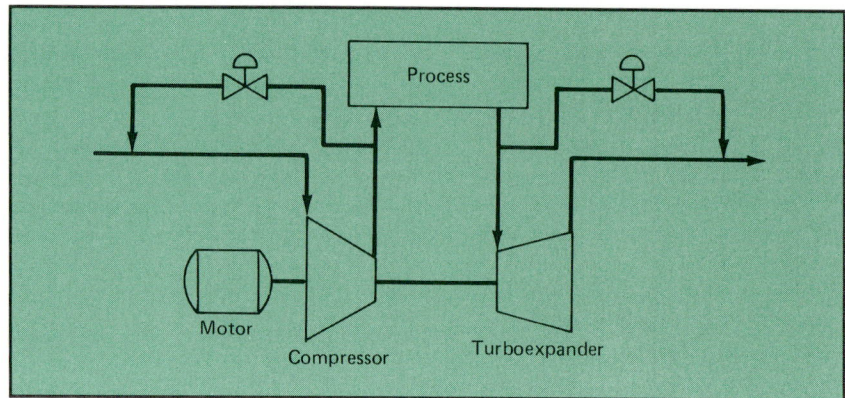

Fig. 10-6. Electrical Motor-Driven Compressor with Turboexpander

mass flow rate, inlet pressure and temperature, discharge pressure, molecular weight, and shaft speed. These state variables are typically combined into the following term:

$$\frac{\text{Tip Speed}}{\text{Spouting Velocity}}$$

The calculation of expander efficiency from field data generally requires the use of a computer thermodynamics program. The expansion process is shown in Fig. 10-7. The computer must calculate very accurately the enthalpy for multicomponent gas mixtures. Then the expander efficiency is[5]

$$\nu_e = \frac{\Delta h_o}{\Delta h'} \qquad (10\text{-}1)$$

where:

Δh_o = actual energy removed, Btu/lb
$\Delta h'$ = ideal amount of energy in the gas stream that could be removed by the expander for a 100% efficiency, Btu/lb

Some expanders are equipped with variable inlet vanes or nozzles that direct the flow into the wheel and control inlet flow and pressure. The volumetric flow rate at the inlet is controlled by varying the area between the vanes. The area is proportional to the following flow parameter, which can be approximately calculated from measured state variables:

$$\text{Flow parameter} = \frac{w}{p}\sqrt{\frac{T}{MW}} \qquad (10\text{-}2)$$

Figure 10-8 shows the effect of the nozzle position (or flow parameter) on the expander efficiency.

Most expanders are equipped with bypass valves, as shown in Fig. 10-6. The valve is used to control the expander inlet pressure. The valve is opened to lower the inlet pressure by allowing part of the gas to flow to the lower pressure at the expander discharge. Equation (10-2) shows that lowering the inlet pressure increases the flow parameter. Figure 10-8 indicates that increasing the flow parameter increases efficiency.

The most desirable operating strategy is to adjust the inlet nozzles to a greater area and close the bypass valve. This strategy has the primary benefit of minimizing the power loss represented by the pressure drop across the control valve and

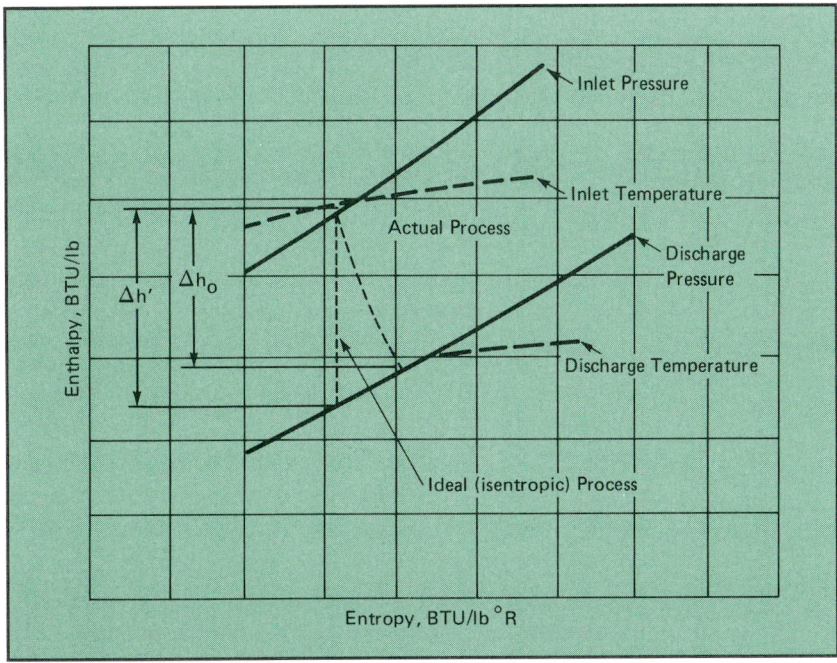

Fig. 10-7. Thermodynamic Diagram of Gas Expansion Showing an Ideal Reversible Process and a Real Irreversible Process

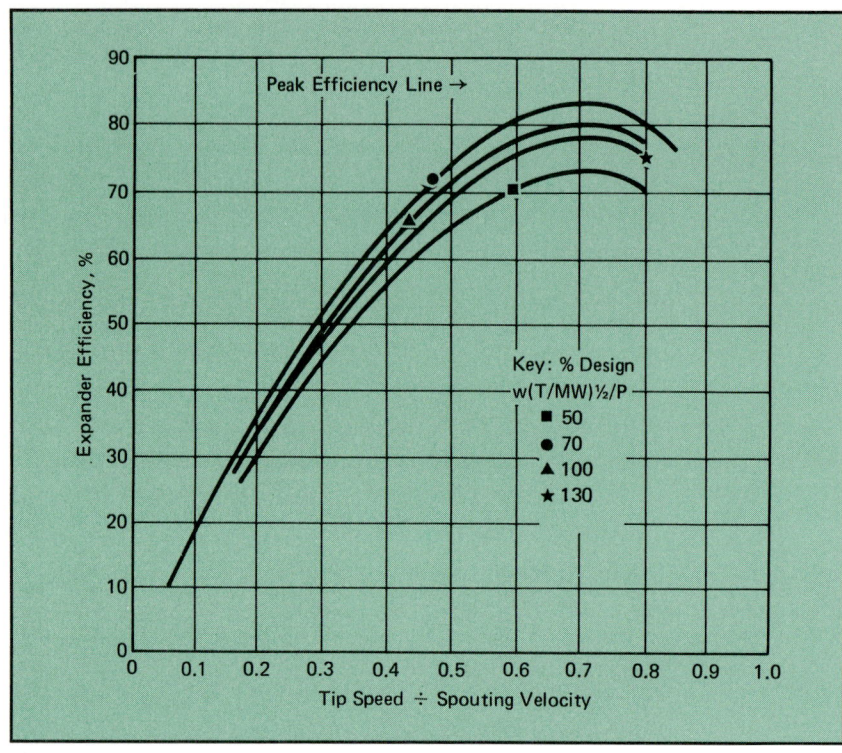

Fig. 10-8. Effect of Nozzle Movement on Expander Efficiency

a secondary benefit of speeding up the compressor-expander unit, making it operate more efficiently.

Other unit operations in the process also contribute to efficiency. For example, consider the heat exchangers in Fig. 10-3. Their performance affects overall system efficiency. Any increase in pressure drop across a heat exchanger is equivalent to lost efficiency in the expander or compressor.

The optimizing program will consider system performance in terms of the variable q/s. This variable, as formerly discussed, comes from the Fan Law, and reduces the curves on the compressor map to single lines. It is also approximately proportional to the x-axis variable of Fig. 10-8. Thus, it can be simply programmed to describe the efficiency of both compressor and expander and is easily measured. This information is combined with the openings of the anti-surge and bypass valves and the heat exchanger fouling to calculate the optimum operating point to meet demand.

10-6. Optimization of Multiple Compressors[1]

Units 8 and 9 have described the S-Criterion concept as a technique for causing multiple compressors to operate equidistant from their surge lines. Thus the compressors are optimized in the sense that no compressor will open its anti-surge valve while another one is overloaded. Figure 10-9 shows an alternative optimization system that is applicable when demand will be sufficiently constant so that a compressor can be shut down in the infrequent event of a change in demand.

The optimization system shown in Fig. 10-9 loads the compressors shown in the order of their efficiencies. Load shifting is accomplished by adjusting the ratio settings of FRIC-01 and 02. The system can be computer-optimized by calculating compressor efficiencies in terms of flow per unit power and having the computer load the compressors.

The pressure controller (PIC-01) in Fig. 10-9 directly fixes the set point of the speed controllers (SIC-01 and SIC-02). The balancing controllers (FRIC-01 and FRIC-02) slowly bias the speed set points to drive each compressor toward maximum efficiency for the measured total load. Total flow demand (load) is computed by block 9 and transmitted to the ratio controllers (FRIC) that regulate to the adjusted ratio. Surge controls are not shown for simplicity.

A compressor is automatically shut down when one of the anti-surge valves is open to a predetermined stem position, as sensed by surge limit delay FSL-03. Likewise, an additional compressor is started up when relay PSH-04 senses that one of the compressors has reached full speed. The output of the pressure controller (PIC-01) is corrected as compressors are started or stopped. The summed speed signals (block 2) and the controller output are combined in the high-speed integrator. Equal inputs will result in a constant integrator output. Thus, the shutting down of one compressor will cause the summed speed signal to be less negative, and the integrator output will increase to speed up the remaining compressors. The integrator responds in a fraction of a second to avoid degrading the response of the pressure control system.

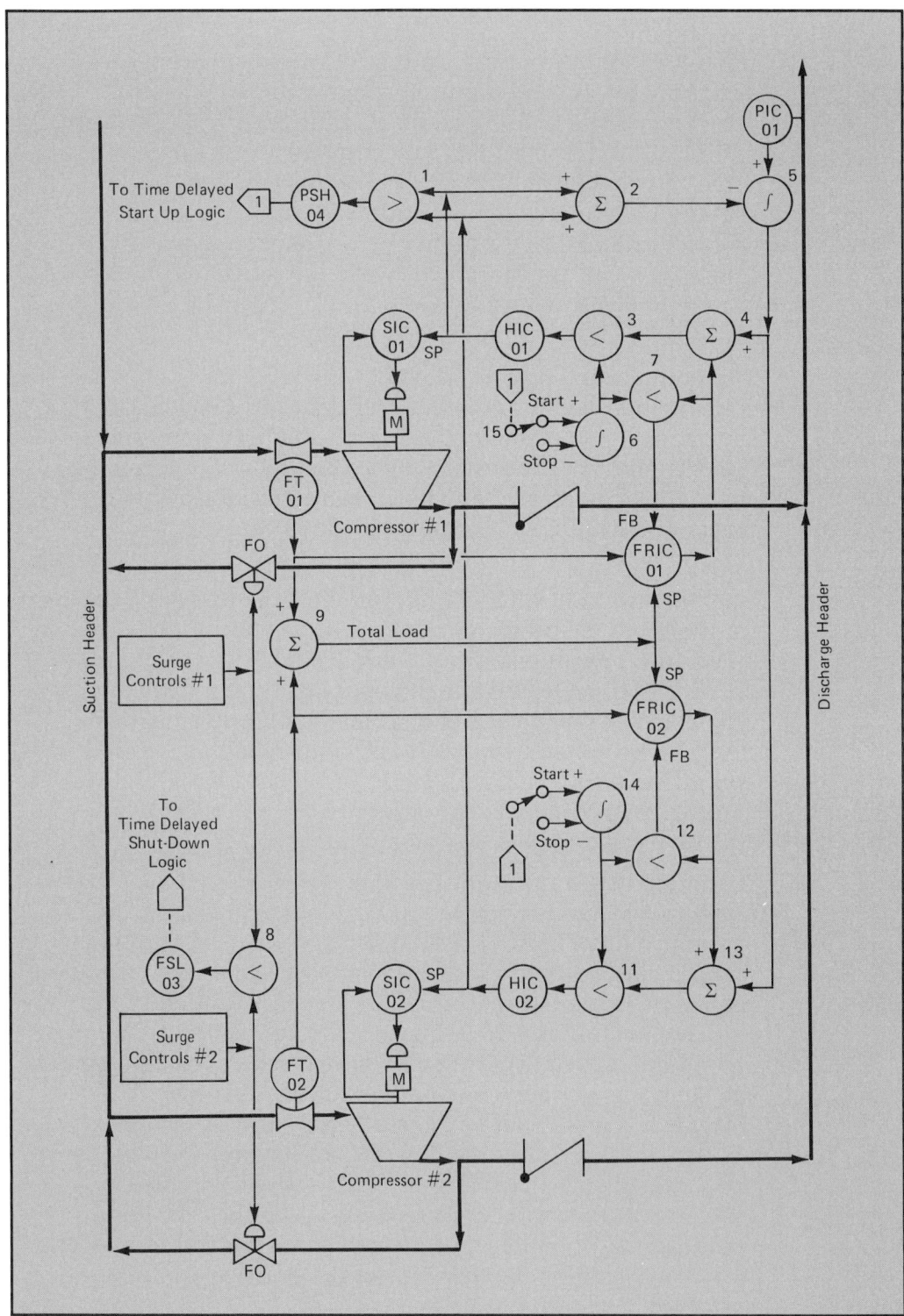

Fig. 10-9. Efficiency Optimization of Compressors in Parallel

10-7. Optimization of Pumps[1]

The operation of fluid processing plants involves the transportation of liquids. Centrifugal pumps (liquids) and centrifugal compressors (gases) have been shown to have similar operating characteristics, so the foregoing efficiency optimization configuration (Sections 10-3 and 10-6) could be readily adapted to pumps and pumping stations.

Energy conservation control configurations emphasize minimizing pressure drop across control valves. Throttled control valves waste power. The following rule of thumb can be used to calculate the cost of power lost in a control valve:

$$\text{Cost of power loss} = 0.5 \, \frac{\$/\text{year}}{\text{gpm} - \text{psi}} \qquad (10\text{-}3)$$

The cost of the power lost in control valves is significant. Figure 10-10 shows a configuration that optimizes power

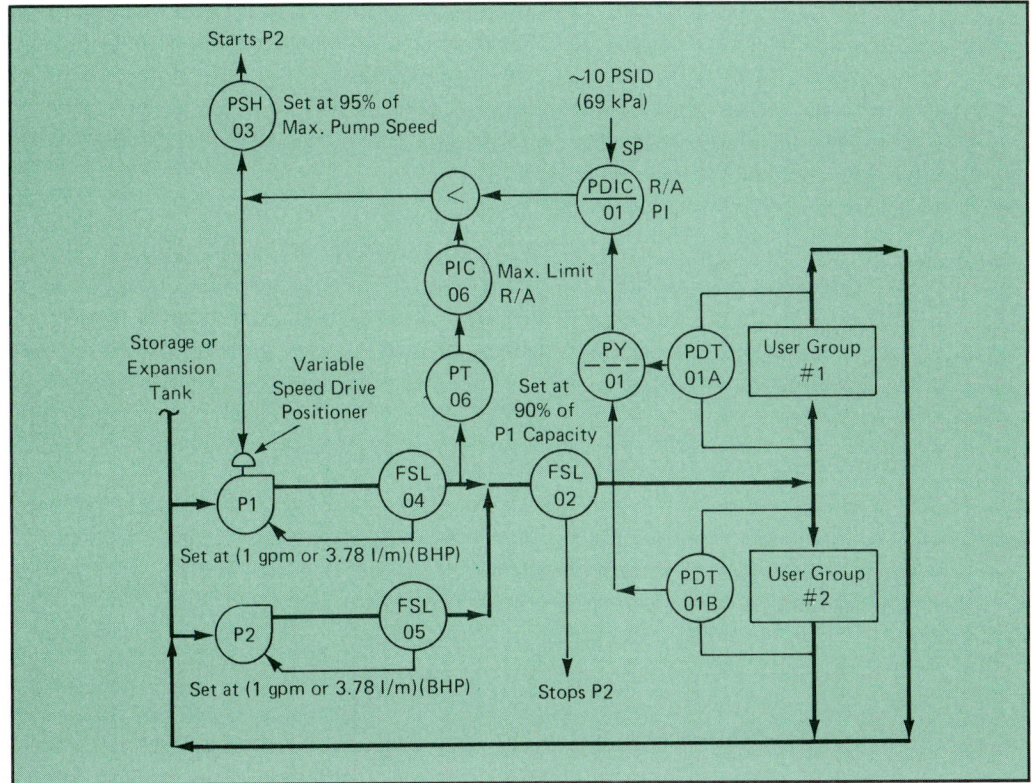

Fig. 10-10. Pump Station Optimizing Control System

usage by eliminating control valves from the system. This system meets the process demand for liquid at minimum pumping cost. It also minimizes installation costs by providing one variable speed pump (very expensive) and one constant speed pump.

The pressure differential controller (PDIC-01) maintains a minimum pressure difference of 10 psi between supply and return from each user. Flow demand by each user varies. Should the user control valve open to meet the demand of more fluid, the differential pressure decreases, and the pressure controller (PDIC-01) speeds up the pump (P1) to provide more flow. Similarly, a decrease in demand closes the control valve, the differential pressure increases, and the pressure control system reduces pump speed.

The limit switches shown by Fig. 10-10 are also a part of the automated system. As the variable speed pump approaches maximum speed, pressure switch PSH-03 starts another pump. Upon decreasing flow demand load, the flow switch shuts off one of the pumps. Low pump throughput, which can cause pump overheating or cavitation, is detected by flow switches FSL-04 and 05, which stop the pumps. The pumps are protected from high discharge pressure by the override actuated by pressure controller PIC-06.

While the use of variable speed pumps to response control valves is not the most efficient installation in every case because of the cost and inefficiency of the variable speed drive, operating costs will be reduced if the system is largely friction and the yearly average loading is less than 80 percent of pump capacity.

10-8. Optimization by Pressure Modulation[13]

Optimization concepts are in sharp contrast to the common compressor control practices of operating at the most stringent pressure or temperature requirements (even if these requirements are seldom encountered) and the use of manual adjustment of compressor operational parameters (based largely on operator availability, judgment or experience). Optimization strategies provide significant energy use savings and peak demand reductions (on the order of 20 to 40%) for a wide range of electrically driven compressor systems. One strategy is the real-time modulation of compressor discharge and suction pressures as required to meet instantaneous operation requirements.[13]

The magnitude of the electrical energy savings and peak demand reductions provided by compressor capacity control modulation techniques are dependent on three primary application considerations: the size (or capacity) of the compressor system; the magnitude and duration of load fluctuations; and the need for year-round operation. In large compressor systems, the magnitude of compressor system load variations (indicated by discharge pressure or capacity requirements) is a primary indicator of the potential for optimization controls. Large load variations represent potential savings by applying suction pressure modulation techniques to achieve energy savings and peak demand reductions. Minimal load variations offer little potential for savings. Table 10-2 shows some of the potential savings.

Figure 10-11 is a diagram of the air conditioning system in a large office building complex. The system consumes 9.9 billion kWh/year for operation. Data is required on the compressors, the cooling system equipment, building occupancy, equipment operational hours, solar heat gain, and

Facility Type	Application Addressed	Compressor System Size	Number of Sites	Energy Savings (1)	Peak Demand Reductions (1)	Projected Cost Savings (2)
Office buildings and office complexes	Commercial space conditioning	164 to 780 horsepower	3	16 to 22%	10 to 20%	$5,000 to $70,000/yr.
Large tire fabric manufacturer	Industrial space conditioning	2180 horsepower	1	16%	Not Calculated	$9,000/yr.
Food processors: - Soft drink bottler - Cheese processor - Bakery - Meat packing	Process cooling and refrigeration	300 to 3100 horsepower	5	10 to 34%	9 to 28%	$3,500 to $235,000/yr.
Cold storage warehousing	Refrigeration	825 horsepower	1	46%	41%	$35,800/yr.
Manufacturers: - Plastics - Glass - Steel	Industrial compressed air	500 to 12,750 horsepower	3	3 to 15%	2 to 10%	$8,000 to $129,000/yr.
Electrical utility boiler soot blowing	Industrial compressed air	11,060 horsepower	1	30%	Approx. 30%	$448,000/yr.

(1) Energy savings and peak demand reductions shown are for compressors only. They do not include energy savings for auxiliaries (pumps and fans.)
(2) Projected cost savings were calculated based upon actual energy and peak demand charges in effect at time of study.

Table 10-2. Typical Energy Savings and Peak Demand Reductions Available from Compressor Capacity Modulation Control Techniques[13]

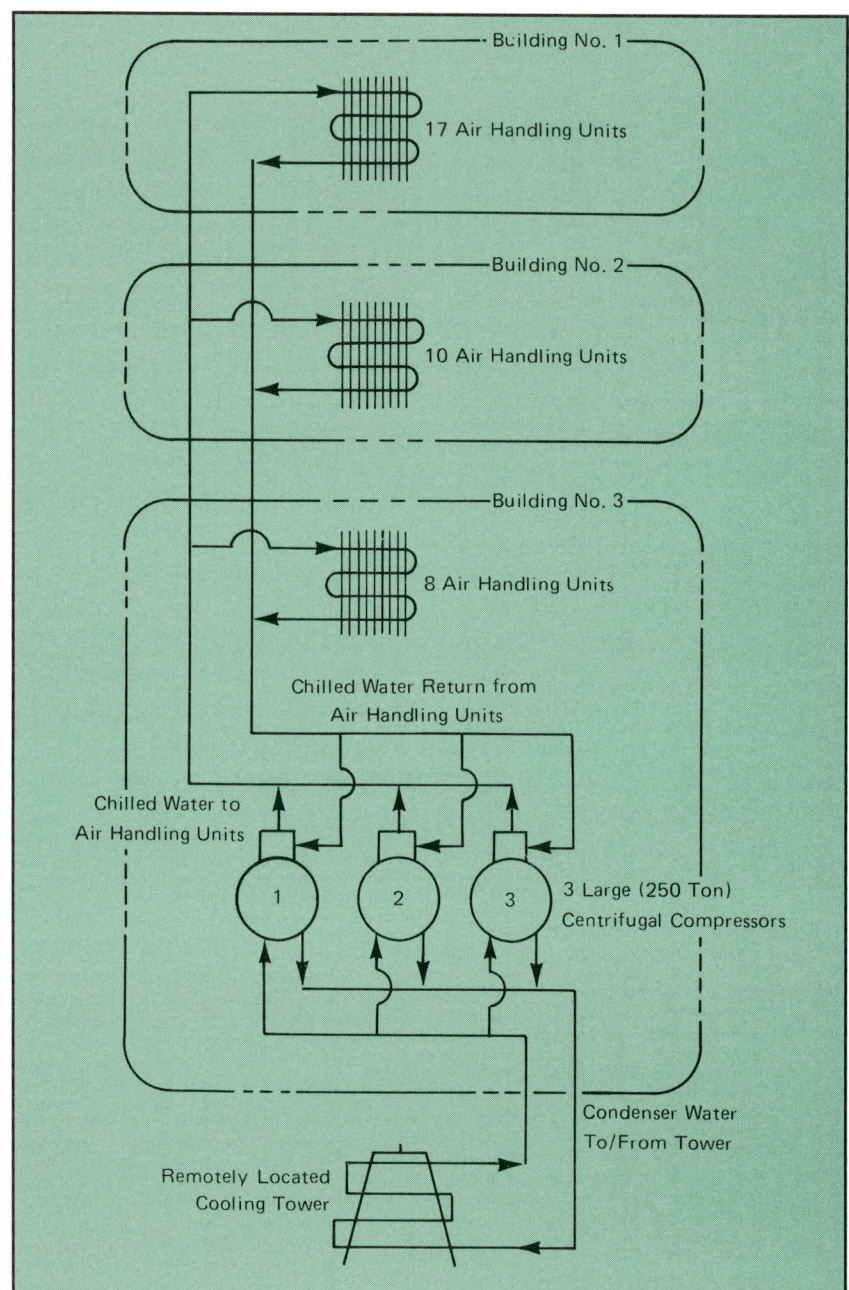

Fig. 10-11. Space Conditioning System in a Large Office Building Complex

other factors determining the demand of the cooling system. Compressor suction pressure is controlled by regulating compressor throughput. Suction pressure set point is determined as the pressure of the saturated refrigerant that will produce the warmest value of the chilled water supply temperature to meet the cooling requirements. Discharge pressure is regulated to maintain a constant approach of the cooling tower, leaving water temperature to the ambient wet bulb temperature. Savings include 380,000 kWh/year for the suction pressure modulation and 770,000 kWh/year for discharge pressure modulation.[13]

10-9. Optimization of Pipeline Compressors[9]

Compressors are used to force flow through transcontinental natural gas pipelines. Since the lines are hundreds of miles long, substantial power is required to overcome pipeline friction. Load allocation or other criteria independent of fuel consumption will generally determine the compressors on-line at any time. Therefore, fuel consumption is expressed as a function of flow for each compressor.

The optimization program allocates flow through parallel or head among tandem compressors in operation. The allocation should yield minimum fuel usage within the constraints on total throughput and operating limits on such compressor parameters as speed, torque, and surge flow. The program starts with initial flow share estimates that satisfy any constraints, then computes the actual flows and the corresponding fuel consumptions. The flow shares are then varied in the direction that lowers total fuel consumption. The procedure is then iterated until no changes in fuel consumption can be found that yield further reductions. Dynamic programming lends itself well to the iterative procedure.[14] The strategy for centrifugal compressors would be as follows:

- Select a flow ratio.
- Compute the head across the unit under current suction and discharge pressure, temperature, and compressibility conditions.
- Determine the speed required to achieve the computed head.
- If speed is outside the limits, select a new ratio and repeat the calculation.
- Compute the flow.
- If the computed flow is smaller than the surge limit, select a new ratio and repeat the procedure.
- Compute the horsepower as a function of flow.
- Compute the fuel consumption necessary to achieve this horsepower.

Optimization calculations involve mathematical models. Basic data and initial parameters are typically obtained from manufacturers, as formerly discussed. The model of the compressor is modified periodically using measurements of the various parameters. The flow shares demanded by the optimization procedure may be achieved by adjusting pressure, flow, or temperature set points. Control is typically achieved by compressor speed control.

10-10. Power House Optimization[10,11]

Compressors used to provide plant air and instrument air on an operating plant site are commonly clustered in an area of the power house sometimes called the engine room. The compressors are a part of a much larger conglomerate of equipment that can be optimized to satisfy steam, electrical, compressed air, and refrigeration demands while spending minimum total dollars for fuel and purchased electricity. Fuel can include the choice of waste chemicals or hydrocarbons from the plant, and gas, fuel oil or coal. Optimum compressor performance does not necessarily yield optimum power house performance. Figure 10-12 is a diagram of a typical plant energy system.[11]

The compressors provide 100-psi air by any combination of three turbine-driven compressors. Air at 210 psi can be produced by boosting 100-psi air and/or running any combination of two electrical motor-driven and one turbine-driven compressors.

Optimization is attained by linear programming concepts combined with mixed integer programming extension on a computer. Linear programming is a mathematical model made up of linear constraints describing the equipment piece, e.g., the compressor. An "objective function" is a mathematical statement of the optimization criterion. The mixed integer extension enables consideration of the on/off state of equipment, i.e., an integer independent variable of "one" means that a compressor is running; "zero" means that it is shut down. With this programming, the computer:

- continuously receives information on flow, kilowatts, pressures, temperatures, and other variables from sensors in the process.

- calculates input energy versus output for each piece of equipment.
- receives current fuel costs and equipment availability from the operator.
- displays to the operator the output of each piece of equipment that will satisfy all demands while resulting in the lowest total cost of fuel and purchased electricity. This is a "local" optimum.
- performs "global" optimum calculation described above. Informs operators of new equipment line-up when reduced purchased energy costs justify starting up and shutting down necessary equipment.
- compares actual performance to standard performance for each piece of equipment, determines reason for substandard performance, and provides operating and maintenance advice.
- automatically makes small adjustments in loading boilers, refrigeration machines, and compressors to meet load demands.
- validates data by closing mass and energy balances.

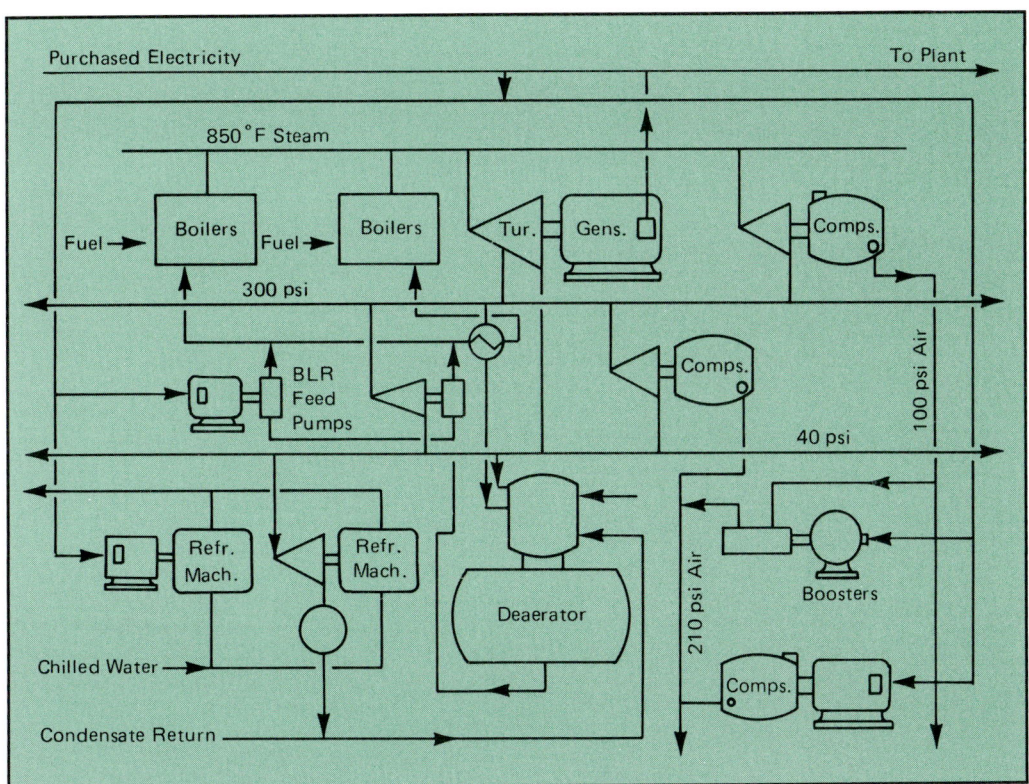

Fig. 10-12. Typical Plant Energy System

An energy system that contains combinations of two or more devices that deliver the same service can be loaded in a unique way to satisfy the demand for least cost. Linear plus mixed integer programming provides a calculation of the unique loading required.

Exercises

1. What would happen, in Fig. 10-2, if the widest open valve were to go past 90% open?

2. If lowering inlet pressure on an expander increases the flow parameter (eq. (10-2)) and increasing the flow parameter increases efficiency (Fig. 10-8), why not always operate with the bypass valve open?

3. A control valve in a liquid line passes 600 gpm and reduces pressure from 90 psig to 40 psig. What is the annual cost of operation?

4. Describe the action of the high discharge pressure override (PIC-06) shown in Fig. 10-10.

5. Why is the pressure modulation technique (Section 10-9) useful only for (a) large installations, and (b) installations that have large changes in demand?

References

1. Liptak, B. G., *Optimization of Unit Operations*, Chilton Book Co., Radnor, PA., 1987.
2. Bierenbaum, H. S., C. R. Zielke, and I. L. Harry, "Use of Floating Pressure Setpoint Capacity Modulation Controls for Improving of Compressor End-Use Efficiency," Instrument Society of America, Paper No. 85-0673, International Conference and Exhibit, Philadelphia, PA., 1985.
3. Moore, R. L., *The Dynamic Analysis of Automatic Process Control*, Instrument Society of America, Research Triangle Park, NC, 1985.
4. Usher, J. M. and A. B. Corripio, "Self-tuning Control of an Industrial Centrifugal Compressor," Instrument Society of America, Paper No. 85-0722, International Conference and Exhibit, Philadelphia, PA, 1985.
5. McIntire, R., "Performance Monitoring Can Boost Turboexpander Efficiency," *Oil and Gas Journal*, July 5, 1982, p. 95.
6. McIntire, R., "Examples show System Performance Upgrade," *Oil and Gas Journal*, July 12, 1982, p. 125.
7. Bozeman, H. C. "Analog Computer Prevents Surge in Centrifugal Compressor," *Oil and Gas Journal*, December 1, 1964, p. 87.
8. Hatton, C., "Controlling a Gas Compressor Station Electronically," *Instruments and Control Systems*, May, 1967, p. 145.

9. Baqui, A., "Optimizing Compressor Operation with Dynamic Programming," *InTech*, September, 1982, p. 105.
10. Robnett, J. D., "Computer-Based Power Advisory and Control System," ASME Paper No. 81-IPC-PWR-2, 1981.
11. Robnett, J. D., "Control and Instrument Features of An Energy Optimization System," Instrument Society of America Paper No. C.I.82-813, Annual Conference and Exhibit, Philadelphia, PA., 1982.
12. Davis, F. T., and A. B. Corripio, "Dynamic Simulation of Variable Speed Centrifugal Compressors," *Instrumentation in the Chemical and Petroleum Industries*, Volume 10, Instrument Society of America, Research Triangle Park, NC, 1974.
13. Bierenbaum, H. S., C. R. Zielke, and I. L. Harry, "Use of Floating Pressure Setpoint Capacity Modulation Controls for Improving of Compressor End-Use Efficiency," Instrument Society of America, Paper No. 85-0673, International Conference and Exhibit, Philadelphia, PA., 1985.
14. Nemhauser, G. L., *Introduction to Dynamic Programming*, John Wiley & Sons, New York, NY., 1966.

Appendix A: Nomenclature

Appendix A
Nomenclature

- A — Area, sq ft
- A — Torque, ft-lb
- C_p — Specific heat at constant pressure, ft lb/lb °R
- C_v — Specific heat at constant volume, ft lb/lb °R
- h — Enthalpy, ft lb/min
- H — Head, feet
- hp — Horsepower
- i — Amperage, coulombs/sec
- k — Ratio of specific heats, C_v/C_p, dimensionless
- kW — Kilowattage
- K — Constant, any required dimension
- K_r — Radius of gyration, ft
- m — Mass flow, lb/hr
- mV — Millivoltage
- M — Molecular weight, lb/lb-mol
- p — Pressure, psi
- q — Volumetric flow, cfm
- Q — Heat transfer, ft lb/min
- r — Radius, ft
- R — Gas constant, 1545/M ft-lb/lb-°R
- R_c — Compression ratio, dimensionless
- s — Speed, rpm
- S — Entropy, ft lb/hr-°R
- T — Temperature, absolute, °R
- u — Internal energy
- v — Specific volume, cu ft/lb
- V — Volume, cu ft
- W — Work, ft lb/lb
- Z — Compressibility, dimensionless
- Δp — Pressure rise across the compressor, psi
- ν — Velocity, ft/sec
- ρ — Density, lb/cu ft
- ϕ — Angle of torsional displacement, radians
- ω — Rotational speed, rad/sec (equals 0.105s)

Appendix B: Suggested Readings and Study Materials

Appendix B
Suggested Readings and Study Materials

In addition to the references listed at the end of each chapter, the following are also recommended:

"Centrifugal Operation and Control," Tutorial papers presented by the Education Committee, Oct. 11–14, 1976, at Houston, TX, Instrument Society of America

Liptak, Bela G., and Vencel, Kriszta, eds., *Instrument Engineers' Handbook*, Revised Edition, Chapters 5 and 8 (Chilton Book Co., Radnor, PA, 1985).

Spitzer, David W., *The Application of Variable Speed Drives* (Instrument Society of America, 1987).

Periodicals

Chemical Engineering, McGraw-Hill, New York, NY.

Chemical Engineering Progress, American Institute of Chemical Engineers, New York, NY.

Compressed Air, Ingersoll Rand Co., Phillipsburg, NJ.

Control Engineering, Dun-Donnelly Publishing Corp., New York, NY.

Hydrocarbon Processing, Gulf Publishing Co., Houston, TX.

InTech, Instrument Society of America, Research Triangle Park, NC.

ISA Transactions, Instrument Society of America, Research Triangle Park, NC.

Oil and Gas Journal, Penwell Publishing Co., Tulsa, OK.

Pipeline and Gas Journal, Harcourt, Brace, Jovanovich, Orlando, FL.

ISA Publications

The Instrument Society of America is a technical, application-oriented society. Its primary goal is providing educational materials and services to its members. If you are seriously interested in the study of process control, you will find membership in the ISA invaluable.

Independent Learning Modules:

One of your best sources of material for further reading and study of process control and instrumentation are the ILMs published by ISA. They are custom designed and created for this exact purpose. Place a *standing purchase order* to receive new ILMs as they are published.

Appendix C:
Solutions to All Exercises

Appendix C
Solutions to All Exercises

Unit 2

1. Pumping is the addition of energy to a fluid.

2. Pumps pump liquids, compressors pump compressible fluids.

3.

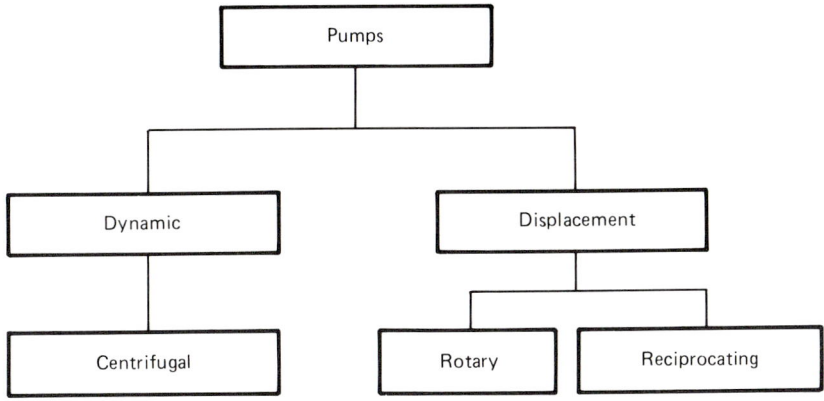

4. $q = K_1 n \quad H = K_2 n^2 \quad hp = K_3 n^3$

5. $p(lb/in.^2) = \dfrac{H(ft)\rho(lb/ft^3)}{144(in.^2/ft^2)}$

6. 40 gpm at 40-ft head

7. The narrow range of pump head would make proper pump choice virtually impossible, because field conditions always vary from design conditions.

8. (1) Inadequate start-up head, (2) Instability, (3) Unequal division of flow.

Unit 3

1. $n = 1.3$

2. 1245 ft-lb/lb

3. 795°F

4. Increase

5. Discharge pressure increases; driving force across the load increases; flow increases; and discharge pressure decreases.

6. The velocity of the fluid flowing through the compressor reaches a Mach number of one somewhere in the machine.

7. As load line OA' moves to OA" due to increasing system resistance, differential pressure must increase to maintain foward flow. It cannot increase past discharge pressure p_{max}, and in fact decreases according to the characteristic. Thus flow will reverse while pressure stays at the maximum.

8. Exactly

9. (1) The "stonewall" point is not identified, and thus compressor rangeability is unknown, and (2) the low flow surge curve is not specified for designing protection at start-up and shutdown.

Unit 4

1. 2 minutes

2. 1 minute

3. Turbines and reciprocating engines are inherently variable speed, while electrical motors must have expensive equipment added to make them variable speed.

4. Advantages include speed of response, simplicity, reliability, and the ability to transmit power; disadvantages include the need to protect oil from dirt and water contamination, and leakage.

5. The danger is that it cannot be started because the high-temperature interlock will shut off power to the motor before it reaches operating speed.

Appendix C: Solutions to All Exercises

6. The danger is that the compressor will stop before the discharge volume is exhausted, and the residual pressure will cause the compressor to run backwards.

7. They are inherently variable speed devices.

8. The seal oil system provides high-pressure oil to balance the pressure of the compressor lube oil and prevent it from leaking.

Unit 5

1. a. Simplicity. Loss of horsepower.
   b. Efficiency. Complexity.

2. Diffuser guide vanes provide a wider span of vane operation for the flow range.

3. No. A 2% change in speed causes a (92% minus 40%) 52% change in throughput. The performance curve is too flat, and the control would be too sensitive.

4. Since the compressor responds to actual inlet volumetric flow, the ideal location for the flow element is between valve and compressor, where that flow occurs.

5. Disadvantages include (a) running at higher pressure, closer to surge, to provide pressure drop, (b) inefficiency, where the pressure drop of denser gas represents more horsepower, and (c) the flow element should be corrected to inlet conditions to be meaningful.

6. A static plus friction system.

7. Friction or kinetic.

8. Figure 5-13 shows a typical surge line. Figure 5-7 has a different surge line for each performance curve, and the "surge" line shown is the locus of the points where the respective surge lines cross their performance curves.

Unit 6

1. Surge will (a) extensively damage the compressor, and (b) cause the compressor to lose capacity, representing a possible hazard to downstream equipment.

2. Anti-surge control (venting, recycling, operating at off-specification conditions) all represent inefficiencies, and incur economic penalties.

3. Venting is always used for air. Other gases and vapors are not vented because (a) they are too valuable to throw away, and (b) toxic gases must be contained.

4. Limitations are inherent in the assumptions that (a) Fan Laws can be applied to a high-pressure compressure and (b) the pressure ratio term can be linearized.

5. Rangeability and energy savings.

6. No. Pressure and temperature are compensated for in the control configuration itself.

Unit 7

1. The anti-surge valve is closed, and the surge control system is not responding to the flow through the compressor.

2. Increased pressure upstream of the anti-surge control valve results in more flow through the valve.

3. $o_i = \sum_{j=1}^{m} P_{ij} m_j \quad i = 1, 2, \ldots, m$

4. With the anti-surge valve closed, the surge control system is open loop and cannot interact with the pressure control system.

5. $\lambda_{w_1, x_1} = 1.01$

6. $\lambda_{w_1, x_1} = 0.72$. Minimize the inlet pressure drop to reduce interaction and to minimize energy consumption.

7. The RGA terms are linearized approximations (slopes) of nonlinear functions. As operation takes place at different points on the nonlinear curve, the slope (gain) changes.

Unit 8

1. Surge is evident by its sound, a decreased forward flow through the compressor, and the location of the operating point on the compressor map.

2. 0.25 sec. From Fig. 3-11, 0.5 sec. Close agreement.

3. Large and fast disturbances.

4. While some savings accrue from deletion of the communications link and configuring the controllers, the major savings is in the anti-surge control valve. These valves are large, low-noise types that cost in the neighborhood of $25,000.

5. Upon start-up, the latter compressor stages have not yet reached operating pressure, and thus are not carrying away adequate volumetric flow from the first stage. Because of this the first stage tends to surge. No means exist for increasing the recycle around the first stage during the transient period (see Fig. 6-18).

6.

Time	Action
H	Turn off electrical motor
H	Anti-surge valve full open
H–J	Pressure controller ramps down at preselected rate
I	Check valve closes
J	Inlet valve at idle position
K	Compressor stopped

Unit 9

1. The booster compressor is recycled to save energy.

2. Increase the speed of compressor B.

3. Process conditions force installation of the check valve at a distance from the compressor. The pressure of the fluid trapped in the discharge piping upon an emergency shutdown decays more slowly than the decrease in speed of the compressor. Flow tends to take place from the high discharge pressure through the compressor.

4.

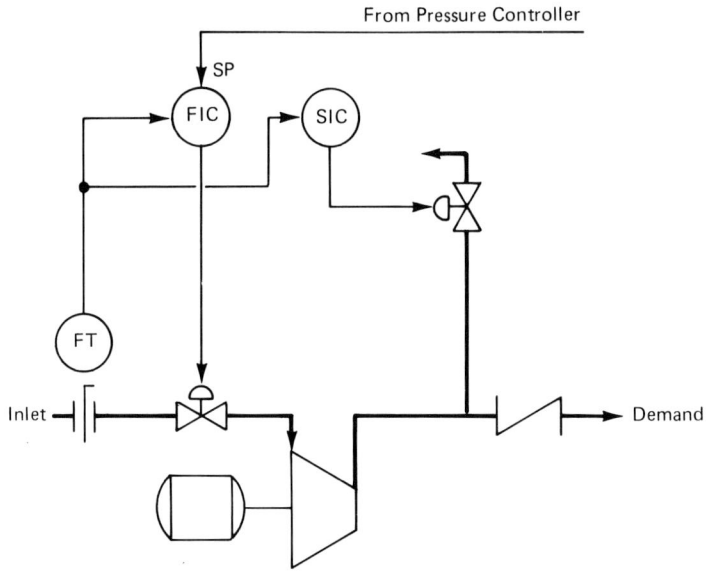

5.

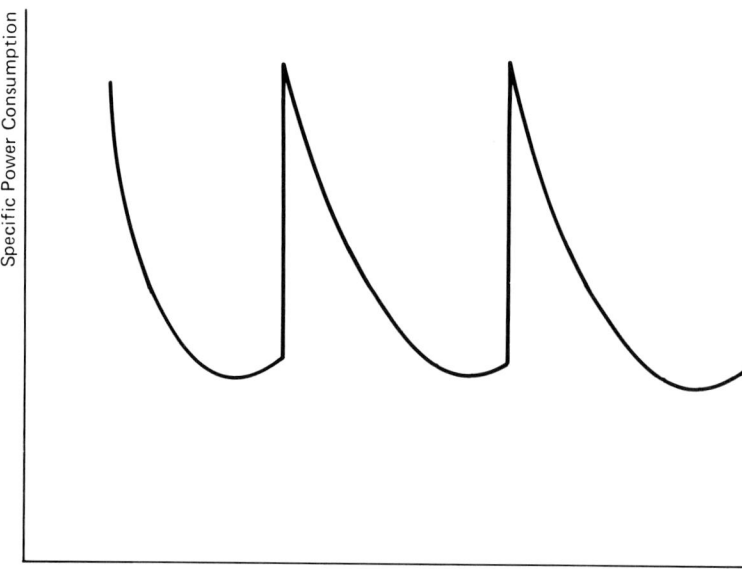

6. Shutting down compressor 2 at discharge pressure point b would cause compressor 1 to open its anti-surge valve to maintain flow down to 50% capacity so that compressor 2 could be shut down.

Unit 10

1. The position controller (VPC) would raise the set point of the pressure controller (PIC) and thus raise the header pressure. This would raise the user valve inlet pressure, causing the valve to close while meeting the same flow demand.

2. Gas flowing through the bypass valve undergoes an isenthalpic expansion, and no energy is removed from the gas stream. Thus, a 10% flow through the bypass valve represents a 10% drop in expander efficiency. This loss must be balanced against the increase in expander operating efficiency.

3. Cost of operation $= \left[0.5 \dfrac{\$/yr}{gpm - psi} \right]$ [600 gpm] [40 psi]

   $= \$12,000$ per year

4. Controller (PID-06) output is at maximum while the pump discharge pressure is less than the controller set point. As discharge pressure rises toward the set point, controller output decreases until it is transmitted by the low selector relay to the variable speed drive. The decreasing signal decreases speed, causing the discharge pressure to stabilize at the controller set point.

5. a. Savings from optimization are a percentage of operating cost. Thus significant costs must be incurred to accumulate worthwhile savings.
   b. Compressors that serve a constant demand can be adjusted to optimum operation manually and left there. No optimization control is required.

Index

Index

Affinity Law (see Fan Law)
Anti-surge control 56, 123, 198
Anti-surge valve 155
   Sizing 156
Axial compressor 184

Block diagrams 170
Blowers 94

Capacity 51
Carnot Cycle 38
Casings, pump
   Diffuser 11
   Turbine 11
   Volute 11
Cavitation 20
Compression
   Adiabatic 32, 35
   Isentropic 35
   Isothermal 35
   Polytropic 35
Compression ratio 51
Compressor map (see Performance curve)
Compressor systems
   Adipic acid 74
   Aeration 233
   Alkylation 111
   Catalytic cracking 111
   Ethylene 67
   Office complex 272
   Power house 274
Continuity equation 29
Controller
   Compressor controls 152
   Microprocessor-based 137, 149, 244
   Surge 158
Cross coupling 167

Demand loads (see System load curve)
Dynamic Interaction 185

Eddy current coupling 22
Electrical motor
   ac 21, 65, 102
   dc 22
Enthalpy 30
Entropy 30
Equation of State (see Gas Law)
Expanders 75, 263

Fans 94
Fan Law 15, 108, 137
Flashing 20

Flow measurement ... 59, 154
    Time constant ... 47

Gas Law ... 30
Governor
    Electronic ... 81
    Fly-ball ... 79
    Hydraulic ... 81
Guide vanes
    Diffuser ... 96, 101
    Inert ... 96, 101

Head-capacity curve
    Compressor (see Performance curve)
    Pump ... 7, 15, 18
Horsepower ... 15, 66, 105, 133

Impeller ... 10
Interaction, compensation ... 188
Intercooling, stage ... 38
Internal combustion engine ... 70

Kinetic energy ... 66
Knockout pots ... 37

Laplace variable ... 180
Load curve (see System load curve)

Mass balance ... 92
Matrices ... 172
Moment of Inertia ... 64

Newton's Laws ... 64
NPSH ... 20

Oil
    Hydraulic ... 85
    Lube ... 83
    Seal ... 85
Optimization ... 16
Overload ... 77, 103
Override ... 95

Parallel operation ... 18, 228, 234
Performance curve
    Compressor ... 40, 93, 94, 109, 146, 152, 238
    Pump ... 13
Performance curves (actual compressors)
    Atlas Copco ... 152
    Dresser Clark ... 127
    G.E. ... 57
    JOY ... 101
    Mitsui ... 114, 146

Pneumatic instruments 143, 154
Power demand 21, 104
Prime mover 63
Pumps
    Centrifugal 9, 10, 269
    Displacement 9
    Dynamic 9
    Reciprocating 9
    Rotary 9

Radius of gyration 64
Rangeability 18, 101, 102
Relative gain array 176
Reset windup 96, 159
Rotor (see Impeller)

Self-regulation 45, 70
Series operation 18, 228
Simulation, compressor 56
Speed 49, 133, 152
    Servomechanism 78
Stage 37, 112
Start-up 77, 117, 147
Static inverters 22, 24
Stonewall 42, 47, 135, 201
Surge 42, 47, 101, 123, 137, 141, 195, 201
    Detection 200
    Margin 56, 206
    Open-loop control 198
    Period 46
    Pressure compensation 129
    Universal control 137, 240
System-head curve (see System load curve)
System-load curve 9, 17, 48
System resistance curve (see System load curve)

Thermal efficiency 39
Thermodynamics
    First Law 29
    Second Law 29
Turbine
    Gas 72
    Steam 73

Variable speed 9, 21
Vector diagrams 10, 99
Velocity, fluid 10